D0782561
CHICAGO PUBLIC LIBRARY
HAROLD WASHINGTON LIBRARY CENTER

DATE			

Farm Power and Tractors

McGraw-Hill Publications in the Agricultural Sciences

Brown: Farm Electrification
Campbell and Lasley: The Science of Animals that Serve Mankind
Campbell and Marshall: The Science of Providing Milk for Man
Christopher: Introductory Horticulture
Edmonds, Senn, Andrews, and Halfacre: Fundamentals of Horticulture
Jones and Aldred: Farm Power and Tractors
Kipps: The Production of Field Crops
Kohnke: Soil Physics
Kohnke and Bertrand: Soil Conservation
Krider and Carroll: Swine Production
Lassey: Planning in Rural Environments
Laurie and Ries: Floriculture: Fundamentals and Practices
Laurie, Kiplinger, and Nelson: Commercial Flower Forcing
Maynard, Loosli, Hintz, and Warner: Animal Nutrition
Metcalf, Flint, and Metcalf: Destructive and Useful Insects
Muzik: Weed Biology and Control
Smith and Wilkes: Farm Machinery and Equipment
Sorenson: Animal Reproduction: Principles and Practices
Thompson and Troeh: Soils and Soil Fertility
Thompson and Kelly: Vegetable Crops
Thorne: Principles of Nematology
Treshow: Environment and Plant Response
Walker: Plant Pathology
Warwick and Legates: Breeding and Improvement of Farm Animals

Farm Power and Tractors

Fifth Edition

Fred R. Jones, M.S.
Professor Emeritus and Former Head of Agricultural Engineering Department
Texas A&M University
Fellow, American Society of Agricultural Engineers

William H. Aldred, M.S.
Associate Professor
Agricultural Engineering Department
Texas A&M University
Member, American Society of Agricultural Engineers

McGraw-Hill Book Company

New York St. Louis San Francisco Auckland Bogotá Hamburg Johannesburg London Madrid Mexico Montreal New Delhi Panama Paris São Paulo Singapore Sydney Tokyo Toronto

This book was set in Times Roman by Bi-Comp, Incorporated.
The editors were Carol Napier, James S. Amar, and Scott Amerman;
the production supervisor was Phil Galea.
New drawings were done by Fine Line Illustrations, Inc.
R. R. Donnelley & Sons Company was printer and binder.

FARM POWER AND TRACTORS

1 2 3 4 5 6 7 8 9 0 DODO 8 9 8 7 6 5 4 3 2 1 0

Library of Congress Cataloging in Publication Data

Jones, Fred Rufus, date
Farm power and tractors

(McGraw-Hill publications in the agricultural sciences)
First-4th. ed. published under title: Farm gas engines and tractors.
Includes index.
1. Farm tractors. 2. Farm engines. I. Aldred, William H., joint author. II. Title.
S711.J63 1980 629.22'5 79-22912
ISBN 0-07-032781-5

Contents

Preface

This book represents the fifth edition of a textbook first published in 1932 and titled, *Farm Gas Engines and Tractors*. The first edition was divided into two sections, the first dealing largely with single-cylinder, slow-speed, stationary power units which supplied power for water pumping, wood sawing, and similar light jobs. Such power units were soon displaced by electric power when it became available for farm use. The second section was devoted to tractors because they were making noticeable headway in supplying power in agricultural production, particularly for such heavy power-consuming operations as plowing, harrowing, grain threshing, and combining.

But animal power still dominated the farm power picture. However, in spite of a seriously depressed economic condition existing at that time, progress in the design and improvement of tractors accelerated with the adoption of rubber-tired wheels; multiple-cylinder, high-speed engines; electric starting; hydraulic controls; and other improvements. World War II, as well as other factors, encouraged and further accelerated a shift from animal to tractor power. By 1950, the changeover was practically 100 percent.

In keeping with this progress in the use of tractor power in agriculture, this book was revised in 1938, 1952, and 1963. By 1970, it was again evident that the American farmer had not only accepted the tractor as an essential and vital element in practically all types of crop production, but that a size and type of tractor was wanted that fitted farming needs from every standpoint including economy, utility, ease of handling, servicing, and dependable performance. During the period 1960 to 1979 manufacturers observed this trend and responded with a number of rather unexpected but exciting developments such as (1) an almost 100 percent changeover to diesel engines in the 20-hp and higher power range; (2) the introduction of large four-wheel-drive tractors ranging from 120 to 250 hp; (3) increasing the number of available sizes per company from two or three to as many as eight to ten models; (4) developing transmissions with as many as twelve or more travel speeds and providing on-the-go gear shifting; (5) adapting and utilizing hydraulic power more extensively to permit easier and faster machine and tractor performance; and (6) providing operator safety and comfort by means of rollover protective devices, canopies, and enclosed dust-, weather-, and sound-proof cabs together with a comfortable adjustable seat and convenient location of all controls.

Another very significant development relative to the manufacture and merchandizing of tractors and other machines is the change which has taken place with respect to repairing and servicing them. The complexity and precision involved in the design and assembly of a tractor is such that all successful dealers or farmer owners of several machines must now have a well-arranged and equipped shop and use only reliable, well-trained, and experienced service mechanics. In this connection also, the manufacturers are providing tractor owners as well as service personnel with excellent well-prepared instruction books and literature. The preceding story points out clearly that a textbook dealing with the development and adoption of the tractor as a major source of power in agriculture must undergo periodic revision and improvement.

The objective of the authors in preparing this revised edition has been to prepare a text for use particularly by college-level students and high school students taking vocational agriculture courses, as well as farm tractor owners and operators who desire a complete, up-to-date, and detailed knowledge of tractors and similar automotive types of equipment.

This book could not have been prepared without the generous assistance of tractor manufacturers and numerous other companies whose products are allied with this industry. This assistance in supplying illustrations and specific information is gratefully acknowledged and appreciated. Special recognition and thanks are extended to Dr. Alan L. Dorris, Corporate Manager, Product Safety, J I Case Company, for his effort and care in preparing Chapter 21, "Tractor Safety and Comfort."

Fred R. Jones
William H. Aldred

Farm Power and Tractors

Chapter 1

Survey of Farm Power

The story of American agricultural progress from the colonial period to the present time is an interesting and spectacular one. At the beginning of the nineteenth century, industrial activities were limited, cities were relatively small, and a large part of the population of working age was engaged in the production of food and fiber products for self-subsistence. There was no appreciable incentive for an excess per capita agricultural production and output, because living standards were low as compared with present-day standards and the many conveniences and luxuries that now seem so commonplace did not exist. But, as the population increased, simple machines and processes were developed, new cities were established, factories were built, railroads and better transportation facilities were constructed, and American industry began to grow.

This slow but certain industrial expansion and a steadily increasing population created a demand for more land and a greater agricultural output, as well as for other basic and essential materials such as coal, iron and other metals, lumber, and oil. Consequently new agricultural areas were opened up, and individual farmers and farm families found themselves operating more acres and producing not only for their own needs and subsistence but for the needs of this constantly increasing number of people engaged in commerce and industry and in other nonagricultural enterprises. As the decades passed, this trend gained momentum. It was accelerated by many new inventions, by the discovery of petroleum and the great development of this industry, by such factors as the Civil War and the two world wars, and by normal technological progress. Not only is the United States today the most highly industrialized nation in the world, but its agricultural production and the efficiency and capacity of its farm workers are much greater than those of any other nation. This condition can be attributed largely to one significant and spectacular development—mechanical power and modern farm machines.

Farm-Power Progress and Transition. Obviously, the first kind of power used in agriculture was human power, and all operations from land preparation through cultivation, harvesting, and processing of the final product were performed more or less by hand. Such was probably the situation in America during the Revolutionary period and the early part of the nineteenth century. But, as the need arose for increased crop production, the development and invention of heavier and more effective field tools and machines became imperative. Hence, heavier and larger plows and harrows, mechanical planters, cultivators, and harvesting devices were designed and introduced. But they required more and better power; therefore, the ox, the horse, and the mule entered the farm-power scene in appreciable numbers about 1840.

During the next 50 years numerous laborsaving machines were developed and introduced. These included the steel walking plow in 1837, the grain reaper in 1831, the thresher in 1842, and riding machines such as planters, cultivators, plows, mowers, and grain binders between 1850 and 1870. All these required one or more animals for power.

By 1850 the settlement of the fertile and expansive prairie area now known as the Middle West had started. Wheat was the predominating crop, and this area was ideal for its production. Hence the acreage of wheat and similar grains expanded rapidly. This expansion stimulated a need for an even better source of power than animals, particularly for harvesting and threshing. Consequently the steam engine came into use about 1870 and proved very popular for nearly half a century as a power medium for such heavy-duty operations as grain threshing, lumber and wood sawing, and corn shelling. The early steam engines were mounted on wheels and pulled about by horses. Later on they were made self-propelling, and with the development of the large wheat farms of the West and Northwest, steam tractors displaced animal power to a certain extent for preparing land and for sowing and harvesting the crop. The steam tractor for field work had its limitations. It was very heavy and slow-moving, the fuel was bulky and difficult to handle, and the matter of boiler-water supply and fueling meant constant attention on the part of one man, with a second man to handle and guide the machine. The manufacture of steam tractors was discontinued in the 1920s, and the steam engine became obsolete as a source of farm power.

Horses and mules provided a large part of the power requirements of American agricultural production from about 1880 to 1920. By this time the gasoline-powered internal combustion engine had been developed and proven successful for automotive power for both transportation and agricultural operations and production. As farm tractor design progressed and improved, the use of animal power in agriculture gradually declined and, eventually, the industry became completely mechanized. This transition has resulted in two pronounced effects on agricultural production and the overall status of the economy of the United States and other important food and fiber-producing countries. First, it has made possible the availability of a greater number of workers in other types of employment and, second, it has resulted in a higher output per worker engaged in agricultural production.

Mechanization and Labor Requirements. The data given in Table 1-1 depict very clearly how mechanization reduces the human-labor requirements for producing the major farm crops. In 1840, 35 work-hours were needed to grow an acre of wheat yielding 15 bushels. By 1900, with the use of binders and threshers, this had been reduced to 15 work-hours per acre. Further mechanization, particularly of the harvesting operation, again reduced this labor requirement to 2.9 work-hours per acre in 1970.

In 1840, corn production required 69 work-hours per acre; in 1900, with the use of horse-drawn machines, the requirements were reduced to 38 work-hours per acre; in 1970, largely as a result of the use of row-crop tractors and mechanical harvesters, the requirements were further reduced to six work-hours per acre.

Cotton has always been considered as a high-labor-consuming crop. But very definite progress has been made in reducing this hand-labor requirement through mechanization, such as improved planting and tillage methods and mechanical harvesting. Again referring to Table 1-1, we find that, in 1840, 135 work-hours of labor were required to produce an acre of cotton. With a yield of 154 lb of lint per acre, the total labor requirement per 500-lb bale was 439 work-hours. In 1900, with the use of improved methods and horse-drawn equipment, the requirements were 112 work-hours per acre, 283 work-hours per 500-lb bale, with a yield of 198 lb per acre. In 1970, with the use of row-crop tractors, mechanical harvesters, and other modern equipment, the requirements dropped to 34 work-hours per acre, 32 work-hours per 500-lb bale, with a yield of 506 lb per acre.

Table 1-1 Estimated Work-hours Used to Produce Wheat, Corn, Cotton, and Hay for Designated Periods, United States Average[1]

Crop and item	Yearly average for							
	1840	1880	1900	1920	1940	1950	1960	1970
Wheat								
Work-hours per acre	35	20	15	12	7.5	4.5	3.8	2.9
Yield per acre, bushels	15	13.2	13.9	13.8	15.9	16.5	22.3	26.6
Work-hours per 100 bushels	233	152	108	87	65	35	17	10
Corn for grain								
Work-hours per acre	69	46	38	32	25	15	10	6
Yield per acre, bushels	25	26	26	28	30	38	50	75
Work-hours per 100 bushels	276	180	147	113	83	40	20	8
Cotton								
Work-hours per acre	135	119	112	90	98	83	66	34
Yield of gross lint per acre, pounds	154	196	198	160	257	269	428	506
Work-hours per bale	439	304	283	281	191	132	74	32
Hay								
Work-hours per acre			12	12	11	8.4	6.0	5.5
Yield, tons per acre			1.15	1.22	1.24	1.35	1.61	1.88
Work-hours per ton			10.3	9.9	9.1	6.2	3.7	3.0

[1] Information obtained from various publications of the United States Department of Agriculture.

Table 1-2 Tractors and Other Machines on Farms, 1930–1978[1]

Year	Tractors, thousands	Garden tractors and motor tillers, thousands	Motor trucks, thousands	Grain combines, thousands	Corn pickers, thousands	Pick-up hay balers, thousands	Field-forage harvesters, thousands
1930	920		900	61	50		
1945	2,354	68	1,490	275	168	42	20
1950	3,399	216	2,209	714	456	196	81
1955	4,345	332	2,675	980	688	448	202
1960	4,688	450	2,834	1,042	792	680	291
1965	4,787	699	3,030	910	690	751	316
1970	4,619	805	2,978	790	635	708	304
1975	4,463	870	3,032	666	615	609	337
1978	4,350	917	3,054	615	600	595	345

[1] *Source:* Reports of the United States Department of Agriculture and the Bureau of the Census, and *Implement and Tractor,* 1978. Market Statistics issue.

Likewise, the mechanization of hay production has reduced the labor requirement for this crop from 12 work-hours per acre in 1900 to 5.5 work-hours per acre in 1970.

Tractor Census and Distribution Table 1-2 shows the number of tractors in the United States during the period from 1930 to 1978. The first large increase in tractor population came after the end of the First World War, when agricultural and industrial production for peacetime consumption were at high levels. Development of the all-purpose type of tractor and its widespread adoption by farmers in the late 1920s were chiefly responsible for maintenance of the rapid upward trend in the number of tractors on farms. In the late 1930s, rubber-tired all-purpose tractors came on the market. These were bought in great numbers by farmers, and the increase in tractor purchases, which had ceased during the Depression years, was resumed.

A number of other more recent developments in tractor design and equipment have contributed to the accelerated adoption of tractor power beginning about 1940. Some of these are (1) greater range of power sizes, (2) improved engine design including smaller diesel engines, (3) power take-off, (4) power steering, (5) hydraulic implement control, (6) improved transmissions including more operating speeds and on-the-go shifting, (7) fully enclosed power operated brakes, and (8) differential lock. Of the items listed, rubber tires, the power take-off, and hydraulic control mechanisms have contributed most to the present wide adaptability and use of tractors in nearly all types of farming and crop production.

PROBLEMS AND QUESTIONS

1 Explain briefly what has happened with respect to the types of farm power and their relation to farming since about the year 1800.
2 Name three leading major field crops produced in the United States and list them in order with respect to the effect of mechanization on the labor required for their production since 1800.
3 Name six developments in the past 40 years which have contributed to the popularity of tractor power for farming.

Chapter 2

Farm Power Sources and Adaptability

Power is required on the farm for doing two kinds of work, namely *tractive work,* requiring pulling or drawing effort, and *stationary work,* usually accomplished by means of a belt, gears, power take-off, or direct drive. Tractive jobs include (1) plowing and land preparation, (2) planting and seeding, (3) crop cultivation, (4) harvesting, and (5) hauling. Stationary jobs include (1) water pumping, (2) processing, (3) ensilage cutting, and other jobs of a like nature. A number of farm operations are performed by machines for which the tractor supplies power in two ways simultaneously, namely (1) for pulling the machine and (2) for operating its mechanism through a power take-off. Examples are hay harvesting machines, forage harvesters, and weed cutters.

SOURCES OF POWER

There are five possible sources of power for doing the various kinds of work. In other words, it can be said that there are five prime movers available for the farmer. These are (1) domestic animals, (2) wind, (3) flowing water, (4) electricity, and (5) heat engines. Some of these are necessarily limited in use, as will be mentioned later. In fact, up to this time only two of the five mentioned, namely, domestic animals and heat engines, have proved practical for supplying tractive power. Thus far, wind, water, and electric power are confined entirely to stationary work.

Animal Power

The use of horses and mules for farm power in the United States reached a peak about 1918 and has shown a rapid decrease since about 1930. All farming operations in most sections of the United States are now carried on without the use or need for a single work animal. On the other hand, animals such as the horse, the mule, the ox, the water buffalo, and even the camel are still the principal source of farm power in certain Latin American, European, and Asiatic countries. There are a number of reasons for this such as (1) size of farms, (2) topography, (3) kinds of crops grown, (4) lack of a suitable fuel at a reasonable cost, (5) high initial cost of mechanical equipment, and (6) plentiful supply of low-cost labor.

Power of Horses and Mules The power and pulling ability of horses and mules are matters that are frequently debated but seldom understood. According to King,[1] a horse working continuously for several hours and walking at the rate of 2½ mph should not be expected to pull more than one-tenth to one-eighth of its weight. On this basis, a 1,000-lb horse can develop 0.67 to 0.83 hp; a 1,200-lb horse, 0.80 to 1.00 hp; and a 1,600-lb horse, 1.07 to 1.33 hp.

Studies and tests made at the Iowa State College[2] demonstrated that:

1 It is possible for horses to exert a tractive effort of one-tenth to one-eighth of their own weight and travel a total of 20 miles per day without undue fatigue.

2 It is possible for horses weighing 1,500 to 1,900 lb or over to pull continuously loads of 1 hp or more for periods of a day or longer.

3 A well-trained horse can exert an overload of over 1,000 percent for a short time.

4 For a period of a few seconds and over a limited distance of perhaps 30 ft or less, a horse can exert a maximum pull of from 60 to 100 percent of its actual weight. Under such conditions one horse may develop as much as 10 hp or more, depending upon its size and pulling ability.

Wind Power The energy of the wind, like that of flowing water, is more or less limited for farm use, chiefly because it cannot be controlled and is seldom available when needed. Consequently, the use of wind

[1] F. H. King, "Physics of Agriculture," published by the author, 1910.

[2] *Iowa State Coll. Agr. Exp. Sta. Bull.* 240.

power on the farm is confined largely to water pumping, because whenever the wind blows, even if but once or twice a week, enough water can be pumped and conveniently stored to last several days, or until the wind blows again. The power of the wind is made available by means of the common windmill.

The power developed by this device depends primarily upon the size of wheel and the wind velocity. However, a number of other factors such as type of wheel, design of wheel and mill, and height of tower affect the performance of a windmill. The theoretical power of a stream of air passing through a circular area perpendicular to the direction of travel of the air is represented by the formula

$$\text{hp} = 0.00000525 D^2 W^3$$

where D = maximum diameter of wind wheel or circle, ft
W = wind velocity, mph.

Owing to certain reactions between the wind and the revolving wheel and to mechanical imperfections, it is usually considered that the actual efficiency of the common multisail type of farm windmill based upon this formula will be approximately 30 percent for wind velocities up to 10 mph, 20 percent for wind velocities of 15 to 20 mph, and 15 percent for wind velocities of 25 mph or more. Table 2-1 gives the approximate horsepower developed by windmills of various sizes and at different wind velocities, based upon the above assumptions.

The airplane-type wind wheel, commonly used where high speed is desirable, as in the case of wind-driven electric plants, probably has a higher efficiency than the ordinary farm-windmill type of wheel.

Water Power The power developed by flowing water depends upon two factors, namely, the volume of water flowing per minute and the head or vertical distance the water drops at the point where the power installa-

Table 2-1 Power of Windmills for Wheels of Different Sizes and for Different Wind Velocities

Wind velocity, mph	Diameter of wind wheel, ft					
	6	8	10	12	14	16
6	0.01	0.02	0.03	0.05	0.07	0.09
10	0.06	0.10	0.16	0.23	0.31	0.40
15	0.13	0.23	0.35	0.51	0.70	0.91
20	0.30	0.54	0.84	1.21	1.65	2.15
25	0.44	0.79	1.23	1.77	2.42	3.15
30	0.77	1.36	2.12	3.06	4.16	5.45

Figure 2-1 Measuring the stream flow with a rectangular weir.

tion is located. The former can be measured either by the float method or by a weir (Fig. 2-1). The head is determined by measuring the difference in surface level before and after the water falls.

For example, suppose that the following stream measurements have been made:

	Ft
Average width	12
Average depth	2
Velocity per minute	15
Head	4

Knowing that water weighs 62.4 lb/ft^3 and that 33,000 ft · lb/min is equal to 1 hp, the theoretical power available from the stream is

$$\frac{12 \times 2 \times 62.4 \times 15 \times 4}{33{,}000} = 2.7 \text{ hp}$$

Owing to frictional losses in the water wheel or other means used, the actual available horsepower would probably be somewhat less than 2.7.

Electric Power The use of electricity on farms has increased tremendously in the past 50 years, and it is now playing a major part in farm mechanization. Electricity contributes directly to agricultural production by supplying heat, light, and power for lighting buildings and heating water, and for operating brooders, water pumps, and dairy and refrigeration equipment.

Electric motors for stationary power have a number of distinct advantages, such as: (1) they are relatively simple and compact in construction, (2) they are light in weight per horsepower, (3) they require little attention and limited care and servicing, (4) they start easily and readily, (5) they operate quietly, (6) they produce smooth, uniform power, and (7) they are adapted to uniform or variable loads.

Heat Engines Fortunately, when the demand arose for larger power-producing units for operating such stationary farm machines as the thresher, the wood saw, and the ensilage cutter, at about that time the steam engine and, later, the gas engine were invented. These engines are known to the engineer as *heat engines.* In either case, whether it is a steam engine or a gas engine, some kind of combustible material known as a fuel—particularly wood, coal, oil, or natural gas—is ignited, combustion takes place either slowly or rapidly, heat is produced and utilized in such a way as to create pressure, and the latter, when applied to certain movable parts of the apparatus, produces motion and therefore energy and power.

The steam engine (Fig. 2-2) and the ordinary gas engine (Fig. 2-3) are the two common types of heat engines. In the former, the heat of the burning fuel is applied to water in a closed receptacle called a boiler. As the water becomes heated, it is converted into steam. As the heating continues, more steam is formed and high pressure results. This steam, under pressure, when conducted by a pipe into a cylinder behind the

Figure 2-2 A complete stationary steam-engine layout.

Figure 2-3 A stationary diesel-engine power unit. (*Courtesy of Deere and Company.*)

piston, places the latter in motion and thus generates power. Since the fuel is ignited and burned outside the cylinder and its heat energy applied indirectly to the piston by an intermediate medium, namely, water vapor, the steam engine is called an *external-combustion engine*.

The gas engine resembles the steam engine in that the pressure is applied to a piston sliding back and forth within a cylinder, but it differs greatly in the combustion of the fuel and the application of the pressure resulting from the heat produced. In the case of the common gas engine and engines of a similar type, the combustible fuel mixture is first placed inside the cylinder in a gaseous condition and compressed before it is ignited. The ignition of this compressed mixture causes very rapid combustion and an instantaneous application of pressure to the piston, more commonly known as an explosion. The piston is consequently set in motion and power is generated. Since the fuel is ignited and burned inside the cylinder, the gas engine and all engines that operate in a similar manner are known as *internal-combustion engines*.

Internal Combustion Engine Versus Steam Engine The internal-combustion engine has completely replaced the steam engine for all types of farm power applications for the following reasons:

1 It is more efficient; that is, a greater percentage of the heat and energy value of the fuel is converted into useful power. The efficiency of the internal-combustion engine varies from 15 to 30 percent, whereas that of the external-combustion engine is often as low as 3 and seldom exceeds 10 percent.

2 Weighs less per horsepower.

3 More compact.

4 Original cost less per horsepower.

5 Less time and work necessary preliminary to starting.

6 Less time and attention required while in operation.

7 Can be made in a greater variety of sizes and types and adapted to many special uses; that is, it has a greater range of adaptability.

PROBLEMS AND QUESTIONS

1 Name the possible sources or types of power for doing farm work and state their relative value and use in agricultural production today.

2 If all five types of power listed in this chapter were available on a certain farm, select which ones you would use to perform the following operations.

- **a** Operating a hay baler
- **b** Irrigating 100 acres of cotton
- **c** Filling a 1,000 gallon water tank from a 50 ft (depth) well to care for 10 cows in a pasture
- **d** Grind and mix the feed needed for a 50-cow dairy.

Chapter 3

History and Development of the Internal-Combustion Engine and Farm Tractor

The first ideas concerning the operation and construction of an internal-combustion engine were based upon the action of the ordinary rifle or cannon; that is, the barrel served as a cylinder and the bullet or cannon ball acted as a piston. The difficulty encountered, however, was in getting the piston to return to its original position, thus producing a continuous back-and-forth movement to ensure a continuous generation of power.

Early Ideas and Inventions Nothing of consequence was accomplished in this field before the seventeenth century. In 1678, Abbe Jean de Hautefeuille, a Frenchman, proposed the use of an explosive powder to obtain power. He is said to have been the first man to design an engine using heat as a motive force and capable of producing a definite quantity of continuous work. Christian Huyghens, a Dutchman, is credited with having been the first man actually to construct an engine having a cylinder and piston. This device used explosive powder as fuel and was exhibited to the French minister of finance in 1680.

None of these early attempts was successful, however, and further efforts in the construction of an internal-combustion engine were abandoned for about a hundred years. During the eighteenth century the possibilities of utilizing steam for power were recognized and developed, and the energies of the engineers of that time were turned almost entirely toward applications of the steam engine. About 1800 the thoughts of these investigators were again directed toward the possible design of a gas engine. During the period from 1800 to 1860, a number of engines were constructed, none of which was really successful. Some notable steps were made, however, among which were the use of compression and an improved system of flame ignition by Barnett in 1838 and the actual construction and manufacture of an internal-combustion engine on a commercial scale by Jean Joseph Lenoir in 1860. The Lenoir engine later proved to be impractical.

Beau de Rochas Perhaps the individual making the first really important contribution toward the development of the present-day types of internal-combustion engines was Beau de Rochas, a French engineer. In 1862 this man advanced the actual theory of operation of all modern types of internal-combustion engines. He first stated that there were four conditions that were essential for efficient operation. These were as follows:

1 The greatest possible cylinder volume with the least possible cooling surface.
2 The greatest possible piston speed.
3 The highest possible compression at the beginning of expansion.
4 The greatest possible expansion.

These principles are still considered as fundamental and extremely important in gas-engine design. Beau de Rochas proposed further that a successful engine embodying these principles must consist of a single cylinder and a piston that made a stroke for each of the four distinct events constituting a cycle, as follows:

1 Drawing in of the combustible fuel mixture on an outward stroke.
2 Compression of the mixture on an inward stroke.
3 Ignition of the mixture at maximum compression producing an outward power or expansion stroke.
4 Discharge of the products of combustion on a fourth or inward stroke.

Otto and Clerk Beau de Rochas never succeeded in constructing an engine based upon his theories, but they were promptly accepted as being essential. Although considerable effort was expended in the design of an engine during the next few years, it was not until 1876 that Dr. N. A. Otto, a German, patented the first really successful engine operating on this four-stroke-cycle principle. The engine was first exhibited in 1878. This cycle, although originally proposed by Beau de Rochas, is commonly known as the *Otto cycle*.

The invention of the four-stroke-cycle engine by Otto was soon followed by the issue of a patent, in 1878, to Dugald Clerk, an Englishman, on the first two-stroke-cycle engine, that is, an engine producing one power impulse for every revolution instead of for every two revolutions. This particular engine was not marketed at once, however, and was not really perfected until 1881.

Diesel Another notable contribution to the development and varied application of the internal-combustion engine was the work of Dr. Rudolph Diesel, a German engineer, who conceived the idea of utilizing the heat produced by high compression for igniting the fuel charge in the cylinder. In 1892 he secured a patent on an engine designed to operate in this manner. This first machine proved unsatisfactory, however, and it was not until about 1898 that the first successful diesel-type engines were produced. During the past 35 years, rapid strides have been made in the development and utilization of the diesel principle in internal-combustion engines for both stationary and tractor applications.

We thus observe that the invention of the internal-combustion engine is a comparatively recent one and that the many finely constructed modern types of single- and multiple-cylinder engines have been designed and developed almost overnight. But when it is considered that petroleum, now the almost universal source of fuels and lubricants for these engines, was first discovered about 1858 and that very little was known concerning the application of electricity in the operation of internal-combustion engines previous to the latter part of the past century, we readily perceive the explanation of the retardation of this invention—now considered an indispensable device to modern life throughout the world.

HISTORY AND DEVELOPMENT OF THE FARM TRACTOR

A *tractor* is defined specifically as a self-propelled machine that can be used for supplying power for (1) pulling mobile machines and (2) operating the mechanisms of either stationary or mobile machines by means of a belt pulley or a power take-off. The two common types are the steam tractor, in which an external-combustion or steam engine supplies the power, and the gas tractor, in which an internal-combustion engine serves as the source of power.

Steam Tractor The invention and development of the steam engine preceded those of the internal-combustion engine by 100 years or more. Consequently, the earliest known tractors were of the steam type. They first came into general use for operating threshers (Fig. 3-1) in the wheat-

Figure 3-1 Steam tractor operating a grain thresher.

and grain-growing sections of the country during the last two or three decades of the nineteenth century. Their self-propelling feature was utilized primarily for moving about from one threshing job to another. Later on, with the opening up of the large wheat farms of the West and Northwest, steam tractors displaced animal power to a certain extent for preparing the land and sowing and harvesting the crop.

The steam tractor for field work had its limitations. It was very heavy and slow-moving, the fuel was bulky and difficult to handle, and the matter of boiler-water supply and fueling meant constant attention on the part of one man, with a second man to handle and guide the machine.

Early Gas Tractors Certain manufacturers, foreseeing a greater future demand for suitable mechanical power for field work, particularly in the wheat-growing sections of the country, started the construction of gas tractors even before the close of the nineteenth century. For example, Fig. 3-2 shows a machine that is said to have been built in 1892. Another tractor (Fig. 3-3) is reported to have been put into use in North Dakota in 1897, and Fig. 3-4 shows the first Hart Parr tractor, which was sold in 1902. These heavy, cumbersome-appearing machines were the forerunners of the present-day tractor industry, which started soon after the opening of the present century and began to gain momentum about 1905.

According to L. W. Ellis and E. A. Rumely,[1]

> By the spring of 1908 the builders of the first successful tractor had about 300 machines in the field and the sales of that year equaled those of the 5 years

[1] Ellis, L. W., and E. A. Rumely, *Power and the Plow,* Doubleday & Company, Inc., New York, 1911.

Figure 3-2 A gas tractor built in 1892.

preceding. The following year the number in the field was again doubled and by the close of the year 1910 over 2,000 of these tractors were said to be in active service. Another company began to produce a small tractor in 1907 and by the close of the decade was selling several thousand yearly. Dozens of gas tractor factories sprang up and practically every manufacturer of steam traction engines either went out of business or added an internal-combustion engine to his line.

These early tractors consisted usually of a large one-cylinder gas engine mounted on a heavy frame placed on four wheels. The two rear wheels were connected by a train of heavy, exposed cast-iron gears to the crankshaft of the engine, thus making the machine self-propelling. Like the steam tractors, they were heavy, cumbersome, and powerful, and in fact they seemed to be designed as a mere substitute for the former. They possessed certain advantages, however. The fuel was easier to handle, there was less water to haul, and less time and attention were required for starting and during operation. One man was usually able to handle the largest outfit.

The Lightweight Tractor About 1910 the designers turned their attention toward the possibilities of a lighter weight gas tractor to meet the approaching demand of the smaller grain and livestock farmer for mechanical power. Consequently, about 1913, there began to appear on the market a number of machines, comparatively light in weight and differing

Figure 3-3 A gas tractor built in 1897.

greatly in construction and appearance. In most cases, they were equipped with two- and four-cylinder engines.

By 1915, the farmers were presented with an amazing array of types, models, and sizes, ranging from the giant one- or two-cylinder four-wheelers to a tractor attachment for a small automobile. No two machines resembled each other. Some had two wheels, some three, and some four.

Figure 3-4 A gas tractor built in 1902.

Some were driven from the front and others from the rear. Some had the plows attached under the frame, while others pulled them in the usual manner. Competition became very marked, and large sums of money were spent for public demonstrations. Many of these freak machines were sold. Some proved more or less successful, while others gave unsatisfactory results and tended to destroy the faith of their owners in the future value of the tractor. True it was that a few of these first lightweight tractors were unreasonable in design and weak in construction. On the other hand, many of them were not. Their failure in the hands of the farmer could be attributed largely to the lack of knowledge of their operation and care.

The period of the freak tractor was short. By 1917, many of them had disappeared from the market, and the more farsighted designer observed that there were certain fundamentals of tractor design that had to be adhered to.

The First World War and the Tractor The First World War had a very pronounced effect upon farm-tractor development and use. Maximum agricultural production was urged. There was a shortage of labor and prices became abnormally high. All these factors meant an increased demand for laborsaving and timesaving machinery, especially small farm tractors. As a consequence, a large number of small tractor manufacturers sprang up overnight, so to speak, and placed upon the market between 200 and 300 different makes and models of machines. These ranged from the small two-wheel garden tractor to the larger four-wheel types. The average and most popular size seemed to be a machine rated at about 10 to 15 drawbar hp and 20 to 30 belt hp. The tendency in design was toward a four-wheel rear-drive tractor with a four-cylinder engine. Most of these tractors seemed to give much better satisfaction and service than those of earlier manufacture.

Thus the First World War proved to be a great stimulus to the adoption of mechanical power by the farmer. Farm tractors were being used successfully in every nook and corner of the country.

Effect of the Agricultural Depression of 1920 The tractor industry, like many others that were dependent upon the prosperity of the farmer, received a severe setback as the result of the unexpected agricultural depression in 1920. At this time more than 100 different companies were offering about 250 sizes, models, and types of machines. Many of these concerns were small and lacked the capital and the organization necessary to enable them to compete with the older, larger, and better established companies. They attempted to struggle along but, with little or no surplus to draw upon, soon fell by the wayside. Within 2 years' time practically every tractor company that had been organized during the 5 or 10 years preceding for the primary purpose of manufacturing and selling farm gas tractors was compelled to quit. Even some of the older and better established farm-equipment manufacturers, who had entered the tractor business, found little sale for their machines.

Those companies that were able to remain in the business realized that it was no longer a problem of convincing the small or average farmer that mechanical power, in the form of the gas tractor, was practical and economical. They saw that he was willing to pay the price for a well-built, sensibly designed machine that would actually do the work for which it was recommended. Consequently, in spite of the generally depressed and unprosperous condition of agriculture throughout the United States during the years 1921 to 1926, the demand for farm tractors continued to grow, as shown by the fact that the number in use on farms increased from about 250,000 in 1920 to over 500,000 in 1925. An analysis of the figures by states shows further that the greatest increase during this period was in the Middle Western, Eastern, and Southern states, indicating beyond a doubt that the small or average farmer, as well as the large grain raiser, was now a tractor convert.

The agricultural depression of 1930 to 1935 was even more pronounced than that of 1920. The prices of agricultural products dropped to the lowest point reached in many years. Consequently, there was little demand for farm tractors and equipment and sales were relatively small, particularly in 1932 and 1933. However, the small, low-priced, all-purpose tractor was introduced at this time, and as commodity prices improved, an immediate, widespread demand developed for such a tractor. As a result, tractor sales increased steadily and manufacturers were operating at capacity by 1935.

Design Standardized It is true that very few of the machines sold during this period, 1920 to 1926, were adapted to doing everything about the farm; that is, they were not of the all-purpose type. A really successful machine of this type was yet to appear on the market. Most of the machines were lightweight, two- or three-plow tractors that were used largely for plowing and land preparation and for belt work. The owners were satisfied to have something that would prepare the land more quickly and better and likewise relieve the horse of this heavy work.

Another outstanding fact was that, during this period, the design of these tractors, as a whole, seemed more stable and uniform. In other words, there was less variation in design and construction than formerly, indicating that the manufacturers had now, after years of costly experience, settled upon many of the essentials of a successful farm tractor.

All-Purpose Tractors The next and most logical step seemed to be the design of a lightweight, low-priced, all-purpose tractor that would do any kind of field or stationary work on the average farm, including plowing, harrowing, planting, cultivating, harvesting, threshing, or anything requiring similar power. Several machines of this type were designed and sold during the period 1917 to 1921, but few of them ever proved really successful. In other words, it appeared to be a difficult problem to construct a tractor that would be heavy and rugged enough to plow and harrow the heaviest soil and still be practical for such lighter jobs as planting and cultivating of row crops, mowing hay, and so on. Efforts were continued, however, by certain manufacturers to build a really successful and practical all-purpose farm tractor, and in 1924 one of these concerns introduced a machine that came nearer to meeting the requirements of such a tractor than any that had been previously built. Within a few years all the established manufacturers had designed and introduced one or more models of all-purpose tractors.

Power Take-off—Diesel Tractors—Pneumatic Tires—Power Lifts and Hydraulic Controls—Small Tractors Numerous other significant developments have taken place since 1925 which have accelerated the adoption of tractors for agricultural power, extended their range of utility, and improved their convenience and ease of operation. The first tractors supplying power directly to the mechanism of a field machine by means of a power take-off attachment appeared about 1927 and this device is now considered as standard equipment on all farm tractors and quite indispensable for many farm applications.

Tractors equipped with diesel engines were introduced in 1931. These early engines were of the heavy-duty type and their use was confined largely to track-type tractors for earth-moving and various types of construction operations. Sometime about 1950, smaller diesel engines were developed and introduced, and by 1960 were gradually replacing gasoline-burning electric-ignition engines in the power range of 3 to 150 hp and higher.

Low-pressure pneumatic tires first appeared on farm tractors about 1932. In spite of their higher cost compared with steel-wheel tractors, it was soon discovered that rubber-tired tractors offered a number of definite advantages. By 1940 the demand for this type of equipment was almost 100 percent. Many improvements have been made in rubber tires and tractor-wheel equipment in recent years.

During the late 1930s a few manufacturers introduced small, one-row all-purpose tractors. Others followed with similar or even smaller tractors during or immediately following the end of the Second World War.

Other developments that have contributed to the popularity and more effective utilization of tractors, particularly the all-purpose types, are (1) improved hydraulic lifting and control devices, (2) special, direct-connected, and quickly attached implements and tools, (3) improved transmissions providing more travel speed and "on-the-go" shifting, (4) better power take-off operation and speed control, (5) power steering, (6) electric starting and lighting, (7) better brakes, and (8) improved operator comfort and safety.

PROBLEMS AND QUESTIONS

1 Name what you consider as six important changes and improvements which have taken place with respect to overall design and operation between the earliest and the latest types of tractors.
2 What effect, if any, did World War I have on the development and use of tractors?

Chapter 4

Types of Farm Tractors

Classification of tractor types. Tractors as now manufactured can be classified as follows:

A According to method of securing traction and self-propulsion:
- **1** Wheel tractors
  - **a** Tricycle with single or double front wheels.
  - **b** Four wheels with single or dual rear-drive wheels and standard- or high-clearance front axle.
  - **c** Heavy-duty two- or four-wheel drive.
- **2** Track-type tractors

B According to Utility
- **1** General purpose or utility
- **2** All-purpose or row-crop type
- **3** Orchard
- **4** Industrial
- **5** Garden and lawn

WHEEL TRACTORS

The wheel-type tractor, Figs. 4-1 and 4-2, is the predominating type of machine, particularly for agricultural purposes. As previously mentioned,

Figure 4-1 A tricycle row-crop tractor. (*Courtesy of Deere and Company.*)

Figure 4-2 A four-wheel general-purpose tractor. (*Courtesy of International Harvester Company.*)

track-type tractors make up less than four percent of the total number of tractors on farms of the United States.

For several years following its introduction, the three-wheel or tricycle arrangement, Fig. 4-1, was used almost exclusively for row-crop operations and production. However, the four-wheel tractor with an adjustable front axle for both wheel tread and clearance height, Fig. 4-2, is available for the more popular tractor sizes and has noticeably displaced tricycle types. The steering member of a tricycle-type of tractor may consist of a single wheel, or of two wheels placed very close together, Fig. 4-1. This double steering-wheel arrangement provides somewhat better front-load support and stability, better traction, and more positive steering action under most conditions.

Track-Type Tractors The traction mechanism in the track-type tractor (Fig. 4-3) consists essentially of two heavy, endless, metal-linked devices known as tracks. Each runs on two iron wheels, one of which bears sprockets and acts as a driver. The other serves as an idler. Steering is accomplished through the tracks themselves by reducing the movement of one track below the speed of the other.

The use of the track-type tractor was quite limited up to the First World War period. The tanks that were developed at that time and proved so successful in traveling over the battle-swept areas did more perhaps than anything else to demonstrate the adaptability and utility of such a traction arrangement. Consequently, during the succeeding years, the de-

Figure 4-3 Track-type tractor pulling a disk harrow. (*Courtesy of Caterpillar Tractor Company.*)

Figure 4-4 Industrial tractor. *(Courtesy of Deere and Company.)*

velopment of successful track-type tractors in both small and large sizes for every known use was rather rapid. In fact, they have practically supplanted the wheel tractor for many heavy-duty, earth-moving, and industrial jobs requiring tractor power.

Track-type tractors have a limited use in agriculture. They are well adapted for commercial orchard cultivation and maintenance; for grain and other crop-production operations in some hilly sections; for terracing, road maintenance, and special earth-moving jobs on large farms, particularly in irrigated areas; and for land clearing.

General-Purpose or Utility Tractor A general-purpose tractor is one of more or less conventional design such as an ordinary four-wheel machine (Fig. 4-2). It is made to perform practically all tractor jobs such as plowing, harrowing, road grading, combining, hay baling, and the like.

Orchard and Industrial Tractors Orchard tractors are small- or medium-size, general-purpose machines of either the wheel or crawler type, so constructed and equipped as to be operated to better advantage around trees. Such tractors are often built lower, with as few projecting parts as possible and with special fenders.

Industrial tractors are machines of any size or type specially constructed for various industrial operations and for heavy hauling about factories, airports, and so on. They may be equipped with hoisting, excavating, power-loading, and similar attachments built on them as shown in Fig. 4-4.

ALL-PURPOSE TRACTORS

An all-purpose or row-crop type tractor (Figs. 4-1, 4-5, 4-6, and 4-7) is a tractor designed to handle practically all the field jobs on the average farm, including the planting and intertillage of row crops. The most important requirements of such a tractor are (1) greater clearance, both vertical and horizontal; (2) adaptation to the usual row widths; (3) quick, short-turning ability; (4) convenient and easy handling; (5) quick and easy attachment and removal of field implements; and (6) essential accessories such as hydraulic controls and power take-off. All-purpose tractors are made in several types and sizes to adapt them to the many kinds of crops and varying field and farm sizes and conditions.

Steering and Control Ease of handling and control of an all-purpose tractor and its attached field equipment are essential. The steering mechanism should (1) permit short, quick turning; (2) require the minimum of effort in its operation, regardless of whether the machine is moving or stationary; and (3) permit precise and accurate control of the attached units, particularly planters and cultivators. Tractors equipped with a hy-

Figure 4-5 Single-row four-wheel tractor. (*Courtesy of Deere and Company.*)

Figure 4-6 Four-wheel general-purpose tractor building a terrace. (*Courtesy of J I Case Company.*)

draulically operated steering mechanism have a definite advantage in these respects.

Power Take-off and Hydraulic Controls A power take-off drive and shaft and hydraulic controls and lifts for the various attachments and machines operated by the tractor are essential accessories for any tractor.

Figure 4-7 Four-wheel general-purpose tractor with eight-row cultivator. (*Courtesy of Deere and Company.*)

The power take-off is needed to operate mowers, combines, hay balers, corn pickers, stalk shredders, forage harvesters, and other field machines. Positive, conveniently operated hydraulic mechanisms are required for lifting or other adjustment and control of such machines as plows, harrows, cultivators, planters, and other similar machines.

Trend in Farm Tractor Sizes It was explained in Chap. 1 that there has been a pronounced transition in agricultural production methods in the United States since about 1940 because of the adoption of mechanical power and the development of new and larger types of machines for crop production and processing. As a consequence the trend has been toward larger farms operated with less labor. Other factors contributing to the situation are (1) the development of higher-yielding crop varieties; (2) the more extensive use of commercial fertilizers; (3) better seedbed preparation, planting, and cultivation practices; (4) more effective disease and insect control; and (5) faster and more efficient harvesting and processing methods.

A concurrent but significant effect of this revolution in American agriculture has been the shift from smaller to larger sizes of tractors and power machines. This is clearly shown in Tables 4-1 and 4-2, according to the U.S. Department of Agriculture.[1] Change in tractor size illustrates the change in mechanical input capability of all types of new farm machinery. Farm tractors averaged about 33 maximum belt hp in 1960, but averaged about 44 hp in 1971. Total available horsepower from tractors increased 38 percent during the decade, with only a 2 percent increase in tractor numbers. Sales of tractors with 120 or more horsepower rose from 3 percent of the total in 1968 to 7 percent in 1970. In 1970, there were 1,330 tractors with 140 or more horsepower sold to farmers or about one percent of total sales. This compares with 879 units of this size in 1969.

Larger machinery, combined with new types of agricultural technology, has contributed to a reduction in the farm labor input. In 1950, 28 hp was available from tractors per 100 acres of cropland harvested. In 1970, with 71 horsepower available per 100 acres of cropland, the labor input was down to 11 hours per acre. Similarly, tractor horsepower per farm worker rose from 9 in 1950 to 46 in 1970, and reached 60 in 1975.

The U.S. Department of Agriculture[2] states further that:

The trend toward larger farms has been a factor in the shift toward larger tractors. (The larger tractors tend to be on the larger farms.) In 1956, for example, farms of 220 acres and over had more than 40 percent of the large-wheel tractors (35 horsepower and over), whereas farms of less than 100 acres had only about 11 percent of these large tractors.

[1] U.S. Dept. Agr., Eco. Res. Service, Stat. Bull. 233.
[2] U.S. Dept. Agr., Agr. Info. Bull. 231, ARS.

Table 4-1 Tractor Sizes and General Specifications

Specification	1	2	3	4	5	6	7
Maximum engine hp range	10–25	30–40	41–50	51–65	66–75	80–120	120–250
Maximum drawbar hp range	8–22	27–35	35–45	46–59	60–69	70–110	110–135
Number of 14–16 in moldboard plow bottoms	1–2	2–3	3–4	4–5	5–6	6–7	8–10
Transmission type[1]	SG	SG CM	SG CM PL HS	SG CM PL HS	SG CM PL HS	SG CM HS	SG
Number of forward speeds	3–4	4–8	4–12	4–12	4–23	6–12	6–12
Travel speeds, range in mph[2]	2–12	1.5–15	1.2–18	1.2–20	1.2–20	1.2–20	1.2–20
Approximate weight, lb	1,600–2,400	3,000–4,000	4,000–5,000	5,500–7,000	6,500–8,000	8,000–12,000	12,000–15,000

[1] SG = sliding gear
CM = constant mesh
PL = planetary
HS = hydrostatic

[2] Varies with engine speed and tire size.

Table 4-2 1975 Retail Sales in the United States of Farm Wheel-Tractors by Horsepower

Horsepower	Number
Under 35 hp	7,762
35–49 hp	27,274
50–69 hp	29,752
60–99 hp	20,791
100–129 hp	40,026
130–149 hp	9,424
150 hp and over	15,181
Four-wheel drive	
150–200 hp and over	10,605

Figure 4-8 Large four-wheel-drive tractor pulling eight-bottom plow. (*Courtesy of J I Case Company.*)

Figure 4-9 Large four-wheel-drive tractor pulling a disk harrow. (*Courtesy of J I Case Company.*)

The type of farming operation also influences tractor size. For example, in 1954 the average cash-grain farm contained 380 acres. Approximately one-half the tractors on these farms in 1956 were large units. Dairy farms, however, averaged under 180 acres in 1954, and in 1956 only one-fourth of the tractors on these farms were 35 belt hp or more. Figs. 4-8 and 4-9 show very large heavy-duty, four-wheel-drive tractors that are particularly adapted to large-grain farming and other similar field operations.

SMALL TRACTORS—GARDEN TRACTORS

For many years the smallest tractors available were largely lightweight, two-wheel machines powered by a 2- to 5-hp engine and guided with two handle bars by the operator walking behind it as shown by Fig. 4-10.

Figure 4-10 A medium-size garden tractor. (*Courtesy of Bolens Product Division, Food Machinery and Chemical Corporation.*)

Figure 4-11 Small garden cultivator and tiller. (*Courtesy of Deere and Company.*)

These machines were designed and utilized almost exclusively for gardening and vegetable production. Today the only available machine resembling these early garden tractors is the rotary tiller, Fig. 4-11. The modern tractor, Figs. 4-12 and 4-13, is a four-wheel riding machine adapted to many more uses such as lawn mowing, earth moving, trench digging, and sawing, as well as the common garden and field operations such as plowing, harrowing, planting, cultivating, and so on. Most manufacturers make a number of sizes and refer to them as compact or garden and lawn tractors. Table 4-3 gives the general sizes and specifications of these machines.

Power Lawn Mowers The compact tractors just described are used extensively for mowing lawns, golf courses, and the like. The principal advantage of these is that the operator rides the machine and can cover a greater area with less effort, particularly in the case of large lawns. For the average residential lawn, the walking type of power mower is widely

Figure 4-12 Large-size riding garden tractor. (*Courtesy of Deere and Company.*)

Figure 4-13 Large-size garden tractor and lawn mower. (*Courtesy of Deere and Company.*)

Table 4-3 Garden- and Small-Tractor Specifications

	Size designation				
Specification	**1**	**2**	**3**	**4**	**5**
Engine hp range	6 to 7	8 to 9	10	12	14
Transmission type[1]	A-B-C	A-B-C	A-B-C	A-B-C	A-B-C
Number of forward speeds	3 to 4	3 to 4	3 to 4	3 to 4	3 to 4
Travel speeds	One to eight miles per hour				
Approximate weight, lbs	300–400	400–600	600–800	600–800	600–800

[1] A = Gear
B = Belt
C = Hydrostatic (infinite speed range)

used. These machines are powered with a small, air-cooled, single-cylinder engine that drives a rotating horizontal blade. This blade may be driven directly by the engine or by a V belt. The smaller lawn mowers are mounted on small wheels and pushed and guided by the operator. Some larger sizes are equipped in such a way that the engine power also propels the machine, and the operator merely guides it.

PROBLEMS AND QUESTIONS

1 What has been the trend in the sale and use of farm tractors in recent years with respect to size and power rating? What are the reasons, if any, for this trend?
2 If you were growing only three field crops, namely hay, soybeans, and corn, would you purchase three-wheel or four-wheel machines, or both? Explain.
3 There are three distinct types of garden tractors with respect to size and utility. How would you classify them?
4 Under what conditions would the purchase and use of a track-type tractor by a farmer be desirable?

Chapter 5

Power and Its Measurement—Fuel Consumption—Engine Efficiency—Nebraska Tractor Tests

In the study of farm power it is important to have a clear and definite understanding of the exact technical meaning of such terms as horsepower, energy, efficiency, inertia, and other terms as applied to mechanical devices. To obtain such an understanding, one must consider, first of all, certain fundamental physical terms, definitions, and units.

Mass Mass is defined as the quantity of matter a body contains irrespective of the kind of material of which it is composed. The unit of mass in the metric system is the gram or kilogram and in the English system it is the pound.

Inertia Inertia is defined as that property of matter by virtue of which a body tends to remain at rest or continue in motion in a straight line.

Momentum Momentum is a term applied to a body which is in motion and is defined as the product of its mass and the velocity at any instant. For example, the momentum of a mass of 1 kg moving with a velocity of 10 cm/s would be 10 kg · cm/sec.

Force A *force* is an action, exerted upon a body, that changes or tends to change its natural state of rest or uniform motion in a straight line. It is thus observed that a force may or may not be effective in producing motion in the body acted upon. Likewise, if a body is in motion, a force may be applied that may or may not change its direction of movement. The unit of measurement of a force is the pound weight.

The force required to move a body may not be the same as the weight of the body. Only when the latter is moved vertically, with respect to the earth's surface, will this be true. Under certain conditions the force required to move an object might be greater than its weight.

Work If a force is applied to a body changing its state of motion—that is, it is made to move from a condition of rest or, if it is already in motion, its rate or direction of travel is changed—then *work* is done. In other words, if a force acts on a stationary body but does not produce motion, no work is done. Work is measured by determining the force in pounds and the distance through which it acts in feet. The product of the two gives the work done in foot-pounds. That is, the unit of work is the foot-pound (ft·lb).

Energy The energy possessed by a body is defined as its capacity for doing work. The energy possessed by an object or body by virtue of its position is known as *potential energy*. The energy possessed by the body by virtue of its motion is called *kinetic energy*. For example, the water stored in an elevated tank possesses potential energy. If this water is now discharged through a pipe, the water in motion in the pipe possesses kinetic energy.

Energy, like work, is measured in foot-pounds. The potential energy of a body, with respect to a given point or surface, is equal to the product of the weight in pounds and the vertical height in feet through which it has been lifted above this point or surface.

The kinetic energy of a body is dependent upon its weight and its velocity or rate of travel. For a body moving at a uniform velocity,

$$\text{Kinetic energy} = \frac{WV^2}{2g}$$

where W = weight of body, lb
V = velocity, ft/s
g = acceleration of a freely falling body
= 32.2 ft/s²

Torque Torque is defined as a turning or twisting effort or action. For example, when a wrench is attached to a bolt or nut and the bolt is turned by a pull on the handle of the wrench, a torque action is produced. Likewise, turning a machine with a hand crank requires torque and when the pistons and connecting rods of an engine react on the crankshaft, the torque effect produces rotation. The unit of torque is the pound-foot; that is, torque is equal to the applied force in pounds and its perpendicular distance in feet from the axis of rotation of the crank or lever arm.

Power Power is defined as the rate at which work is done. In other words, power involves the time element. For example, if a force of 100 lb acts through a distance of 50 ft, 5,000 ft·lb of work is done. If, in one case, this 100-lb force requires 1 min to move the object 50 ft, and, in another case, the same force consumes 2 min to move the object this distance, then twice as much power is required in the first case as in the second because the same work is done in one-half the time.

Horsepower The term *horsepower* is defined as a unit of measurement of power, and 1 hp is equal to doing work at the rate of 33,000 ft·lb/min or 550 ft·lb/s. There is no real reason why this unit should have this particular value. However, it was fixed some time during the eighteenth century as a result of observations made of the work done by a horse in England in hoisting freight. It was estimated from these observations that the average horse was able to lift vertically a load of 150 lb when traveling at the rate of 2½ mph. Calculate as follows:

$$\frac{150 \times 2.5 \times 5{,}280}{60} = 33{,}000 \text{ ft}\cdot\text{lb/min}$$

Therefore, 33,000 ft·lb/min was chosen as the rate at which the average horse could work and, consequently, was termed 1 hp.

In calculating the horsepower developed or required by a machine, it is only necessary to determine the total foot-pounds of work done or required per minute and divide this total by 33,000.

RELATION BETWEEN MECHANICAL AND ELECTRIC POWER UNITS

Electrical Work Electrical work is measured in joules, a *joule* (J) being defined as the amount of work done by a current of 1 ampere (A) flowing for 1 s under a pressure of 1 volt (V), that is,

$$\text{Electrical work} = \text{volts} \times \text{amperes} \times \text{seconds} = \text{joules}$$

Electric Power–The Watt Since power is the rate of doing work, the power of an electric current would be the electrical work it is capable of doing per time unit (second). In other words,

$$\text{Electric power} = \frac{\text{electrical work}}{\text{time}}$$

$$\text{Electric power} = \frac{\text{joules}}{\text{seconds}}$$

$$= \frac{\text{volts} \times \text{amperes} \times \text{seconds}}{\text{seconds}}$$

The unit of electric power, known as the watt (W), is the power required to do 1 joule of electrical work per second; that is,

$$1 \text{ W} = 1 \text{ J/s}$$

$$= \frac{1 \text{ A} \times 1 \text{ V} \times 1 \text{ s}}{1 \text{ s}}$$

$$= 1 \text{ A} \times 1 \text{ V}$$

or

$$\text{Watts} = \frac{\text{joules}}{\text{seconds}}$$

$$= \frac{\text{amperes} \times \text{volts} \times \text{seconds}}{\text{seconds}}$$

$$= \text{amperes} \times \text{volts}$$

It has been found by experiment that if mechanical work is done at the rate of 1 ft·lb/s, than 1.356 W of electric power will be required to do the same work; that is,

$$1 \text{ ft} \cdot \text{lb/s} = 1.356 \text{ W}$$

but

$$1 \text{ hp} = 550 \text{ ft} \cdot \text{lb/s}$$

Therefore

$$1 \text{ hp} = 500 \times 1.356$$

$$= 746 \text{ W}$$

and

$$\text{Horsepower} = \frac{\text{watts}}{746}$$

$$= \frac{\text{volts} \times \text{amperes}}{746}$$

Since

$$1 \text{ kW} = 1{,}000 \text{ W}$$

then

$$1 \text{ kW} = \frac{1{,}000}{746}$$

$$= 1.35 \text{ hp}$$

HORSEPOWER OF ENGINES

Indicated Horsepower The indicated horsepower (ihp) of an engine is the power generated in the cylinder and received by the piston.

Belt or Brake Horsepower The belt or brake horsepower (bhp) of an engine is the power generated at the belt pulley and available for useful work. Several methods are used for measuring brake horsepower, as described later.

Friction Horsepower The friction horsepower of an engine is the power that it consumes in operating itself at a given speed without any load. That is, it is the power required to overcome friction in the moving parts of the engine plus pumping losses on the intake and compression strokes.

$$\text{ihp} - \text{bhp} = \text{friction hp}$$

Drawbar Horsepower The drawbar horsepower of a pulling machine such as a tractor is the power developed at the hitch or drawbar and available for pulling, dragging, or similar tractive effort. In a tractor, for example, the drawbar horsepower would be equal to the belt horsepower less the power consumed in moving the tractor itself. Methods of measurement follow.

Measurement of Horsepower In the measurement of the power of an engine it must be kept in mind that the fundamental problem is to determine specific values for the physical factors involved, namely, the force acting, the distance through which the force acts, and the time it is acting. Knowing these, the rate of power generation in foot-pounds per minute can be calculated, and by dividing by 33,000, the generated horsepower is obtained.

Measurement of Indicated Horsepower The power generated in the cylinder of an engine owing to the explosion pressure acting on the piston is termed the indicated horsepower, because a device known as an indicator is necessary in determining it. The indicator is so equipped and attached to the engine that it records a diagram (Fig. 5-1) from which can be determined the mean effective pressure for a single engine cycle. By measuring the cylinder bore and the length of the piston stroke and knowing the engine rpm, the indicated horsepower of the engine can be calculated by means of the following formula:

$$\text{ihp} = \frac{PLANn}{33{,}000 \times 2} \quad \text{for four-stroke-cycle engine}$$

or $$\text{ihp} = \frac{PLANn}{33{,}000} \quad \text{for two-stroke-cycle engine}$$

where P = mep, lbs/in²
L = length of piston stroke, ft
A = area of cylinder, in²
= (bore)² × 0.7854
N = rpm
n = number of cylinders

Measurement of Belt Horsepower Devices used for the determination of the belt or brake horsepower of engines are known as *dynamometers*. *Dyna* or *dynamo* means force of power, and a *meter* is a measuring device; therefore, a dynamometer is a power-measuring device. There are a number of different types of dynamometers, such as the prony brake (Fig. 5-2), the electric type (Fig. 5-3), and the hydraulic type (Fig. 5-6).

Prony-Brake Dynamometer To explain its operation more clearly, let us assume that the engine crankshaft is locked so that it cannot turn and that a pair of spring scales is fastened to the outer end of the brake arm (Fig. 5-2). Now suppose that the brake band is tightened on the pulley but that the brake is rotated on the latter by grasping the scales and applying the rotative force at the outer end of the brake arm. The scales will register a certain number of pounds, depending upon the tension of the

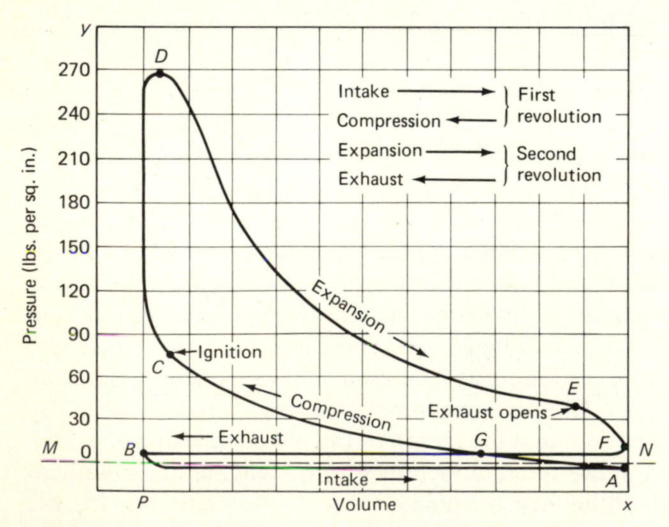

Figure 5-1 Typical indicator diagram for a four-stroke-cycle gas engine.

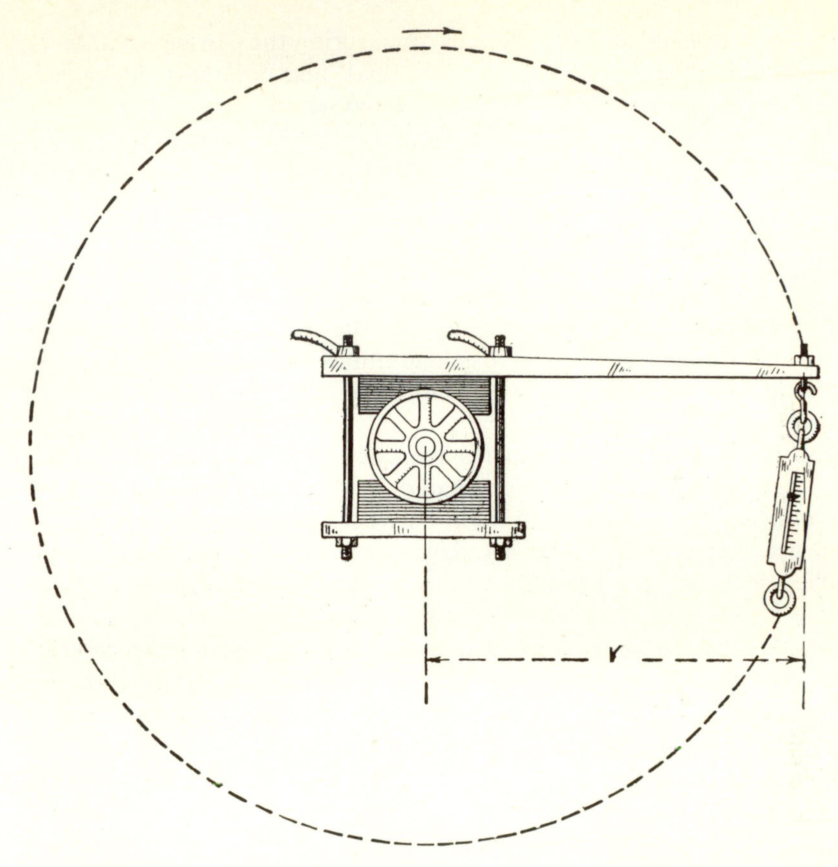

Figure 5-2 Sketch showing principle of operation of a prony brake.

brake in addition to the weight of the arm itself. Likewise, as the arm rotates, the point of application of the rotating force will describe a circle whose circumference is equal to $2\pi r$ (Fig. 5-2). The work done in one turn in foot-pounds is equal to the product of the force in pounds registered by the scales (less the weight of the brake arm) and the circumference in feet of the circle described. By counting the total revolutions made in 1 min, the total foot-pounds of work per minute can be calculated. Dividing by 33,000 gives the horsepower developed. In other words,

$$\text{hp} = \frac{\text{force on scales due to friction only (lb)} \times 2\pi r \text{ (ft)} \times \text{rpm}}{33{,}000}$$

In actual practice, the brake arm is held in a fixed position and the engine pulley turns in the band. This will not alter conditions for the reason that the friction contact will still be the same and will, therefore, create a like pressure on the scales. By measuring the distance r in feet, better known as the length of brake arm (lba), and obtaining the engine rpm, the horsepower output can be calculated as follows:

$$\text{hp} = \frac{\text{net load (lb)} \times \text{lba (ft)} \times 2\pi \times \text{rpm}}{33{,}000}$$

The net load on the scales is determined by first weighing the brake arm, with the brake loose and engine not running, and then subtracting this so-called tare load from the total load on the scales when the engine is running under test; that is, the gross load minus the weight of arm (tare load) equals the net or friction load. The crankshaft rpm is determined by means of a speed counter.

The prony brake is an inexpensive apparatus to construct and is simple and easy to operate compared with other types of brake dynamometers. However, the friction between the brake and the pulley surface generates considerable heat, depending upon the size and speed of the pulley. This heat has a tendency to cause the frictional contact and the resulting brake load to increase or vary. To eliminate or remedy this undesirable load variation it is often necessary to apply some cooling medium such as water or oil to the surfaces.

Figure 5-3 Electric-type dynamometer for determining belt horsepower.

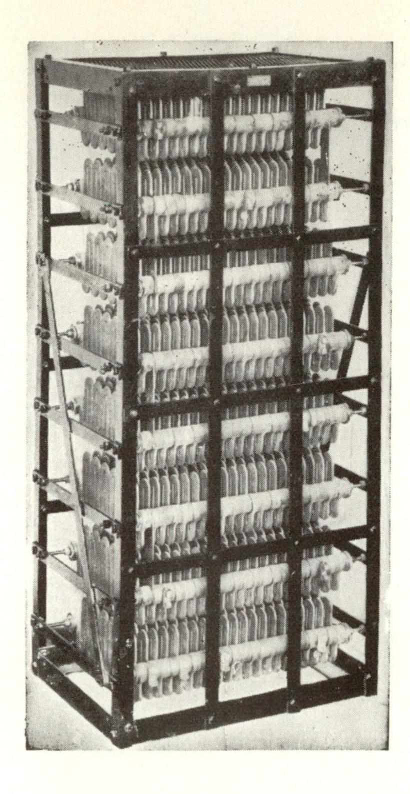

Figure 5-4 Current absorber or resistor for electric dynamometer.

Electric dynamometer For precise power measurements of large, or multiple-cylinder, variable-speed engines, the electric dynamometer is preferred. The complete outfit consists of the generating unit (Fig. 5-3), the resistor unit (Fig. 5-4), and the control and instrument board (Fig. 5-5).

The outer field frame of the direct-current generator is mounted with ball bearings on two heavy iron pedestals as shown, so that it is possible for the entire generator unit to rotate. This rotation is prevented, however, by the torque arm and scales. The generator armature rotates independently in the field frame as in any other generator. The engine to be tested is mounted beside the generator and its crankshaft connected directly to the armature; or, where that is impractical, a pulley and belt drive may be used as in testing a tractor. The load or pressure on the scales is created by the electromagnetic action between the field frame and the armature, which exists in any such machine. That is, as the armature rotates, the electromagnetic field set up tends to cause the field frame to rotate with the armature. This pulling or torque action, as it is called, does make the frame rotate a certain amount and creates a pressure on the scales attached to the torque arm. The electromagnetic field can be closely adjusted so as to increase or decrease the load on the engine by any desired amount. In other words, the stronger the current supplied to the field

Figure 5-5 Control unit for electric dynamometer.

windings, the greater the load on the engine and the resulting pull or torque. In other respects this device is very much like the prony brake. That is, by observing the rpm of the armature, the length of the torque arm, and the net load on the scales, the power output can be calculated by the same formula, namely,

$$\text{hp} = \frac{\text{net load (lb)} \times \text{lba (ft)} \times 2\pi \times \text{rpm}}{33{,}000}$$

The electric dynamometer is much more expensive to install than the prony brake and requires more experience and care in operation. However, where a large amount of accurate and precise engine testing is done, it is the most desirable apparatus to use.

Hydraulic Dynamometer The dynamometer (Fig. 5-6) operates very much like the electric eddy-current type except that the torque action on the outer housing or stator is produced by a hydraulic reaction created by water contained in the housing and surrounding the rotors. The rotation of the rotors imparts motion to the water, and this water movement in turn

Figure 5-6 Hydraulic-type dynamometer. (*Courtesy of Taylor Dynamometer and Machine Company.*)

tends to cause rotation of the stator. The energy applied to the rotors and the resultant action on the water heats the latter. Hence, the load is adjusted and controlled by controlling the temperature and the rate of flow of the water through the stator. (See also Fig. 5-7.)

Determination of Drawbar Horsepower It is often desirable to know the pulling power of a tractor under various conditions. The usual procedure is to hitch the machine to some heavy object or load to be pulled or dragged, which will require not more than the maximum pulling power of the tractor. Between the tractor and the load is placed the dynamometer, which must at least record or indicate the pulling effort in pounds required to drag the load at a certain rate of travel. The rate of travel can be determined by observing the time required to cover some definite measured distance, such as 500 ft. Then, the horsepower developed can be calculated as follows:

$$\text{hp} = \frac{\text{average pull (lb)} \times \text{rate of travel (ft/min)}}{33{,}000}$$

Figure 5-7 Portable hydraulic power take-off driven dynamometer. (*Courtesy of AW Dynamometer, Inc.*)

Drawbar dynamometers may be classified as (1) spring type; (2) distortion type; (3) hydraulic type; or (4) electric, strain-gauge type. The spring type utilizes either a heavy spiral spring or an elliptic spring to absorb the pull. The depression or elongation of the spring actuates an indicating needle or a dial showing the magnitude of the load. This type of dynamometer has a number of disadvantages that limit its use.

The distortion dynamometer consists of a special alloy steel beam that deflects in some manner under tension. The limited movement is multiplied and transmitted by a special linkage to an indicating dial. (See Fig. 5-8.)

The hydraulic dynamometer uses an oil-filled cylinder and piston arrangement by which the tension or pull creates hydraulic pressure. This pressure is read on a dial indicator and converted to pounds of draft. This type of dynamometer is well adapted to drawbar and similar kinds of draft tests. Fig. 5-11 shows a drawbar test being made at the Nebraska tractor testing laboratory. A description of the dynamometer being used is also given.

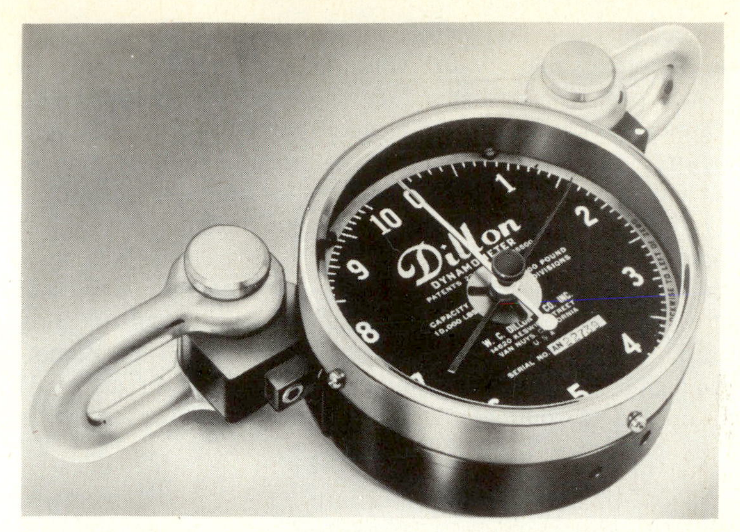

Figure 5-8 Distortion-type indicating drawbar dynamometer. (*Courtesy of W. C. Dillon and Company.*)

Strain-gauge types of dynamometers have limited application and are not well adapted to tractor testing.

ENGINE OPERATING CHARACTERISTICS

Mechanical Efficiency The mechanical efficiency of an engine is the ratio of its brake horsepower to its indicated horsepower, that is,

$$\text{Mechanical efficiency (percent)} = \frac{\text{bhp}}{\text{ihp}} \times 100$$

The principal factors affecting the mechanical efficiency of an engine are losses due to friction in the moving parts such as the crankshaft and connecting-rod bearings, pistons and cylinders, valve mechanism, and cooling fan and pump, and losses involved in the induction of the fuel mixture and the exhaust of the residue. The most practical method of measuring these losses, that is, the engine friction horsepower, is to motor the engine under normal operating conditions with an electric dynamometer or an electric motor and thereby check the actual power requirement. The mechanical efficiency of an internal-combustion engine varies from 75 to 90 percent, depending upon the load, speed, and other factors.

Volumetric Efficiency Volumetric efficiency refers to the relationship of the quantity of fuel mixture actually drawn into the cylinder by the piston on the intake stroke and producing power to the actual volume of the space displaced by the piston on this stroke and, therefore, the theoretical amount of combustible mixture which would have been taken in under ideal conditions. The principal factors affecting volumetric efficiency are (1) atmospheric pressure and temperature; (2) manifold design, such as length, size, and smoothness; (3) intake manifold temperature; (4) air-filter design and operation; (5) fuel characteristics; (6) piston speed; (7) compression ratio; (8) valve size, opening, and timing; (9) engine-operating temperature.

The volumetric efficiency of automotive-type engines should fall within a range of 75 to 85 percent, depending on their design and existing operating conditions and factors. Two methods of increasing the volumetric efficiency of high-speed multiple-cylinder engines are by the use of (1) the supercharger and (2) multiple-barrel carburetors, as described in Chap. 11.

Thermal Efficiency The thermal efficiency of an engine is the ratio of the output in the form of useful mechanical power to the power value of the fuel consumed; that is,

$$\text{Thermal efficiency (percent)} = \frac{\text{bhp}}{\text{power value of fuel}} \times 100$$

In order to determine the thermal efficiency of an engine, the quantity of fuel consumed and the power generated in a given time must be measured. Then this power and fuel must be converted into a common form; that is, the power must be converted into heat-energy units, or the fuel into mechanical-power units. The heat unit used is the British thermal unit (Btu). It has been determined that 1 Btu of heat is equivalent to 778 ft·lb of work. This is known as the mechanical equivalent of heat. Since 1 hp equals 33,000 ft·lb/min,

$$1 \text{ hp} = \frac{33{,}000}{778}$$
$$= 42.42 \text{ Btu/min}$$

and 1 hp generated for 1 h is equal to 60 × 42.42, or 2,545 Btu.

Again the heat value in Btu has been determined for the various engine fuels. Gasoline, kerosene, and other petroleum fuels contain approximately 20,000 Btu/lb (see Table 10-2). A gallon of gasoline (6.2 lb) contains 6.2 × 20,000 or 124,000 Btu. Therefore, if an engine were 100 percent efficient in its operation, that is, if the entire heat value of the fuel burned were converted into useful power, then 1 gal of gasoline would produce 124,000/2,545 or 48.7 hp for 1 h.

Such power production, however, is obviously impossible, because every engine wastes a large quantity of heat through the exhaust, through the cylinder walls and the piston, by friction, and so on.

The thermal efficiency of internal-combustion engines varies from 15 to 35 percent, depending upon the type of engine, speed, load, design, and other factors.

For example, the ordinary stationary gasoline engine uses about 0.7 lb of fuel/hp·h. The input is

$$0.7 \times 20{,}000 = 14{,}000 \text{ Btu/h}$$

The output is

$$1 \text{ hp}\cdot\text{h} = 2{,}545 \text{ Btu}$$

Then

$$\begin{aligned}\text{Thermal efficiency} &= \frac{\text{output}}{\text{input}}\\ &= \frac{2{,}545}{14{,}000} \times 100\\ &= 18.2 \text{ percent}\end{aligned}$$

Certain diesel-type engines often burn as low as 0.45 lb of fuel/hp·h. Calculating as above,

$$\begin{aligned}0.45 \times 19{,}000 &= 8550 \text{ Btu/h}\\ \text{Thermal efficiency} &= \frac{2545}{8550} \times 100\\ &= 29.8 \text{ percent}\end{aligned}$$

Torque The torque of an engine refers to the turning or twisting effort developed through its crankshaft and applied to the belt pulley or power-transmitting mechanism. Figure 5-9 shows torque curves for heavy-duty engines—one a diesel type and the other using electric ignition and gasoline. It will be noted that the torque varies considerably with engine speed and falls off appreciably at high speeds. This is probably due to less effective fuel-mixture induction and reduced volumetric efficiency. It is also noted that the diesel-torque curve is higher and somewhat more uniform than the spark-ignition curve. This explains the better lugging characteristic of a diesel engine.

Factors Affecting Fuel Consumption and Efficiency One of the fundamental objectives in the design, construction, and operation of any internal-combustion engine is to secure the greatest possible effi-

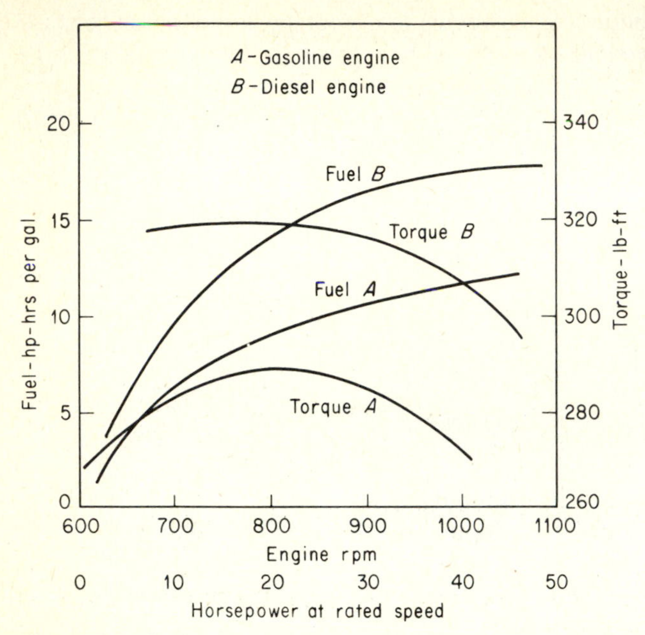

Figure 5-9 Fuel-consumption and torque curves for gasoline and diesel engines.

ciency without interfering with other considerations involved in practical adaptability to a particular purpose. As already explained, efficiency in an engine means obtaining the greatest possible power out of it with the lowest possible fuel cost—not necessarily the lowest fuel consumption. For example, an engine might burn either of two fuels satisfactorily but use slightly less of one than of the other, indicating that the first fuel is the better to use. However, if the second fuel costs considerably less per gallon, it might be the more economical one to use in the engine.

The actual fuel consumption of engines is usually expressed in pounds per horsepower-hour or in horsepower-hours per gallon. The total fuel consumption of engines varies, of course, according to the size, the power generated, and the length of time in operation. However, when reduced to the basis of pounds per horsepower-hour, the fuel consumption of two engines of entirely different size and type may be very nearly the same.

In general, the fuel consumption of all gasoline-burning engines, such as automobile, truck, and tractor engines, is about the same when reduced to the basis of pounds per horsepower-hour, provided they are all operated under a one-half to full load. Such engines seldom burn less than 0.55 lb of fuel/hp · h. The average is around 0.60 lb, and the rate may run as high as 0.70 lb. Diesel-type and similar high-compression heavy-duty oil-

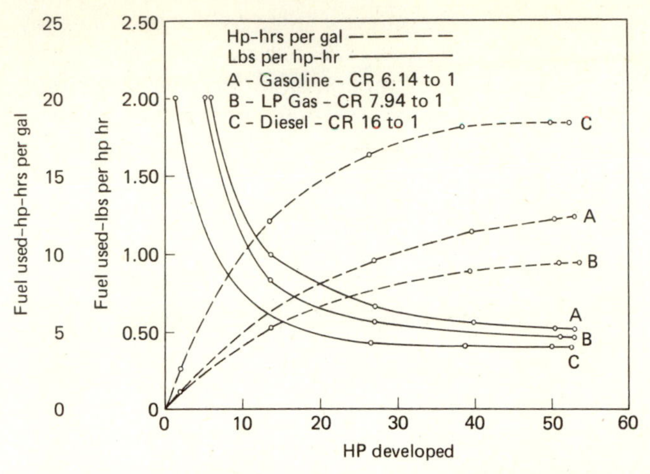

Figure 5-10 Fuel-consumption curves for different kinds of fuels.

burning engines often show a fuel consumption of 0.4 to 0.5 lb hp·h and seldom use more than 0.55 lb (Fig. 5-10).

Given a certain type of engine, burning a given fuel, the most important factors affecting its economical and efficient operation are:

1 Normal operating compression pressure
2 Operating load: light, medium, or heavy
3 Mechanical condition:
   a Ignition correctly timed
   b Valves correctly timed
   c Fuel mixture properly adjusted
   d Piston rings and cylinder not badly worn
   e Bearings properly adjusted
   f Properly lubricated

Compression and Efficiency Referring to Table 6-1 and Fig. 6-9, it is observed that, theoretically, the greater the compression pressure, the higher the thermal efficiency. It is difficult and impractical to take advantage of this, however, in the carbureting type of engine, such as the farm, automobile, or tractor engine, for the reason that too high a compression causes detonation of the fuel mixture. In other words, the high compression produces a higher cylinder temperature and the mixture ignites and burns rapidly and explodes too early. In the diesel and similar types of heavy-duty oil engines, high compression is practical because the fuel charge does not enter the cylinder until the piston is ready to receive the explosion. Engines of this type, therefore, are somewhat more efficient than carbureting engines.

Effect of Load on Efficiency The curves (Fig. 5-9) show that any engine, when operating at a very light load, will use more fuel/per horsepower-hour. As the load increases, the fuel consumption decreases until, at about nine-tenths of the maximum power developed by the engine, it gives the most economical results. It is obvious that a certain amount of fuel is required to operate the engine itself, that is, to supply the power necessary to overcome friction. Furthermore, this friction and the amount of fuel required to overcome it remain practically constant, regardless of the load on the engine. Therefore, as the load increases, this quantity of fuel required to overcome friction becomes less and less in proportion to the total amount burned.

This characteristic of engines to give the best fuel economy at medium to heavy loads means that it is important always to use an engine that fits the job. That is, a 20-hp engine should not be used to operate a machine requiring only 5 hp. On the other hand, it is not good practice to use an engine that is too small and will be overloaded.

TRACTOR POWER RATING—NEBRASKA TRACTOR TESTS

The present-day farm tractor is designed to deliver power in three distinct ways, namely, (1) by a pulley and belt, (2) by pulling effort at the drawbar, and (3) by means of a power take-off.

Power Rating As a rule, tractor sizes are designated according to the power capable of being generated at the power take-off and at the drawbar. Obviously, the maximum power that can be generated by the power take-off will be practically the same as the belt power. However, when a tractor is operating a machine through the power take-off, the machine is usually being pulled. Therefore, only a certain fraction of the engine power is available at each point.

The usual practice of tractor manufacturers is to designate their various sizes by numbers or letters such as Model 520, D-12, and so on, rather than by the actual power output. The specific available power take-off and drawbar power of any model is included in the detailed specifications. In general, tractors deliver 75 to 90 percent of their power take-off power at the drawbar.

Nebraska Tractor Law and Tests

A law known as the "Nebraska tractor law," which was enacted in that state and put into effect in 1919, has been of far-reaching effect, particularly in the design, construction, and operation of the farm tractor. The provisions of the law, the tests involved, and the results are outlined in detail in the discussion that follows.

Purpose of the Law The Nebraska tractor law, as stated in the Revised Statutes of Nebraska, 1943, Chapter 75, Article 9, Sections 75-901 to 75-911, inclusive, as amended in 1949, was enacted to encourage the manufacture and sale of improved tractors and to contribute to a more successful use of the tractor for farming.

It was thought that the best method of accomplishing these objectives would be to require a tractor of each model sold in the state to be tested at the University of Nebraska and to have the results of these tests made public.

Provisions of the Law Stated briefly, the provisions of the law are:

1 That a stock tractor of each model sold in the state shall be tested and passed upon by a board of three engineers under state university management.

2 That each company, dealer, or individual offering a tractor for sale in Nebraska shall have a permit issued by the state railway commission. The permit for any model of tractor shall be issued after a stock tractor of that model has been tested at the University and the performance of the tractor compared with the claims made for it by the manufacturer.

3 That a service station with full supply of replacement parts for each model of tractor shall be maintained within the confines of the state and within reasonable shipping distance of customers.

Test Procedure

General Tractors are tested at the University of Nebraska according to the Agricultural Tractor Test Code approved by the American Society of Agricultural Engineers (ASAE) and the Society of Automotive Engineers (SAE) or official Nebraska test procedure.

The manufacturer selects the tractor to be tested and certifies that it is a stock model. Each tractor is equipped with the usual power consuming accessories such as power steering, power lift pump, generator, etc., if available. Power consuming accessories may be disconnected only when the means for disconnecting can be reached from the operating station. An official representative of the company is present during the test to see that the tractor gives its optimum performance. Additional weight may be added to the tractor as ballast if the manufacturer recommends use of such ballast. The static tire loads and the tire inflation pressures must conform to the Tire Standards published by the ASAE and SAE.

Preparation for Test The engine crankcase is drained and refilled with new oil conforming to specifications in the operator's manual. The operator's manual is also used as the guide for selecting the proper fuel and for routine lubrication and maintenance operations.

The tractor is limbered up for 12 hours on the drawbar, using each gear with light to heavy loads during the limber-up period. Preliminary adjustment of the tractor is permitted at this time. Any parts added or replaced during the limber-up run, or any subsequent runs, are mentioned in the individual test reports. The tractor is equipped with approximately the amount of added ballast that is to be used during the drawbar runs.

Power Take-off (PTO) Performance Power take-off performance runs are made by connecting the power take-off (or the belt pulley if no power take-off is available) to a dynamometer. During a preliminary power take-off run the manufacturer's representative may make adjustments for the fuel, ignition or injection timing, and governor control settings. These settings must be maintained for the remainder of the test. The manually operated governor control mechanism is set to provide the high-idle speed specified by the manufacturer. During the power take-off runs an ambient air temperature of approximately 75°F is maintained.

Maximum power is obtained at the rated engine speed specified by the manufacturer, with the governor control lever set for maximum power. This same setting is used for all subsequent PTO runs. Time of the run is two hours. Whenever the power take-off speed during the maximum power run differs from the speeds set forth in the ASAE and SAE standards, an additional run is made at either 540 or 1,000 rpm of the power take-off shaft. Time of this run is one hour.

Drawbar Testing Drawbar pull is determined by means of a hydraulic cylinder inserted in the hitch between the tractor and the load. The drawbar pull exerted on a piston in this hydraulic cylinder creates a pressure in the cylinder. This pressure is carried through flexible tubing to the draft-recording instrument. The essential mechanism of this draft-recording unit is an engine-pressure indicator that has been so modified and calibrated that it leaves on a chart a continuous record of the drawbar pull of the tractor. From this chart, showing the pull in pounds, and from electronic timer readings showing the time of travel over a measured course 500 ft long, the drawbar horsepower developed is calculated. Figure 5-11 shows a tractor at the University of Nebraska being tested for drawbar power.

In determining slippage, the number of revolutions made by the drive wheels is accurately determined while the tractor is driven with no drawbar load over a 500-ft distance. Rotary switches having 10 contact points are attached to each drive wheel and are connected to magnetic counters that record to one-tenth of one revolution the number of revolutions made over the 500-ft distance. The difference between the actual number of revolutions recorded by the counter and the no-load count, divided by the former and the result multiplied by 100, is the percent of slippage. When speaking of rubber-tire tests, some prefer to use the term travel reduction.

Figure 5-11 Testing a tractor for drawbar power at the University of Nebraska.

The engine speed is calculated by multiplying the wheel count by the gear ratio and dividing by the time.

In determining the drawbar horsepower of a tractor, it is necessary to provide a source of load that may be readily varied. The load is obtained by placing the loading unit in gear and operating its engine against compression without fuel or ignition. The load may be increased or decreased by closing or opening valves that have been installed on the exhaust manifolds of the engine. If a light load is desired, the tractor may be placed in gear, the fuel and ignition turned on, and the throttle set to give such assistance as will result in the desired load. Additional loading units consisting of five tractors can be added to the load unit on the test car. The current instrument or test car was built in 1963 but several changes have been made in the power unit and transmission since then. The test car provides protection from the weather and dust for the recording instruments and the operator. Some of the instruments included in the car are the draft-recording instrument, two temperature indicators showing the tractor cooling-medium temperature and the air temperature, the wheel counters, and an engine speed indicator. The rear cab contains the sound-level analyzing equipment.

For all drawbar tests at least two suitable drawbar runs are obtained over a measured distance of 500 ft on the testing course. A suitable run is one in which the operating temperatures are normal, the load is relatively constant, and the average engine speed is at or very close to that specified by the manufacturer. Very seldom is a pull used in which the variation from rated speed is more than 1 percent, and in each test in each gear an attempt is made to obtain an average number of revolutions per minute that does not deviate from rated speed more than one-half of one percent.

Sound Measurement Sound is recorded during each of the varying power and fuel consumption runs as the tractor travels on a straight section of the test course. The dB(A) sound level is obtained with the microphone located near the right ear of the operator. Bystander sound-readings are taken with the microphone placed 25 ft from the line of travel of the tractor. An increase of 10dB(A) will approximately double the loudness to the human ear.

Test Report

The following is a representative test report as prepared and issued upon completion of a test:

NEBRASKA TRACTOR TEST 1295 — CASE 2090 POWERSHIFT DIESEL

POWER TAKE-OFF PERFORMANCE

Hp	Crank-shaft speed rpm	Fuel Consumption Gal per hr	Lb per hp-hr	Hp-hr per gal	Temperature Degrees F Cooling medium	Air wet bulb	Air dry bulb	Barometer inches of Mercury
MAXIMUM POWER AND FUEL CONSUMPTION								
Rated Engine Speed—Two Hours (PTO Speed—998 rpm)								
108.29	2100	7.247	0.463	14.94	204	57	75	28.587
VARYING POWER AND FUEL CONSUMPTION—Two Hours								
95.28	2178	6.618	0.481	14.40	195	57	76	......
0.00	2317	2.520	......	......	179	56	72	......
49.52	2257	4.467	0.624	11.09	186	56	73	......
109.90	2100	7.411	0.467	14.83	203	58	76	......
25.05	2283	3.448	0.952	7.27	181	56	72	......
73.13	2224	5.507	0.521	13.28	18	58	76	......
Av. 58.81	**2226**	**4.995**	**0.588**	**11.77**	**189**	**57**	**74**	**28.980**

DRAWBAR PERFORMANCE

Hp	Draw-bar pull lbs	Speed miles per hr	Crank-shaft speed rpm	Slip of drivers %	Fuel Consumption Gal per hr	Lb per hp-hr	Hp-hr per gal	Temp Degrees F Cool-ing med	Air wet bulb	Air dry bulb	Barometer inches of Mercury
Maximum Available Power—Two Hours 8th (3I) Gear											
89.40	5106	6.57	2101	4.89	7.155	0.554	12.49	198	53	68	28.690
75% of Pull at Maximum Power—Ten Hours 8th (3I) Gear											
74.71	4024	6.96	2199	3.70	6.245	0.578	11.96	190	54	62	28.536
50% of Pull at Maximum Power—Two Hours 8th (3I) Gear											
51.67	2678	7.24	2257	2.36	5.095	0.682	10.14	184	44	49	28.695
50% of Pull at Reduced Engine Speed—Two Hours 10th (4L) Gear											
52.10	2696	7.25	1481	2.41	3.957	0.525	13.17	185	47	54	28.720
MAXIMUM POWER IN SELECTED GEARS											
87.94	11211	2.94	2103	14.84	4th (2L) Gear			194	59	65	28.520
92.05	8191	4.21	2100	8.55	5th (2I) Gear			200	57	71	28.640
93.60	7302	4.81	2099	7.21	6th (3L) Gear			200	56	70	28.640
90.36	6262	5.41	2100	5.99	7th (2H) Gear			201	56	70	28.640
93.95	5369	6.56	2101	4.97	8th (3I) Gear			191	55	70	28.650
91.70	4138	8.31	2101	3.77	9th (3H) Gear			199	56	70	28.640

LUGGING ABILITY IN RATED GEAR (8th) (3I)

Crankshaft Speed rpm	2101	1893	1681	1474	1256	1042
Pounds Pull	5369	5914	6259	6275	6040	5766
Increase in Pull %	0	10	17	17	12	7
Horsepower	93.95	92.76	86.83	76.25	62.61	49.62
Miles Per Hour	6.56	5.88	5.20	4.56	3.89	3.23
Slip of Drivers %	4.97	5.68	5.68	5.99	5.83	5.68

TRACTOR SOUND LEVEL WITH CAB	dB(A)
Maximum Available Power 2 Hours	78.0
75% of Pull at Max. Power 10 Hours	77.5
50% of Pull at Max. Power 2 Hours	78.0
50% of Pull at Reduced Engine Speed 2 Hours	73.0
Bystander in 12th (4H) Gear	88.5

TIRES, BALLAST AND WEIGHT		With Ballast	Without Ballast
Rear Tires	—No., size, ply & psi	Two 20.8-38; 8; 18	Two 20.8-38; 8; 18
Ballast	—Liquid (each)	1465 lb	None
	Cast Iron (each)	None	None
Front Tires	—No., size, ply & psi	Two 11.00-16; 8; 40	Two 11.00-16; 8; 40
Ballast	—Liquid (each)	None	None
	Cast Iron (each)	115 lb	None
Height of drawbar		21 in	21 in
Static weight with operator—rear		11640 lb	8710 lb
front		3550 lb	3320 lb
total		15190 lb	12030 lb

Department of Agricultural Engineering
Dates of Test: November 6-13, 1978.
Manufacturer: J. I. Case, 700 State Street, Racine, Wisconsin 53404.

FUEL, OIL AND TIME: Fuel No. 2 Diesel **Cetane No.** 50.4 (rating taken from oil company's typical inspection data) **Specific gravity converted to 60°/60°** 0.8309 **Fuel weight** 6.918 lbs/gal **Oil SAE** 30 **API service classification** SE-CD **To motor** 4.088 gal **Drained from motor** 3.882 gal **Transmission and final drive lubricant** Case TFD fluid **Total time engine was operated** 36.5 hours.

ENGINE Make Case Diesel **Type** 6 cylinder vertical **Serial No.** 10159810 **Crankshaft** lengthwise **Rated rpm** 2100 **Bore and stroke** 4.625" x 5.00" **Compression ratio** 16.0 to 1 **Displacement** 504 cu in **Cranking system** 12 volt **Lubrication** pressure **Air cleaner** two paper elements **Oil filter** one full flow cartridge **Oil cooler** radiator for hydraulic and transmission oil **Fuel filter** two paper cartridges **Muffler** vertical **Cooling medium temperature control** 2 thermostats.

CHASSIS: Type Standard **Serial No.** 8835687 **Tread width** rear 64" to 88" front 60" to 88" **Wheel base** 104" **Center of gravity** (without operator or ballast, with minimum tread, with fuel tank filled and tractor serviced for operation) Horizontal distance forward from center-line of rear wheels 29.0" Vertical distance above roadway 40.7" Horizontal distance from center of rear wheel tread 0" to the right/left **Hydraulic control system** direct engine drive **Transmission** selective gear fixed ratio with partial (3) range operator controlled power shift **Advertised speeds mph** first 2.0 second 2.6 third 3.3 fourth 3.4 fifth 4.5 sixth 5.0 seventh 5.6 eighth 6.7 ninth 8.4 tenth 10.2 eleventh 13.6 twelfth 18.8 reverse 3.3, 5.6, 8.4 **Clutch** multiple wet disc hydraulically operated by foot pedal **Brakes** multiple disc hydraulically operated by two foot pedals which can be locked together **Steering** hydrostatic **Turning radius** (on concrete surface with brake applied) right 142" left 142" (on concrete surface without brake) right 167" left 167" **Turning space diameter** (on concrete surface with brake applied) right 296" left 296" (on concrete surface without brake) right 347" left 347" **Power take-off** 998 rpm at 2100 engine rpm and 534 rpm at 2100 engine rpm.

REPAIRS AND ADJUSTMENTS: No repairs or adjustments.

REMARKS: All test results were determined from observed data obtained in accordance with SAE and ASAE test code or official Nebraska test procedure. Temperature at injection pump was 191°F. Six gears were chosen between 15% slip and 10 mph.

We, the undersigned, certify that this is a true and correct report of official Tractor Test **1295.**

LOUIS I. LEVITICUS
Engineer-in Charge
G. W. STEINBRUEGGE, Chairman
W. E. SPLINTER
K. VON BARGEN
Board of Tractor Test Engineers

The Agricultural Experiment Station
Institute of Agriculture and Natural Resources
University of Nebraska—Lincoln
H. W. Ottoson, Director

IMPLEMENT & TRACTOR, January 31, 1979 A-177

PROBLEMS AND QUESTIONS

1 Distinguish between the terms inertia, momentum, and kinetic energy.

2 Distinguish between the terms torque and work.

3 Define the term horsepower and explain the origin of the unit's actual value.

4 What size of engine (horsepower rating) would you recommend to operate a 120-volt, direct-current electric generator having a maximum output of 75 A and an operating efficiency of 80 percent?

5 A four-cylinder, four-stroke-cycle engine with 3- by 4-in cylinders develops 19 hp at 1,650 rpm. Assuming a mechanical efficiency of 85 percent, compute the ihp and mep.

6 Compute the thermal efficiency of the engine described in Prob. 5 if it used 47 lb of gasoline in a 4-h test.

7 A tractor with a 9-in pulley is belted to a prony brake having a 24-in pulley. If the engine-pulley speed is 950 rpm, the brake arm length is 54 in, and the net load on the scales is 60 lb, what is the value of the brake constant and what horsepower is developed?

8 Calculate the drawbar horsepower required to pull a plow with three 14-in bottoms at a rate of 3.25 mph if the draft is 8 lb/in^2 of furrow section and the depth of cut is 7 in.

9 A drawbar dynamometer shows that the average pull required for a certain machine is 2,800 lb. If the tractor travels 1,000 ft in 3½ min, what is the horsepower developed? What is the rate of travel in miles per hour?

10 If an engine shows an average output at the crankshaft of 35 hp at 1,650 rpm, what torque is exerted?

11 Referring to the curves on Fig. 5-10, compare the cost per 10-h day of operating a tractor on LP gas, gasoline, and diesel fuel if the average load is 30 hp and the fuel cost per gallon is 55, 90, and 85 cents for the respective fuels.

12 State the reasons for the passage of the Nebraska tractor law and name its basic provisions.

13 Three major kinds of tests are made in compliance with the Nebraska tractor law. What are they?

14 To what extent, if any, do representatives of the manufacturers participate in carrying on the testing of their tractors under the Nebraska tractor law?

Chapter 6

Thermodynamic Principles and Applications—Engine Cycles and Efficiencies

The following symbols are used in this chapter (and throughout the text):

A	area	k	the ratio c_p/c_v
C	a constant	P	pressure or total pressure
c	specific heat	p	unit pressure (lb/in^2)
c_p	specific heat at constant pressure	Q	transferred heat
c_v	specific heat at constant volume	R	specific gas constant (pv/T)
D	distance or diameter	T	absolute temperature (°F.)
e	thermal efficiency	V	volume
F	force or total load	W	work
J	Joule's constant (778)	w	weight of substance

Since an internal-combustion engine generates power from the heat energy of certain gaseous fuels, its operation is based specifically upon certain physical principles and theories as applied to gases. The basic factors entering into the energy and power generated by a combustible gaseous mixture are (1) volume, (2) pressure, and (3) temperature. Scientists have been able to establish certain definite relationships and reactions between these. An understanding of the relationship of these factors to each other is important in establishing a clear understanding of the fundamentals involved in the design and efficient operation of an internal-combustion engine.

Thermodynamics Thermodynamics is defined as that phase of physical science that is related to the conversion of heat into mechanical force or energy, or vice versa, and the specific factors and relationships involved. Two basic principles known as the two laws of thermodynamics have been evolved and must be well understood in any discussion of heat and its relationship to mechanical energy. The first law, basically, is the law of the conversion of energy and states that "when heat energy is transformed into mechanical energy, the work done is equivalent to the quantity of heat involved." Joule, an English scientist, made extensive studies of this hypothesis and established the presently accepted value of 778 ft·lb of work as being equivalent to 1 Btu. This is known as the mechanical equivalent of heat or Joule's constant.

The second law of thermodynamics states that heat will, of itself, pass from a hot to a cold substance, but external work is required to transfer heat from a cold substance to a hot substance. For example, the operation of the heat engine is based upon a flow of heat from a hot to a cold substance, but a mechanical refrigerating machine requires external energy because it transfers heat away from a cold material and into a hotter substance.

Boyle's Law Suppose a cylinder is filled with a gas such as air or natural gas or some common fuel vapor and then a piston is inserted in the cylinder which compresses the gas into one end of the cylinder without leakage or loss. The volume of the gas is decreased and there is an increase in its pressure. Likewise its temperature is increased on account of the energy expended in producing compression.

Not only is there an increase in pressure with a decrease in volume of a gas, but there is a definite pressure-volume relationship for any gas. This relationship was first expressed by Boyle and is known as *Boyle's law* as applied to gases. It states that, if the temperature of a gas is kept constant, its volume will vary inversely as the pressure to which it is subjected. That is, $PV = C$, C being a constant depending upon the kind of gas, initial pressure, temperature, and other factors. Or if V_1 represents the volume of a gas at a pressure P_1 and V_2 represents the volume of the same gas when the pressure is P_2, it follows that

$$P_1V_1 = P_2V_2 = C$$

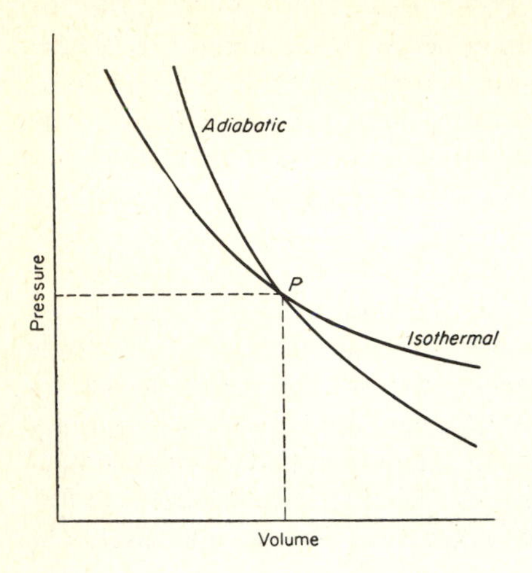

Figure 6-1 Pressure-volume diagram showing adiabatic and isothermal changes and relationships.

The change of a gas with respect to pressure and volume, the temperature remaining constant, is known as an *isothermal change*. If a curve is drawn showing this *PV* relationship of a gas at any time, it is called an *isothermal curve* (Fig. 6-1).

Charles' Law Normally heat is generated when a gas is compressed and heat is released when its volume is increased. Or we can say that the volume of a gas increases when its temperature is raised and decreases with a drop in temperature. Again, let us assume that a free-moving piston is inserted in a tight cylinder containing a gas and that heat is applied to the gas. Increased pressure will be developed and the piston will move outward. Now if the piston is allowed to move in such a manner that there is no actual increased pressure built up behind it, it will be found that there is a definite relationship between the increase in the volume of the gas and the rise in temperature. In fact, experiments show that when a quantity of gas is heated or cooled while the pressure to which it is subjected remains constant, the volume will change $^1/_{492}$ of its volume at 32°F. or $^1/_{273}$ of its volume at 0°C. for each degree of temperature change. Therefore, if a mass of gas were cooled to −273°C. (273 degrees below 0°C.), the gas would contract to zero volume, provided the same law held true to this point. This we may assume to be the case with a perfect gas, although any known gas would change its state before this temperature were reached.

In the study of gases it was also found that if a gas were heated without allowing it to expand, that is, with its volume remaining constant, then the pressure increased $^1/_{273}$ of its value for each degree rise in temperature above 0°C. If the pressure of the gas is measured at 0°C. and this rate of decrease in pressure continues as the temperature is gradually lowered, it is clear that when the temperature has become $-273°$ the pressure will disappear and the gas will no longer be able to exert a pressure. From the standpoint of the kinetic theory of gases, a gas must always exert a pressure as long as the molecules are in motion. Hence, the temperature at which these molecules cease to exert a pressure is the temperature at which these molecules cease to move. The temperature at which the molecules have no motion is known as absolute zero and is $-273°$C. or $-460°$F. Temperatures when measured from these points as zero are known as *absolute temperatures.*

This temperature-volume-pressure relationship of a gas is known as *the law of Charles* and is stated as follows: The pressure remaining constant, the volume of a mass of gas is directly proportional to its absolute temperature. It may also be stated in the following way: The volume remaining constant, the pressure of a gas is directly proportional to its absolute temperature.

From the above, if the volume V_1 of a gas changes to the volume V_2 when the absolute temperature T_1 changes to the value T_2, the pressure remaining constant, Charles' law may be expressed as

$$V_1T_2 = V_2T_1$$

and if the pressure P_1 of a gas changes to the pressure P_2 with a change of absolute temperature from T_1 to T_2, the volume remaining constant, the law may be expressed as

$$P_1T_2 = P_2T_1$$

Boyle's and Charles' Laws Combined We have seen that according to Boyle's law $P_1V_1 = P_2V_2$ (T being constant) and according to Charles' law $V_1T_2 = V_2T_1$ (P being constant). Now if the volume V_1 of a gas changes to a new volume V_2, while at the same time the pressure P_1 and the absolute temperature T_1 change to new values P_2 and T_2 respectively, the final relation between the volume, the pressure, and the absolute temperature of the gas may be determined by conceiving that the change of volume due to the change in pressure and the change in volume due to the change in absolute temperature take place independently and successively. Referring to Fig. 6-2, let us assume that the initial pressure P_1 is

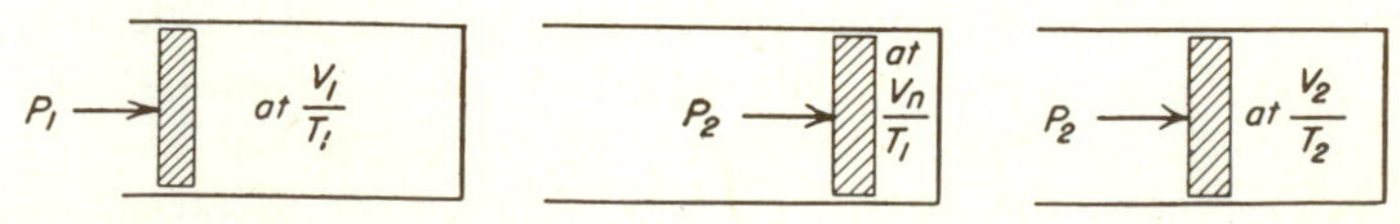

Figure 6-2 Boyle's and Charles' laws functioning independently and successively.

changed to P_2 while the volume V_1 changes to V_n with no change in temperature. Then, according to Boyle's law,

$$P_1V_1 = P_2V_n \qquad (1)$$

But now suppose the temperature changes from T_1 to T_2 and the volume V_n changes to V_2, with no change in pressure. Then, according to Charles' law,

$$V_nT_2 = V_2T_1 \qquad (2)$$

Combining Eqs. (1) and (2), we find that

$$\frac{P_1V_1}{T_1} = \frac{P_2V_2}{T_2}$$

and $\frac{PV}{T} = R$ a constant

or $PV = RT$

That is, the product of the pressure and the volume of a gas is always proportional to its absolute temperature.

If P = absolute pressure, lb/ft²
V = volume of gas, ft³
W = weight of gas, lb/ft³
T = temperature, °F abs
R = gas constant

then, for any gas,

$$PV = WTR$$

and for air, if

$P = 144 \times 14.7$
$V = 1$
$W = 0.0807$
$T = 492$°F. abs

then $R = \frac{144 \times 14.7}{0.0807 \times 492} = 53.3$

Heat and Energy Ordinarily heat is thought of as something created by the rapid combustion or burning of some material. However, it was discovered many years ago that there was a definite relationship between heat and mechanical energy. For example, when friction is developed by two objects rubbing against each other, the work done in breaking down the irregularities between the two surfaces is transformed into an increased kinetic energy of the particles, and the heat thus produced is a measure of the work done against friction. When flint and steel are used to "strike fire," the work done by the flint in tearing away little particles of the steel is so great that these particles are heated to a temperature at which they burn in the air. These and other simple experiments prove that heat is produced as a result of work performed; therefore, it is a form of energy. It is now considered that the heat of a body is the kinetic energy of its ultimate particles, that is, of its electrons and its molecules, and that the temperature of the body is simply a measure of this kinetic energy.

Unit Quantity of Heat—The Calorie and the British Thermal Unit Temperature is a measure of the kinetic energy of the molecules of a body but is entirely independent of the number of these molecules, that is, of the mass of the body. On the other hand, the quantity of heat a body contains denotes the total kinetic energy of its particles and depends upon the mass of the body.

It has been found that the quantity of heat required to raise the temperature of a given mass of one substance to a certain degree will raise the temperature of the same mass of another substance to an entirely different degree. Therefore, in choosing the unit quantity of heat, water was selected as the standard material. The unit is known as a *calorie* and is defined as the quantity of heat that will raise the temperature of 1 g of water 1°C. In the English system, the unit is called the *British thermal unit* (Btu) and is defined as the quantity of heat required to raise the temperature of 1 lb of water 1°F.

Specific Heat The *specific heat* of a substance is defined as the ratio between the quantity of heat required to raise the temperature of any mass of that substance one degree and the quantity required to raise the temperature of an equal mass of water one degree.

In the case of gases, the specific heat is affected by the conditions of pressure and volume when heating takes place. For example, if the gas is confined in a tight vessel and the volume remains constant, the effect of the heat is confined to increasing the kinetic energy of the molecules of the gas, which in turn raises its temperature. The number of Btu required to raise the temperature of 1 lb of gas 1°C at constant volume is called its *specific heat at constant volume* or c_v. If the heat applied to a gas causes expansion and an increase in its volume but the pressure remains constant, external work is done. If 1 lb of gas is thus heated and its temperature is increased 1°C, the amount of heat supplied is known as the *specific heat at constant pressure* c_p. The ratio of these two specific heats, which increases with the pressure, is constant for all diatomic gases at atmospheric pressure. That is, for such gases at atmospheric pressure,

$$\frac{c_p}{c_v} = 1.4$$

The difference between these two specific heats for any given gas must be equal to the heat equivalent of the external work done when a unit weight of the gas is raised one degree at a constant pressure.

Adiabatic Compression Boyle's law states that, the temperature remaining constant, the volume of a gas varies inversely as the pressure to which it is subjected. This is known as *isothermal compression or expansion*. However, as previously explained, according to Charles' law, an increase in the temperature of a gas at a constant volume gives an increase in pressure. Conversely a change in the volume of the gas results in a change of temperature. For example, let us assume that a gas is enclosed in a cylinder provided with a piston and that the walls of the cylinder, as well as the piston, are absolute nonconductors of heat so that, whether the temperature of the gas is high or low, no heat can pass through the walls. If, therefore, the gas is compressed in such a cylinder, the heat produced by the compression remains in the gas itself and so raises its temperature, and if the gas expands, the cold produced by the expansion lowers the temperature of the gas. The compression or expansion of a gas when enclosed in such a cylinder is known as *adiabatic*.

In general, adiabatic changes are possible only when the system is enclosed by a non-heat-conducting material. Rapid changes of condition are approximately adiabatic, since time is required for conduction and radiation of heat; thus the compression of air in an "air-cooled" compressor cylinder is practically adiabatic, as the time is so short that little heat can escape through the cylinder walls.

According to Boyle's law, $PV = C$ (temperature remaining constant). However, if a gas is subjected to adiabatic compression, the change in temperature also affects the pressure produced and an additional factor must be introduced. This factor k depends upon the specific heat of the material, that is, $k = c_p/c_v$. Therefore, for adiabatic compression, $PV^k = C$. The exponent k varies with the initial pressure and the kind of gas.

For example, for air,

If initial pressure is 11½ lb/in² (abs), $k = 1.21$.
If initial pressure is 13 lb/in² (abs), $k = 1.29$.
If initial pressure is 14 lb/in² (abs), $k = 1.34$.
If initial pressure is 14.7 lb/in² (abs), $k = 1.40$.

For other gases, as those used in internal-combustion engines, the value of k may be lower because the ratio of the specific heats is different from what it would be for air and because of losses from leakage past the pistons and valves and heat loss through the cylinder walls. In general practice, the equation of a curve representing the adiabatic change of a gas is

$$P_1V_1^k = P_2V_2^k \qquad \text{etc.}$$

and $$P_2 = P_1\left(\frac{V_1}{V_2}\right)^k$$

Also, if

$$P_1V_1^k = P_2V_2^k$$

then $$TV^{k-1} = C$$

or $$\frac{T_1}{T_2} = \left(\frac{V_2}{V_1}\right)^{k-1}$$

Compression Ratio and Pressure The compression ratio of an engine is equal to the total cylinder volume divided by the clearance volume, or $V_1/V_2 = r$. The greater this ratio the higher the compression pressure in the cylinder; that is $P_2 = P_1V_1/V_2$, or, for adiabatic conditions, $P_2 = P_1(V_1/V_2)^k$. By the use of this formula we may compute the theoretical maximum compression pressure of an engine for certain assumed conditions.

For example, if $P_1 = 14.7$ lb (atmospheric pressure)
$V_1 = 18 \text{ in}^3$
$V_2 = 3 \text{ in}^3$
$k = 1.4$

then $$P_2 = 14.7(18/3)^{1.4} = 180.7 \text{ lb/in}^2 \qquad \text{absolute pressure}$$

or $$180.7 - 14.7 = 166 \text{ lb/in}^2 \qquad \text{gauge pressure}$$

Under most conditions in an engine the initial pressure would be less than the normal atmospheric pressure of 14.7 lb and the exponent k would be about 1.3.

Work and Gas Pressure If a volume of gas is enclosed in a cylinder behind a piston whose area is A and the total gas pressure is P, then $P = Ap$ (p is unit of force per unit area). If the piston is forced inward a distance D and the gas is compressed, the value of p increases. Assuming that the average pressure exerted on the piston during its movement through D is pA and that p is the average unit pressure of the gas, the work done is pAD. But AD represents the decrease in volume of the gas. Letting V equal this volume, $pAD = Vp$; that is, the work done in compressing a gas equals the product of the change in volume and the average unit pressure of the gas during compression, or work done is equal to $p(V_1 - V_2)$.

We have also observed that when a gas is heated at constant volume the increase in internal energy is in proportion to the heat supplied and is equal to $c_v(T_2 - T_1)$ Btu. If the gas is heated at a constant pressure, the total heat supplied is equal to $c_p(T_2 - T_1)$ Btu and includes the heat needed to raise the temperature at a constant volume plus an additional quantity of heat which is expended in doing external work because the pressure remains constant. Therefore, $c_p(T_2 - T_1)$ Btu is greater than $c_v(T_2 - T_1)$ Btu, and the net energy output is equal to

$$(c_p - c_v)(T_2 - T_1) \text{ Btu}$$

Furthermore, for any gas, $778(c_p - c_v) = R$ (gas constant).

The work done by a gas under pressure can also be represented by graphs (Figs. 6-3 and 6-4). In Fig. 6-3 it is assumed that P is constant at all volumes and the volume changes from V_1 to V_2 as represented by line CD. The product of P and CD is represented by area $ABCD$, which, in turn, represents the work done by the gas when it expands and moves the piston.

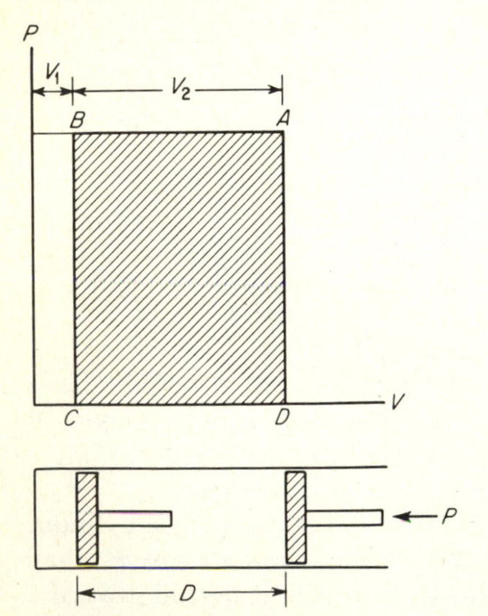

Figure 6-3 Constant-pressure cycle.

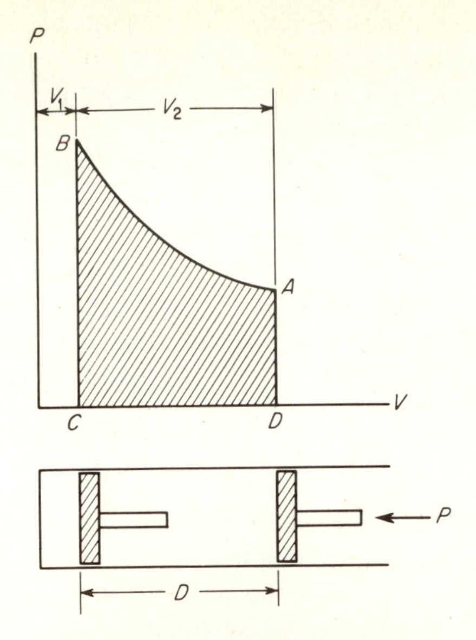

Figure 6-4 Variable-pressure cycle.

In the operation of most heat engines the gas pressure P is variable, as shown by Fig. 6-4; that is, the P curve drops from B to A as the volume increases. Again, the work done during the stroke is represented by the area $ABCD$, which, in turn, is the product of CD and the mean height of curve AB.

Carnot Cycle In 1824, Sadi Carnot, a French physicist, proposed an operating cycle for what may be considered as the ideal heat engine. Referring to Fig. 6-5, let us assume the following:

1 The cylinder and piston assembly C are made of a material that will not conduct heat with the exception of the base which is a perfect conductor.

2 The cylinder contains a perfect gas as the working substance.

3 A and B are heat reservoirs whose temperatures are maintained constant, but T_A is relatively high and T_B relatively low.

4 S is an insulating stand which is completely non-heat-absorbing or transmitting.

5 The engine parts are without friction.

Theoretically, the cycle would function as follows: With cylinder C sitting on heat source A and the piston near the bottom, the gas at temperature T_A occupies a small volume at a relatively high pressure, as indicated on the PV diagram (Fig. 6-6). Now, if the pressure is reduced slightly so that the gas can expand, a quantity of heat Q_1 flows into it from A and maintains a constant temperature. Therefore, isothermal expansion oc-

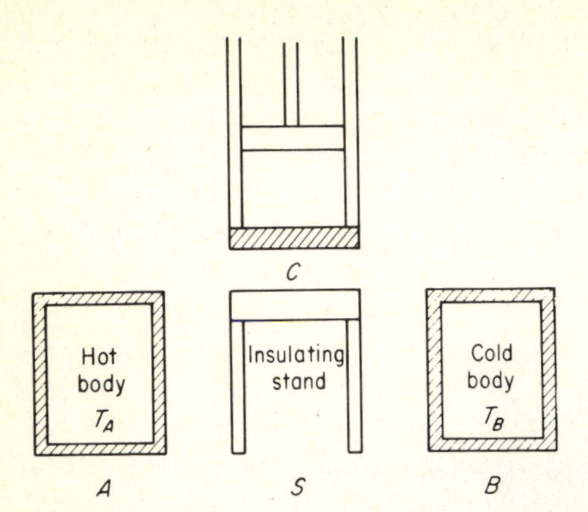

Figure 6-5 Carnot-engine principles.

curs, as indicated by curve *AB*. Now, if the cylinder is placed on insulating stand *S*, the gas will expand further adiabatically, as indicated by curve *BC*, *C* being the point at which the gas temperature reaches the value T_B. At this instant, if the cylinder is placed on *B* and the gas pressure is increased by a slight pressure increase on the piston, the resultant heat Q_2 will flow into the cold body *B*. Again, the gas temperature remains constant and the pressure as indicated by curve *CD* is isothermal. If at point *D* the cylinder is again placed on stand *S*, there is a further increase in pressure which is adiabatic because the heat cannot escape. This pressure is indicated by curve *DA*, and the cycle is completed.

As explained previously, the areas under the *PV* curves represent work. Therefore, referring to Fig. 6-6, areas ABB_1A_1 and BCC_1B_1 represent the work done by the expansion of the gas (positive work) and areas CDD_1C_1 and DAA_1D_1 represent the work required for compression (nega-

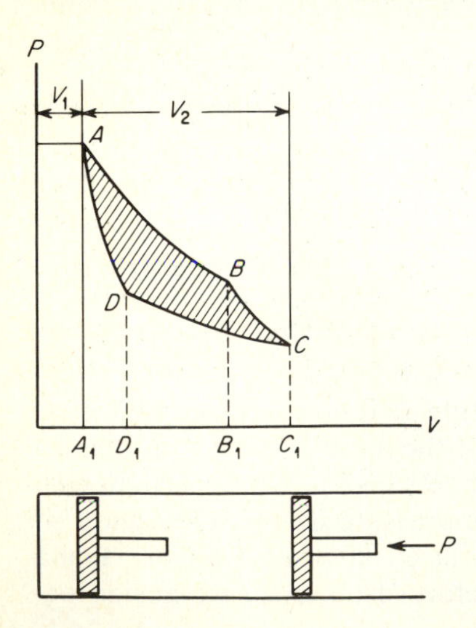

Figure 6-6 Carnot-engine cycle.

tive work). Hence, the net work output of the cycle is represented by area $ABCC_1A_1$ – area $ADCC_1A_1$, or $ABCD$.

During the cycle a quantity of heat Q_1 was absorbed by the gas during isothermal expansion along curve AB at T_A and a quantity of heat Q_2 was rejected by isothermal compression along line CD at T_B. Therefore, during the cycle, $Q_1 - Q_2$ heat units have been transformed into mechanical energy (W) and

$$W = J(Q_1 - Q_2)$$

where J is the number of units of work produced from one unit of heat.

The efficiency of the Carnot cycle depends only upon the temperatures between which it operates. Since the efficiency of any engine has been defined as the ratio of the heat applied to it and the heat equivalent of its energy output, we can say

$$\text{Efficiency} = \frac{Q_1 - Q_2}{Q_1}$$

Furthermore, in the case of the Carnot engine, it may be shown that

$$e = \frac{T_A - T_B}{T_A}$$

where T_A is the temperature (absolute) of the hot body supplying the heat and T_B is the temperature (absolute) of the cold body which absorbs the heat not converted into useful work.

ENGINE CYCLES AND EFFICIENCIES

The operation of any internal-combustion engine is based upon the theoretical thermodynamic principles previously discussed. However, actual operating results may vary somewhat from theoretical values because of such factors as (1) variation in the specific heats of the gases at different temperatures; (2) variation in air-fuel ratio; (3) exchanges and losses of heat to the cylinder walls and piston; (4) leakage and resistance to gas flow; (5) lapse of time between ignition and complete combustion of mixture, and others. It is possible to analyze some of the common engine cycles by using a gas such as air and making certain assumptions, namely, that the specific heat remains constant and that there is no loss or transfer of heat to the working parts. Such analyses will aid in showing the effect of changing operating conditions and provide a better understanding of the basic factors involved in engine design and the part that they play in operating efficiency.

The two cycles used in present-day engines are known as the *Otto cycle* and the *diesel cycle*. Either may be subdivided into two- and four-stroke-cycle types, depending upon the number of piston strokes required to complete a cycle of events in the cylinder.

Ideal Otto Cycle (Air Standard) Figure 6-7 is a *PV* diagram of the Otto cycle as it would function under ideal and air-standard conditions. V_1 represents the total cylinder volume, V_2 the clearance or compression volume, and $V_1 - V_2$ the piston displacement.

The line *MN* represents atmospheric pressure, and *AB* represents the admission of the gas, with the pressure dropping slightly below atmospheric. *BGC* represents compression under adiabatic conditions, with ignition occurring at *C*. It is assumed that combustion is instantaneous and that the rise in pressure *CD* occurs at a constant volume V_2. *DE* represents adiabatic expansion or the working stroke. Exhaust takes place at *E*, and the pressure drops nearly to atmospheric and remains so during the exhaust stroke *FA*.

In examining Fig. 6-7, it will be observed that areas represent work because horizontal distances are volumes in cubic feet and vertical distances are pressures in pounds per square inch. Since work equals *FD*, vertical distances multiplied by horizontal distances equal some area which, by means of suitable units and values, can be converted into units of work. Furthermore, the area *ABG* is relatively very small and can be neglected. The area *ADEB* represents the gross work of the expansion stroke, and area *ACB* represents the negative work done during the com-

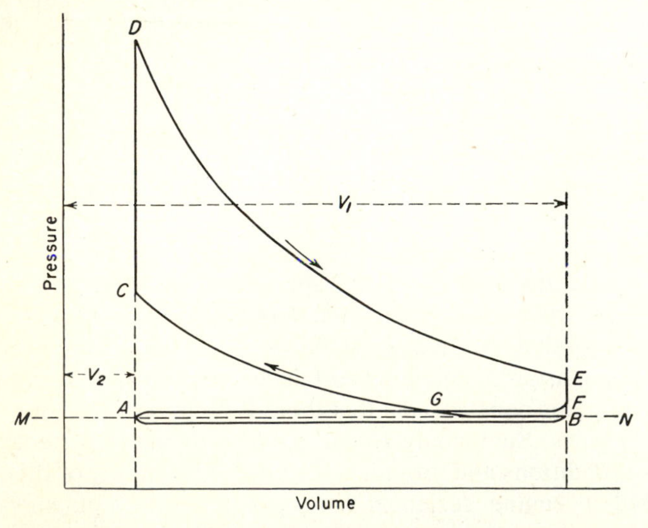

Figure 6-7 Ideal *PV* diagram for a four-stroke-cycle ignition engine (air standard).

pression stroke. The difference between these two areas is *CDEB* and represents the net work of the cycle.

Analyzing the heat and energy reactions of the cycle, we find that heat is added along *CD* and lost along *EB;* that is,

$$Q_1 \text{ (heat added)} = c_V(T_D - T_C) \tag{1}$$
$$Q_2 \text{ (heat lost)} = c_V(T_E - T_B) \tag{2}$$

The theoretical thermal efficiency e of the cycle is given by the equation

$$e = \frac{Q_1 - Q_2}{Q_1} \tag{3}$$

Substituting,

$$e = \frac{c_V(T_D - T_C) - c_V(T_E - T_B)}{c_V(T_D - T_C)} \tag{4}$$

or $$e = 1 - \frac{T_E - T_B}{T_D - T_C} \tag{5}$$

From the laws of gases,

$$\frac{P_B V_B}{T_B} = \frac{P_C V_C}{T_C} = \frac{P_D V_D}{T_D} = \frac{P_E V_E}{T_E} \tag{6}$$

and $$P_B V_B{}^k = P_C V_C{}^k \quad \text{and} \quad P_D V_D{}^k = P_E V_E{}^k \tag{7}$$

Referring to Fig. 6-7 and combining Eqs. (6) and (7),

$$\frac{T_B}{T_C} = \left(\frac{P_B}{P_C}\right)^{(k-1)/k} \tag{8}$$

$$\frac{T_B}{T_C} = \left(\frac{V_C}{V_B}\right)^{k-1} \tag{9}$$

and $$\frac{T_E}{T_D} = \left(\frac{V_D}{V_E}\right)^{k-1} \tag{10}$$

but (Fig. 6-7)

$$V_D = V_C \quad \text{and} \quad V_E = V_B$$

Hence $$\frac{T_B}{T_C} = \left(\frac{V_C}{V_B}\right)^{k-1} \quad \text{and} \quad \frac{T_E}{T_D} = \left(\frac{V_C}{V_B}\right)^{k-1} \tag{11}$$

Therefore $$\frac{T_B}{T_C} = \frac{T_E}{T_D} \quad \text{or} \quad \frac{T_E}{T_B} = \frac{T_D}{T_C} \tag{12}$$

Then $$\frac{T_E}{T_B} - 1 = \frac{T_D}{T_C} - 1 \quad \text{and} \quad \frac{T_E - T_B}{T_B} = \frac{T_D - T_C}{T_C} \tag{13}$$

or $$\frac{T_E - T_B}{T_D - T_C} = \frac{T_B}{T_C} \tag{14}$$

Substituting Eq. (14) in Eq. (5),

$$e = 1 - \frac{T_B}{T_C} \tag{15}$$

From Eq. (8),

$$e = 1 - \left(\frac{P_B}{P_C}\right)^{(k-1)/k} \tag{16}$$

and from Eq. (9),

$$e = 1 - \left(\frac{V_C}{V_B}\right)^{k-1} \tag{17}$$

Equations (15) to (17) show that the air-standard thermal efficiency of an Otto-cycle engine is dependent upon (1) the ratios of the absolute temperatures before and after compression; (2) the increase in the compression pressure; and (3) the relative initial and final volumes.

Ideal Diesel Cycle (Air Standard) The diesel-cycle engine differs from the Otto-cycle one in that the heat of compression instead of an electric spark is utilized to produce ignition. The air and fuel are not premixed by a carburetor and introduced as a gaseous mixture; instead, air alone is taken in and compressed, and the fuel is injected into the highly compressed and heated air at the end of the compression stroke. It is assumed that the rate of admission of the fuel is so regulated that the pressure remains constant during this injection interval and until combustion is complete.

Figure 6-8 is a *PV* diagram of the diesel cycle as it would function under ideal and air-standard conditions. The line *MN* represents atmospheric pressure, and *AB* represents the admission of the air. *BGC* represents adiabatic compression with a compression ratio of about 15 : 1 and a maximum pressure of approximately 500 lb/in^2. Injection of the fuel charge begins at *C* and stops at *D*. Combustion takes place during this

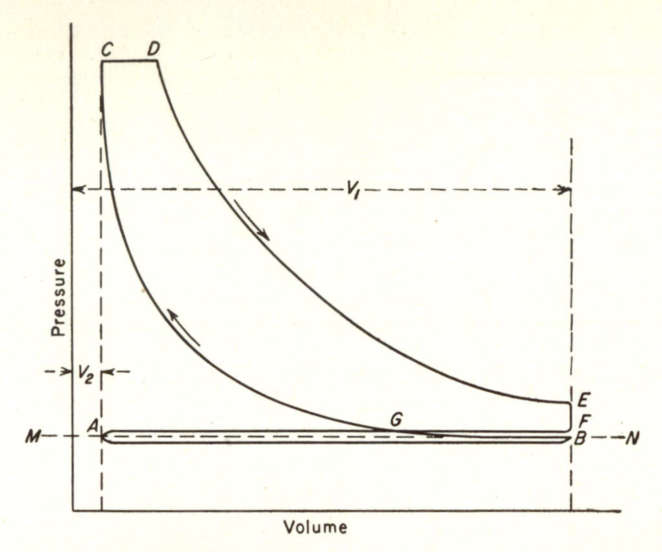

Figure 6-8 Ideal *PV* diagram for a four-stroke-cycle diesel engine (air standard).

interval, and a condition of constant pressure is assumed. The ratio of the volumes at D and C (V_D/V_C) is known as the *cutoff ratio*. Adiabatic expansion occurs from D to E; the exhaust valve opens at E, the pressure dropping nearly to atmospheric and remaining so during the exhaust stroke FA. The area $ACDEB$ represents the total work of the expansion stroke, and area ACB represents the negative work done during the compression stroke. The difference between these two areas is $BCDE$ and represents the net work of the cycle.

In analyzing the heat and energy reactions of the cycle, the following assumptions are made:

1 The air pressure during the stroke AB is atmospheric.

2 Compression is adiabatic and the cylinder walls are nonconductors.

3 Fuel enters the cylinder at the temperature T.

4 Combustion takes place at constant pressure during the interval CD.

5 Expansion is adiabatic.

6 Exhaust takes place at atmospheric pressure.

7 Specific heat remains constant during the entire cycle.

With the above assumptions in mind, we find that heat is added along CD and lost along EB; that is,

$$Q_1 \text{ (heat added)} = c_P(T_D - T_C) \tag{1}$$

$$Q_2 \text{ (heat lost)} = c_V(T_E - T_B) \tag{2}$$

The theoretical thermal efficiency e of the cycle is given by the equation

$$e = \frac{Q_1 - Q_2}{Q_1} \tag{3}$$

Substituting,

$$e = \frac{c_P(T_D - T_C) - c_V(T_E - T_B)}{c_P(T_D - T_C)} \tag{4}$$

or $$e = 1 - \frac{c_V(T_E - T_B)}{c_P(T_D - T_C)} \tag{5}$$

Since $$\frac{c_P}{c_V} = k \tag{6}$$

then $$e = 1 - \frac{T_E - T_B}{k(T_D - T_C)} \tag{7}$$

From the laws of gases,

$$\frac{P_B V_B}{T_B} = \frac{P_C V_C}{T_C} = \frac{P_D V_D}{T_D} = \frac{P_E V_E}{T_E} \tag{8}$$

and $$P_B V_B{}^k = P_C V_C{}^k \quad \text{and} \quad P_D V_D{}^k = P_E V_E{}^k \tag{9}$$

Since CD is at constant pressure and EB is at constant volume, from Eq. (8),

$$T_D = T_C \frac{V_D}{V_C} \tag{10}$$

and $$T_E = T_B \frac{P_E}{P_B} \tag{11}$$

and, from Eq. (9),

$$\frac{P_B}{P_C} = \left(\frac{V_C}{V_B}\right)^k \quad \text{and} \quad \frac{P_D}{P_E} = \left(\frac{V_E}{V_D}\right)^k \tag{12}$$

But $P_C = P_D$ and $V_B = V_E$

Hence $$\frac{P_E}{P_B} = \left(\frac{V_D}{V_C}\right)^k \tag{13}$$

and, substituting Eq. (13) in Eq. (11),

$$T_E = T_B \left(\frac{V_D}{V_C}\right)^k \tag{14}$$

Substituting Eqs. (10) and (14) in Eq. (7),

$$e = 1 - \frac{T_B}{T_C} \frac{[(V_D/V_C)^k - 1]}{k(V_D/V_C - 1)} \tag{15}$$

From Eqs. (8) and (9),

$$\frac{T_B}{T_C} = \left(\frac{V_C}{V_B}\right)^{k-1} = \left(\frac{P_B}{P_C}\right)^{(k-1)/k} \tag{16}$$

Hence

$$e = 1 - \left(\frac{V_C}{V_B}\right)^{k-1} \frac{[(V_D/V_C)^k - 1]}{k(V_D/V_C - 1)} \tag{17}$$

and

$$e = 1 - \left(\frac{P_B}{P_C}\right)^{(k-1)/k} \frac{[(V_D/V_C)^k - 1]}{k(V_D/V_C - 1)} \tag{18}$$

A study of the preceding theoretical analysis of the diesel cycle shows that its efficiency is dependent upon the compression ratio and the cutoff ratio; that is, the efficiency is increased by a high compression ratio and a decrease in the cutoff ratio.

Comparison of Ideal Cycles Table 6-1 and Fig. 6-9 show the comparative efficiency values for ideal Otto and diesel cycles. It will be noted that, at first, the efficiency increases rapidly with the increase in compression but the rate of increase is less as the compression ratio becomes high. It will also be noted that the Otto cycle gives a higher theoretical efficiency

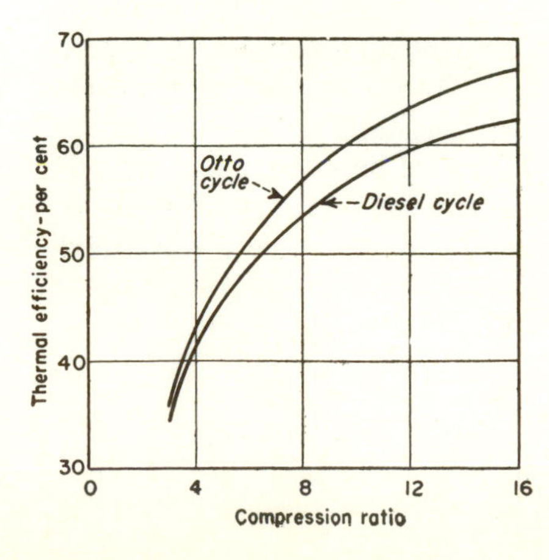

Figure 6-9 Theoretical relationship of compression ratio and thermal efficiency for electric-ignition and diesel-engine cycles.

Table 6-1 Theoretical Thermal-Efficiency Values for Otto and Diesel Cycles (Air Standard)

Compression ratio, V_1/V_2	Clearance vol., percent of piston displacement	P_2, lb/in^2 abs	Thermal efficiency, percent	
			Otto cycle	Diesel cycle
3.0	50.0	69	35.6	34.3
4.0	33.3	102	42.6	41.0
4.5	28.9	121	45.2	43.6
5.0	25.0	140	47.4	45.8
5.5	22.3	160	49.5	47.5
6.0	20.0	180	51.1	48.9
6.5	18.2	202	52.7	50.4
7.0	16.7	224	54.1	51.2
7.5	15.4	247	55.3	52.7
8.0	14.5	270	56.4	53.5
9.0	12.5	318	58.5	55.2
10.0	11.1	369	60.2	56.9
11.0	10.0	422	61.7	58.3
12.0	9.1	477	63.0	59.4
13.0	8.3	533	64.2	60.6
14.0	7.7	590	65.2	61.2
15.0	7.1	651	66.1	62.0
16.0	6.6	713	67.0	62.8

than the diesel cycle. However, in actual practice, it is not higher, because the compression ratio of the Otto-cycle engine is limited by the fuel characteristics, particularly with respect to ignition and detonation. Diesel engines normally use considerably higher compression pressures than Otto-cycle engines and, therefore, give a higher fuel-utilization efficiency.

It should be made clear that engines cannot be made to perform under the ideal cycle conditions just explained. This is true for a number of reasons, namely, (1) compression and expansion cannot be completely adiabatic; (2) the fuel is not a perfect gas; (3) heat cannot be supplied to the gaseous material at either an exact constant volume or constant pressure; and (4) losses from heat, mechanical imperfections, and other sources are unavoidable. A knowledge of the theoretical efficiency and performance of an ideal-cycle engine does make possible a more intelligent analysis of the actual performance of any engine and whether or not it is well designed.

PROBLEMS AND QUESTIONS

1 What is meant by the absolute temperature of a substance?

2 Define calorie, Btu, and specific heat.

3 Distinguish between isothermal and adiabatic expansion.
4 An engine has a compression ratio of 7.5. Compute P_2 (gauge) if P_1 is 13.5 lb/in^2 and k is 1.3.
5 What is the compression ratio of an engine if P_1 is 14.0 lb/in^2, P_2 is 190 lb/in^2, and k is 1.4? What is the percentage change if k is 1.2?
6 Calculate the theoretical thermal efficiency for an ideal Otto-cycle engine if $V_B = 51$ in^3, $V_c = 8.3$ in^3, and $k = 1.35$.
7 Calculate the theoretical thermal efficiency for a diesel-cycle engine if $V_B = 180$ in^3, $V_C = 12$ in^3, $V_D = 14.5$ in^3, and $k = 1.35$.
8 Make sketches of typical PV diagrams for ideal Otto- and diesel-cycle engines (air standard), and explain the P, V, and T reactions for the different events in the cycles.
9 Compare the Otto and diesel cycles as to efficiency from both a theoretical and a practical standpoint.

Chapter 7

Engine Cycles and Principles of Operation

A mechanical device or machine that converts heat and other forms of energy such as wind, flowing water, and electricity into useble power is called an engine or a motor. The basic requirement of such a device is that it must carry out this conversion process in a continuous, efficient and reliable manner consistent with the specific operation involved.

TYPES OF GAS- AND LIQUID-FUELED ENGINES

Four distinct types of internal combustion engines have been developed and are being utilized to a certain extent depending upon their specific characteristics and adaptability. These are (1) the rotary engine, (2) the jet engine, (3) the gas turbine engine, and (4) the reciprocating or piston-type engine.

Rotary Engine The principle of the rotary engine was developed by Felix Wankel, a German, in 1956. As a result of the interest and research efforts of automobile and other manufacturers of internal combustion engines, automobiles equipped with rotary engines are now available, and the future development and adaptability of this principle appears to be promising.

Referring to Fig. 7-1, basically, the rotary engine consists of a triangular and symmetrical rotor that revolves on a large bearing within a figure-eight-shaped or epitrochoidal chamber. This bearing is eccentric with the power output shaft and thus becomes a crank arm for it. An internal or planetary gear concentric with the rotor bearing meshes with a sun gear on the power shaft and thus rotates the latter at thrice the rotor speed. These gears insure that the correct relationship is followed by the engine parts as the rotor travels the epitrochoidal path.

Referring to Fig. 7-1, the apexes of the rotor are in constant sliding contact with the chamber inner wall. During one rotor revolution, each apex completes one power cycle including fuel intake, compression, ignition, and exhaust. The shapes of the chamber and rotor are such that the fuel mixture volume varies in such a manner during a single revolution as to produce the required pressure conditions for each event in a power cycle.

The principal advantages claimed for the rotary engine are (1) fewer parts, and simple construction, (2) quiet running, (3) little or no vibration, and (4) compact and lightweight. The major problem in the design of the rotary engine has been the control of wear between the rotor and the housing and in the apex seals. Any appreciable wear and resulting leakage at these points directly affects and reduces the operating efficiency of the engine. A great amount of research has been carried on to develop reliable seal materials.

Jet Engine—Gas Turbine Engine The operation of these engines is based upon Newton's third law of motion, which states that for every acting force there is an equal and opposite reaction. Referring to Fig. 7-2, the jet engine is basically a heavy cylindrical metal tube whose internal construction is such that as it moves through space, air enters at one end and mixes with fuel injected into a special compression chamber. The ignition and resulting combustion of this mixture results in a highly compressed gas mixture being blown against a multibladed wheel or turbine that spins at a high speed. If the gas is expelled directly from the rear, the

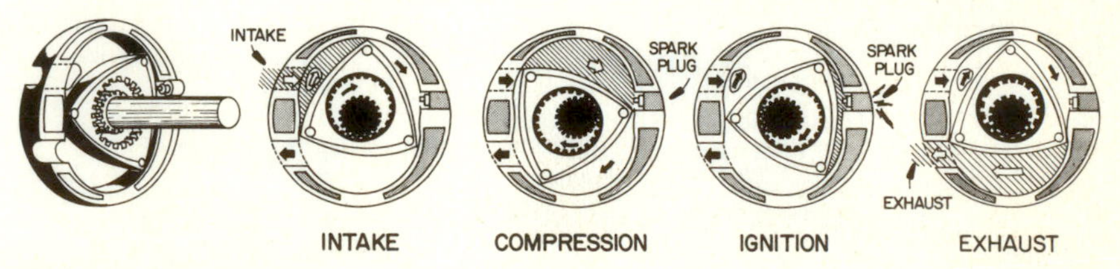

Figure 7-1 Basic design and principles of operation of a rotary engine. (*Courtesy of Mr. Walter Hortens.*)

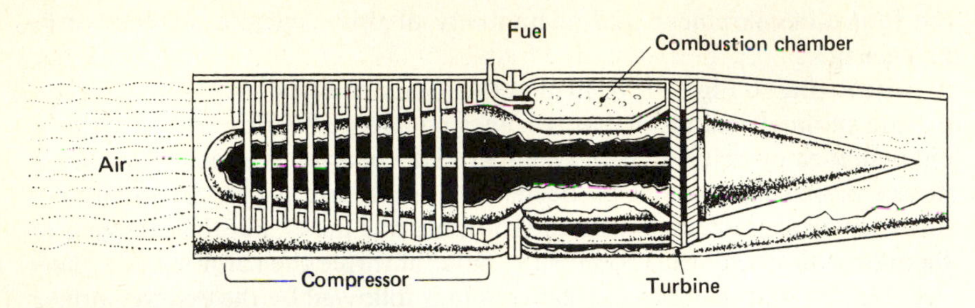

Figure 7-2 Basic design and principle of operation of a jet engine. (*Courtesy of General Motors Corporation.*)

force caused by the difference in the speed of the entering air and that expelled from the rear becomes the propulsion for the vehicle. This is the principle involved in the construction and operation of the jet engine as used in the modern airplane.

Fig. 7-3, shows the basic construction and operation of the turbojet engine. The general arrangement of parts is similar to the plain jet engine except that, in addition to the compressor turbine, a second multistage turbine is connected to a shaft which in turn supplies direct mechanical power to the propulsion device or other mechanism. A common example is the turboprop-driven airplane.

A further development in the possible use of a continuous flow of hot gas under high pressure is the gas turbine engine. Such engines are more complex than the plain turbojet engine but have certain advantages over the common piston-type engines, and much work is being carried on to develop such a power unit for automobiles and trucks. As discussed in Chap. 2, the reciprocating or piston-type engine is a development of the nineteenth century. The first piston-type engine appeared early in the century and eventually became the forerunner of the various types of steam-powered engines that prevailed for many years until the internal-combustion engine entered the mechanical power field at the beginning of the twentieth century (detailed in Chap. 2).

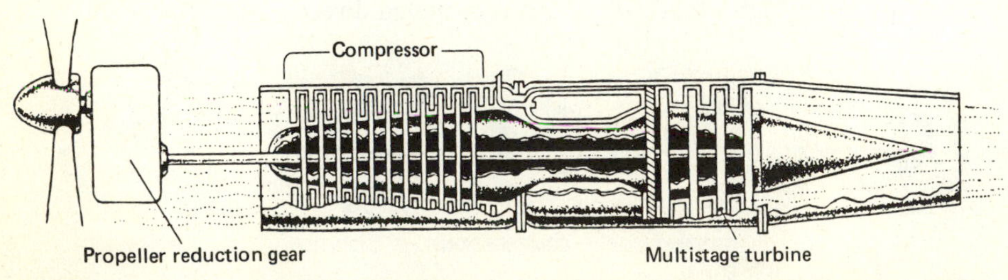

Figure 7-3 Basic design and operating principles of a turboprop engine. (*Courtesy of General Motors Corporation.*)

Cycle of Operations Any piston-type internal-combustion engine, regardless of size, number of cylinders, use, and so on, is of either one or the other of two general types. These are known as the four-stroke-cycle type and the two-stroke-cycle type, or, as more commonly stated, four-cycle and two-cycle. A cycle consists of the events taking place in each cylinder of an engine between two successive explosions in that cylinder. These events or operations, in the order in which they occur, are as follows: (1) the taking in of a combustible mixture; (2) the compression of this mixture; (3) the ignition of the compressed mixture; (4) the expansion of the burned gases producing the power; and (5) the exhaust of the products of combustion.

In the two-stroke-cycle engine, two strokes of the piston or one revolution of the crankshaft is required to complete this cycle. In the four-stroke-cycle type, four strokes of the piston or two complete revolutions of the crankshaft are needed, and the four strokes are frequently referred to as intake, compression, power, and exhaust. It must be kept in mind that, in engines of more than one cylinder, this cycle of events must be carried out in each of the cylinders; that is, one cylinder cannot perform the first event and another perform the second, and so on. Also the events can take place only in the order stated.

Two-Stroke-Cycle Operation Two important characteristics of two-stroke-cycle construction must be noted and kept in mind: (1) that ports or openings in the cylinder walls at some distance below the head serve as intake and exhaust valves and (2) that the open or crank end of the engine cylinder is completely enclosed. Referring to Fig. 7-4A, the piston in its upward travel has closed the ports *e* and *h* and is compressing the charge. At the same time the crankcase volume is being increased and the fuel

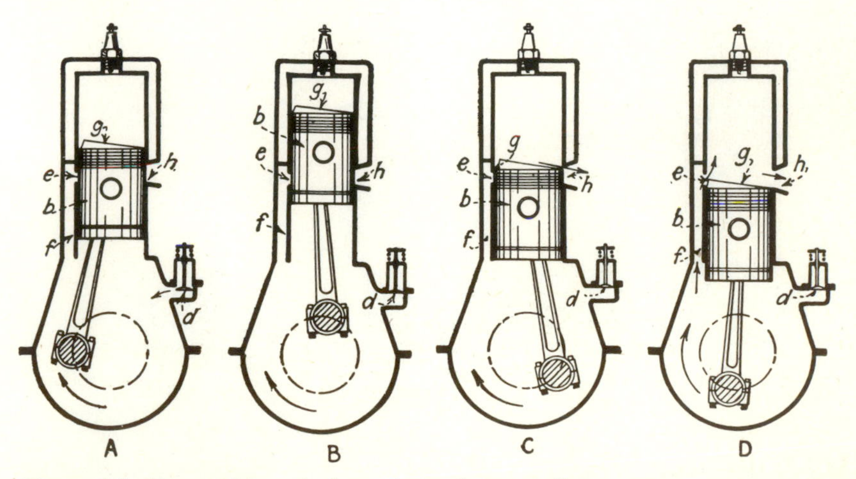

Figure 7-4 Two-port two-stroke-cycle engine operation.

mixture is drawn into the crankcase through an opening *d*, to be compressed on the next downward stroke of the piston and forced through a connecting passage *f* into the combustion space when the inlet port *e* is uncovered. Near the end of the upward or compression stroke of the piston the spark is produced and the compressed charge is fired. The explosion and resulting expansion send the piston downward on its power stroke (Fig. 7-4B), near the end of which the two ports are uncovered as shown by Fig. 7-4c. Even though the piston is nearly at the end of its stroke, considerable pressure remains in the cylinder, thus forcing the products of combustion out through the exhaust port *h*. At the same time, the fresh mixture in the crankcase has been compressed on the downward stroke and passes upward through the intake port to the combustion chamber as shown. The piston has now completed two strokes, and the crankshaft has made one revolution, thus completing the cycle.

Figure 7-5 illustrates the three-port type of two-stroke-cycle engine in which the fuel mixture enters the crankcase through a third port, which is uncovered by the piston on its upward travel.

Two-Stroke-Cycle-Engine Characteristics Two-stroke-cycle engines have the following distinguishing mechanical characteristics: (1) the crankcase is enclosed and must be as airtight as possible; (2) ports or openings in the side of the cylinder, opened and closed by the piston sliding over them, take the place of valves; (3) no valve-operating mechanism of any kind is necessary; and (4) the fuel mixture usually enters and passes through the crankcase on its way to the cylinder.

Some of the important advantages of two-stroke-cycle engines are as follows: (1) they are lighter in weight per horsepower, (2) they are simpler

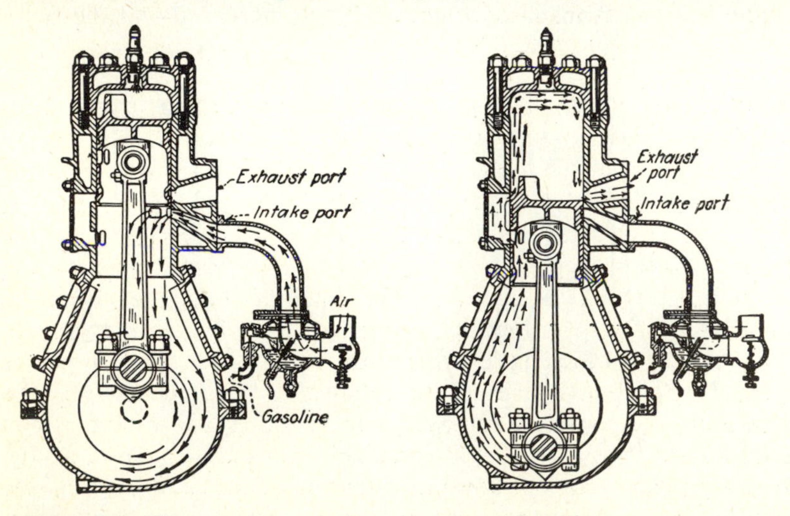

Figure 7-5 Three-port two-stroke-cycle engine operation.

in construction, and (3) they have a greater frequency of working strokes or power impulses. Some disadvantages are as follows: (1) their fuel mixture is controlled with difficulty, (2) they are inefficient in fuel consumption, and (3) they do not operate satisfactorily under fluctuating loads. Two-stroke-cycle engines are usually inefficient in fuel consumption and frequently give considerable trouble in starting and during operation, largely because complete exhaust of the burned fuel residue is extremely difficult. Likewise, the problem of producing the correct fuel mixture and placing it in the cylinder is a difficult one.

Some diesel engines utilize the two-stroke-cycle principle very successfully. Since they use fuel injection rather than carburetion, certain disadvantages and difficulties enumerated above are not encountered.

It might be assumed that a two-stroke-cycle engine of a certain bore, stroke, and speed would develop twice as much power as a four-stroke engine of the same size and speed, because it would have twice as many power strokes in a given time. Such is not the case, however, because it is impossible to get as effective a charge of fuel mixture into the cylinder.

The removal of the burned residue from the cylinder is dependent entirely upon the pressure remaining in the cylinder being somewhat greater than that existing on the outside when the exhaust port is opened. There is not a distinct exhaust stroke, as in the four-stroke-cycle engine, to push this material out; therefore its complete expulsion is unlikely, and the incoming fuel charge may be thus contaminated and its effectiveness reduced. To prevent the possible escape of a portion of the incoming fuel mixture through the exhaust port before the piston has closed the latter, a projection known as a deflector, shown by *g* in Fig. 7-4, is placed on the closed end of the piston at the point where it passes the intake port. As the name implies, this device deflects the charge in an upward direction toward the cylinder head.

Four-Stroke-Cycle Operation Figure 7-6 shows the usual four-stroke-cycle construction and operation. It will be observed, first of all, that valves located in the cylinder head are used instead of ports. Starting with the piston at the head end of the cylinder, it moves toward the crank end, drawing in a fuel mixture through the open inlet valve. Soon after the end of this stroke, the inlet or suction valve closes and the mixture is compressed as the piston returns to the head end of the cylinder. Near the completion of this stroke a spark is produced that ignites the charge, causing an explosion which, in turn, sends the piston on the third or power stroke. Near the end of this stroke the exhaust valve is opened, and the burned residue is completely removed from the cylinder as the piston travels backward toward the cylinder head on what is known as the exhaust stroke. The piston has now passed through four complete strokes, the crankshaft has made two revolutions, and a cycle has been completed.

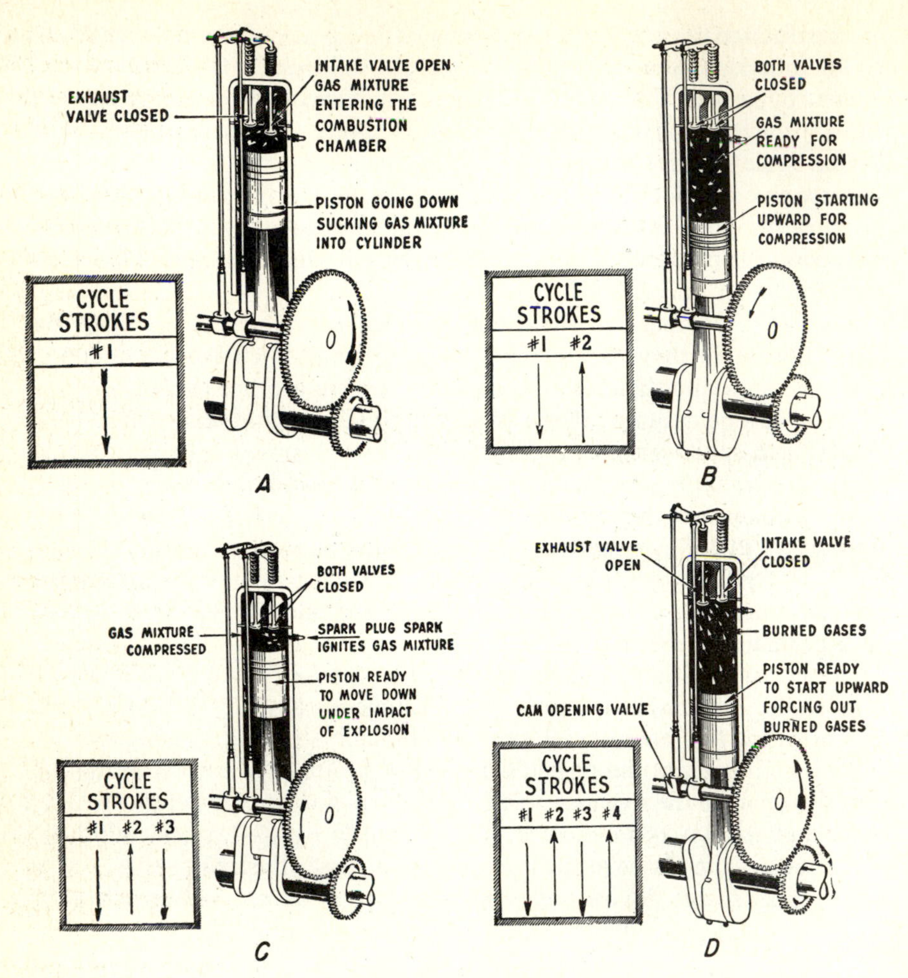

Figure 7-6 Four-stroke-cycle operation: (A) Intake stroke, (B) compression stroke, (C) power stroke, (D) exhaust stroke.

It is thus observed that the two-stroke-cycle single-cylinder engine has a power impulse for each revolution of the crankshaft, and that the four-stroke-cycle single-cylinder engine gives one power impulse in two revolutions. This explains why heavy flywheels are necessary on one-cylinder engines. Were it not for these flywheels, such an engine when under load would have a tendency to lose speed between explosions. These flywheels, owing to their inertia, carrying the piston through these so-called idle strokes in spite of the resistance offered by the load and thus maintain uniformity of speed. A one-cylinder two-stroke-cycle engine will not require so heavy a flywheel as the four-stroke, because it fires twice as frequently.

Multiple-Cylinder Operation As previously stated, these principles of operation likewise apply to multiple-cylinder engines. That is, a four-cylinder four-stroke-cycle engine is nothing more than four single-cylinder four-stroke-cycle engines built into a single unit; or a two-cylinder two-stroke-cycle engine is nothing more than two single-cylinder two-stroke-cycle engines built into a single unit. A better understanding of the application of these principles to multiple-cylinder engines may be obtained by reference to the following illustrations and explanations.

Figure 7-8A illustrates the usual construction of a two-stroke-cycle two-cylinder engine. Note that the crankshaft consists of two opposite cranks; therefore, the pistons will move in opposite directions at all times. A cycle will be completed in each cylinder at every revolution of the crankshaft, but a power impulse or explosion will occur at every half turn, or there will be two explosions per revolution. Figure 7-8B illustrates a typical two-cylinder four-stroke-cycle engine layout. Note that the cranks on the crankshaft are arranged side by side and that the two pistons move together. By reference to the accompanying chart (Fig. 7-7), it will be noted that there will be two power impulses during two revolutions of the crankshaft or one power impulse for each revolution; that is, when piston 1 is on the power stroke, piston 2 is on intake, and so on.

The crank and cylinder arrangement (Fig. 7-8A) also could be used for a four-stroke-cycle, two-cylinder engine, and the chart (Fig. 7-9) shows the power impulse and stroke relationship. It will be noted that the power impulses occur at unequal intervals, namely, 180 and 540°, but this cannot be avoided in this type of engine. It would appear that this uneven power flow would be undesirable. However, if such an engine were running at any appreciable speed—500 or more rpm.—there would not be any noticeable effect. The arrangement shown in Fig. 7-8B gives an equal

Crankshaft has Turned (Degrees)	Cylinder No. 1	Cylinder No 2
180	Power	Intake
360 / 540	Exhaust	Compression
720	Intake	Power
	Compression	Exhaust

Figure 7-7 Chart showing occurrence of events in a two-cylinder four-stroke-cycle engine with crank arrangement shown in Fig. 7-8B.

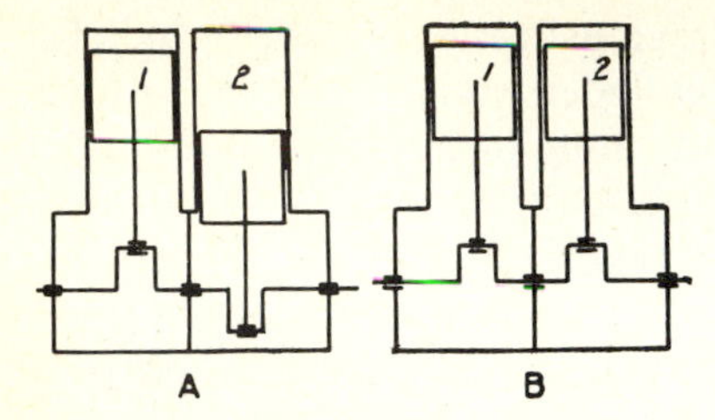

Figure 7-8 Crank arrangements for two-cylinder twin engines: (A) Opposed cranks, (B) twin cranks.

firing interval—360°, according to the diagram (Fig. 7-7)—but is less preferable because of difficulty in securing mechanical balance and therefore the least possible vibration. The opposed-crank arrangement for two-cylinder twin engines seems to be preferable in spite of its unequal firing interval.

Figure 7-10 illustrates the usual four-cylinder four-stroke-cycle construction. Four-cylinder crankshafts are always arranged as shown; that is, cranks 1 and 4 are together and opposite cranks 2 and 3. Referring to the chart (Fig. 7-11) and assuming that piston 1 has just received a power impulse so that it is on the power stroke, piston 4 must necessarily be on the intake stroke. If piston 2 is now placed on the compression stroke, piston 3 must necessarily be on the exhaust stroke. Completing the chart by filling in the strokes for the four pistons in the correct order, the *firing order* becomes 1-2-4-3.

Crankshaft has Turned (Degrees)	Cylinder No. 1	Cylinder No. 2
180	Power	Compression
360 540	Exhaust	Power
720	Intake	Exhaust
	Compression	Intake

Figure 7-9 Chart showing occurrence of events in a two-cylinder four-stroke-cycle engine with opposed cranks.

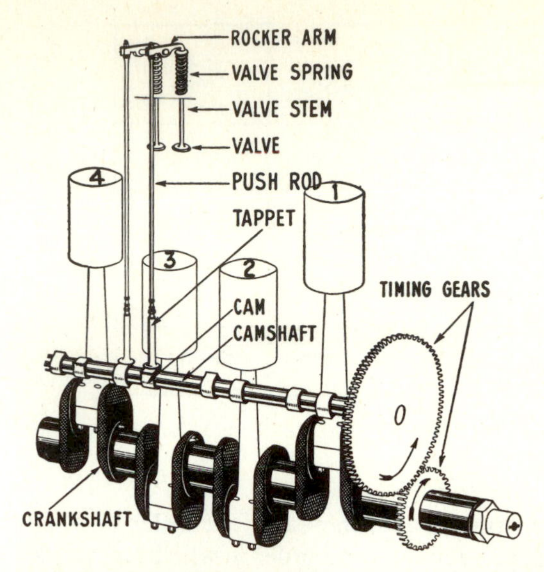

Figure 7-10 Typical four-cylinder four-stroke-cycle engine crankshaft and valve operation.

A second firing order is likewise possible, as shown by the chart (Fig. 7-12); that is, again assuming that piston 1 is on the power stroke and piston 4 on the intake stroke, piston 2 could be on exhaust and piston 3 on compression. Thus the firing order becomes 1-3-4-2. These are the only firing orders possible, however, with such a crankshaft and crank arrangement. Both firing orders are used in four-cylinder four-stroke-cycle engines, and neither has any particular advantage over the other.

Crankshaft has Turned (Degrees)	Cylinder No. 1	Cylinder No 2	Cylinder No. 3	Cylinder No. 4
180	Power	Compression	Exhaust	Intake
360 540	Exhaust	Power	Intake	Compression
720	Intake	Exhaust	Compression	Power
	Compression	Intake	Power	Exhaust

Figure 7-11 Chart showing occurrence of events in a four-cylinder four-stroke-cycle engine having a 1-2-4-3 firing order.

Crankshaft has Turned (Degrees)	Cylinder No. 1	Cylinder No 2	Cylinder No. 3	Cylinder No 4
180	Power	Exhaust	Compression	Intake
360 540	Exhaust	Intake	Power	Compression
720	Intake	Compression	Exhaust	Power
	Compression	Power	Intake	Exhaust

Figure 7-12 Chart showing occurrence of events in a four-cylinder four-stroke-cycle engine having a 1-3-4-2 firing order.

The particular firing order used is determined by the order in which the intake or exhaust valves operate and by the order in which the sparks are delivered to the cylinders. Given two four-cylinder four-stroke-cycle engines, one having a firing order of 1-2-4-3 and the other of 1-3-4-2, the principal mechanical difference will be in the arrangement of the cams on the camshaft (Fig. 7-10) that operate the valves. These principles of opera-

Figure 7-13 Cutaway view of a V-type four-cylinder air-cooled engine. (*Courtesy of Wisconsin Motor Corporation.*)

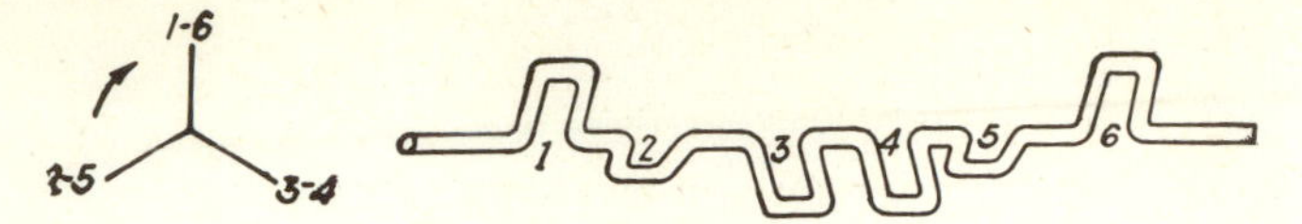

Figure 7-14 Crank arrangement for a six-cylinder engine.

tion likewise apply to engines having six, eight, or any number of cylinders, and the proper distribution of the explosions and the firing order are obtained by the correct crank arrangement and valve operation.

Figure 7-13 illustrates a four-cylinder V-type engine. This gives an extremely compact arrangement, but an equal power-impulse distribution cannot be obtained because the V angle is 90° and at some time or another two opposite cylinders must follow each other; that is, there must be a 90° interval. The crankshaft for this engine has the four cranks arranged in one plane, similar to the standard four-cylinder type. The firing interval is 90-180-270-180°.

Six-cylinder crankshafts are constructed as shown in Fig. 7-14; that is, crank 1 is paired with crank 6, crank 2 with crank 5, and crank 3 with crank 4. The three pairs are then arranged so as to make 120° with each other. With such an arrangement, three cylinders fire during each revolution of the crankshaft, and one common firing order is 1-5-3-6-2-4. The other is 1-4-2-6-3-5.

Eight-cylinder engines may have the cylinders in line or in two sets of four each, forming a 90° V. An eight-in-line crankshaft has four pairs of cranks. They are paired, as shown by Fig. 7-15, giving a firing order of 1-6-2-5-8-3-7-4. The crankshaft for a V-8 engine is similar to a four-cylinder crankshaft; that is, there are only four crank journals, and two connecting rods—one for each cylinder bank—are connected side by side to each crank journal. All cranks may be in the same plane as in a standard four-cylinder crankshaft, or they may be at 90° to each other. The firing order would be determined largely by the crankshaft type. The V-8 type of cylinder arrangement (Fig. 7-16) is used extensively in automobiles and

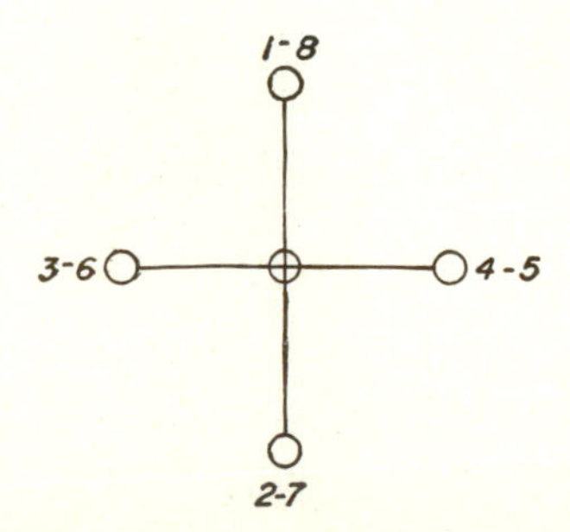

Figure 7-15 Crank arrangement for an eight-cylinder in-line engine.

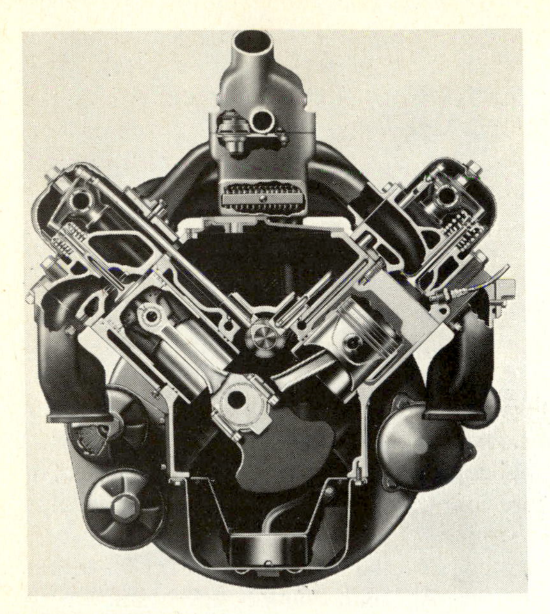

Figure 7-16 Cross section showing construction of a typical V-8 automotive-type engine, (*Courtesy of Reo Division, White Motor Company.*)

trucks because of compactness of design along with good speed and power flexibility.

Firing Interval It should now be clear that the advantage of an engine's having more than one cylinder is that it fires more frequently—that is, there are more explosions in the same number of revolutions of the crankshaft—and consequently the power is said to be more uniform. The greater the number of cylinders, the shorter the distance moved by the crankshaft between two successive explosions in the engine.

The arc of travel of the crankshaft in degrees between successive explosions in an engine is known as the firing interval. For most multiple-cylinder engines it can be determined by the formula

$$\text{Firing interval} = \frac{360^\circ}{\text{no. cylinders}} \quad \text{for a two-stroke-cycle engine}$$

$$\text{Firing interval} = \frac{720^\circ}{\text{no. cylinders}} \quad \text{for a four-stroke-cycle engine}$$

Assuming that energy from any piston is applied to the crankshaft of the conventional four-stroke-cycle engine during 135° of its rotation, it will be observed that, for a four-cylinder engine, there will be a 45° no-energy interval before the next cylinder fires and releases energy to the crankshaft. However, for engines having more than four cylinders, there will be an overlapping of power impulses because the firing interval will be less than 135°. Obviously, this will provide more uniform power and speed, faster acceleration, and better speed flexibility and torque characteristics. A few engines, as previously explained, have an unequal firing interval. For example, most two-cylinder, four-stroke-cycle engines have opposed cranks and fire 180-540°, and a V-4 engine fires 90-180-270-180°. However, this unequal interval does not affect the power and smoothness of operation at normal running speeds.

PROBLEMS AND QUESTIONS

1 In general, what is the operating principle of the Wankel rotary engine and what are its particular advantages and disadvantages, if any?
2 What is the basic principle of operation of a jet engine, and explain the difference between a simple jet engine and a turbojet engine.
3 Compare two- and four-stroke-cycle engines as to simplicity of design, efficiency, power, utility, and general adaptability to various power applications.
4 Compute the firing interval and total number of explosions per minute for a V-12, four-stroke-cycle engine having a crankshaft speed of 3,200 rpm.

Chapter 8

Engine Construction and Design—Tractor Engine Types and Construction

All piston-type internal-combustion engines, regardless of size, type, and number of cylinders, are made up of certain basic and essential parts and assemblies. A thorough knowledge of the correct terminology, function, and specific characteristics of these parts is important.

Principal Engine Parts The parts and systems making up a simple internal-combustion engine are as follows:

1. Cylinder
2. Cylinder head
3. Piston
4. Piston rings
5. Piston pin
6. Connecting rod
7. Crankshaft
8. Flywheel
9. Valve system
   a. Valves—intake and exhaust
   b. Cam gear and cam or camshaft

- c Tappet
- d Push rod
- e Rocker arm

10 Fuel-supply- and carburetion system
- a Fuel tank
- b Fuel line
- c Fuel pump
- d Carburetor
- e Throttle-butterfly and fuel-mixture supply control mechanism
- f Manifold
- g Air cleaner
- h Fuel-injection pump and nozzles, filters, and connections (diesel engines)

11 Ignition and electrical system
- a Battery or magneto
- b Coil
- c Condenser
- d Sparking device
- e Timing mechanism (breaker points, distributor)
- f Generator
- g Starter motor
- h Switch and wire connections

12 Cooling system
- a Radiator
- b Pump
- c Water jacket
- d Fan
- e Thermostat
- f Pipes and connections

13 Lubrication system
- a Oil pump
- b Oil lines
- c Oil gauge
- d Oil filter
- e Oil pan
- f Grease fittings

14 Governing system
- a Weights and springs and speed-control linkage

Cylinder The cylinder and cylinder block (Fig. 8-1) of an engine constitute the principal basic and supporting portion of the engine power unit. Their major function is to provide the space in which the piston operates to draw in the fuel mixture, compress it, and allow it to expand and generate power. The design and construction of the cylinder are dependent upon such factors as power required, exact purpose of engine, compression ratio, valve arrangement, method of cooling, arrangement of

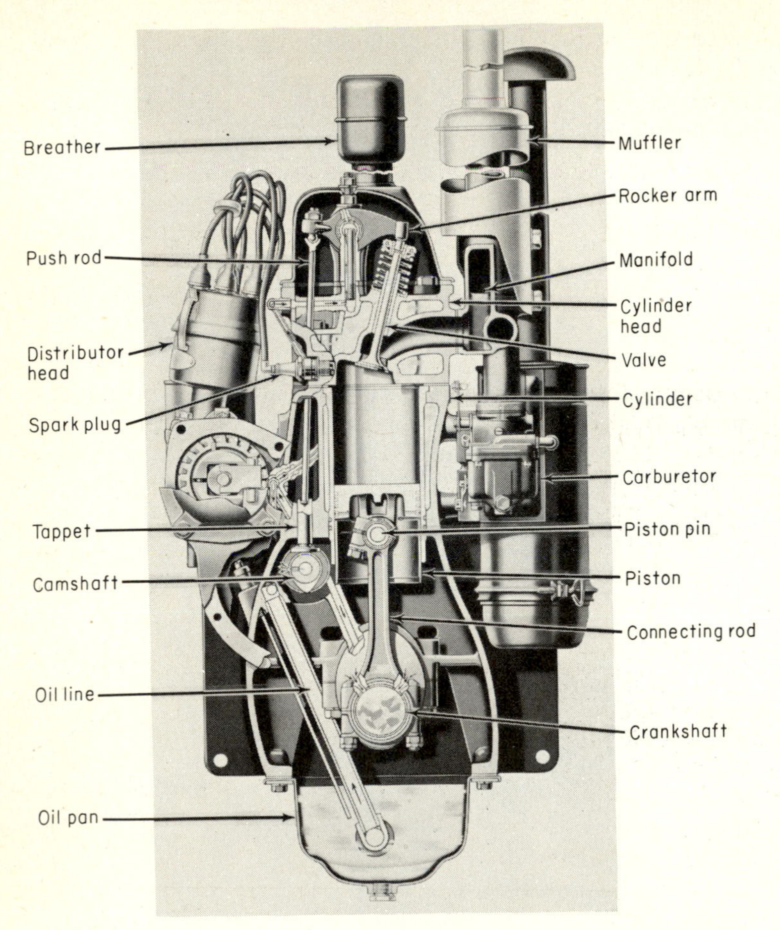

Figure 8-1 Basic design and parts of typical automotive-type engine. (Courtesy of Allis-Chalmers Manufacturing Company.)

cylinders (if multiple-cylinder), and manufacturing and production operations.

Multiple-cylinder, automotive engines usually have all cylinders cast together; that is, the cylinders are arranged side by side or in line in a single unit. Special cylinder arrangements such as the V arrangement and the radial arrangement, in most cases have the cylinders cast singly or in groups. Air-cooled engines usually have the cylinders cast separately and equipped with fins to provide better air circulation and more rapid escape of excess heat.

Casting the cylinders in a single block offers a number of advantages. It gives a more compact and rigid construction, simplifies manufacturing and assembly operations, and provides better enclosure of engine mechanisms. Hollow spaces in the block provide for the circulation of the cooling medium for liquid-cooled engines.

Cylinders are usually made of high-grade cast iron. This material has the necessary strength, machines and grinds readily, withstands extreme pressures and temperatures without distortion, and is low in cost. Usually some scrap steel is mixed with the pig iron in casting cylinder blocks. This provides additional strength and better wearing qualities. In some cases such metals as chromium, nickel, and molybdenum are added to the cast iron to give an alloy that has greater strength and wear resistance, with less weight.

Cylinder Head All engines and particularly liquid-cooled engines have removable cylinder heads. This head is attached to the block with a number of bolts and a copper-asbestos gasket provides a seal between it and the block. The cylinder head contains the pocket or space above the cylinder and piston known as the *combustion chamber.* It contains passages that match those of the cylinder block and allow the coolant to circulate to provide effective cooling. If the engine is of the valve-in-head type, all valves with their guides, springs, and retainers are mounted in the head. Cylinder heads are usually made of the same material as the block.

Cylinder Manufacture In manufacturing cylinders and cylinder blocks, a number of distinct steps are involved, as follows:

1 Melting and casting
2 Cleaning the rough casting
3 Machining: planing and milling
4 Boring, drilling, and tapping
5 Grinding the wall
6 Honing and polishing the wall

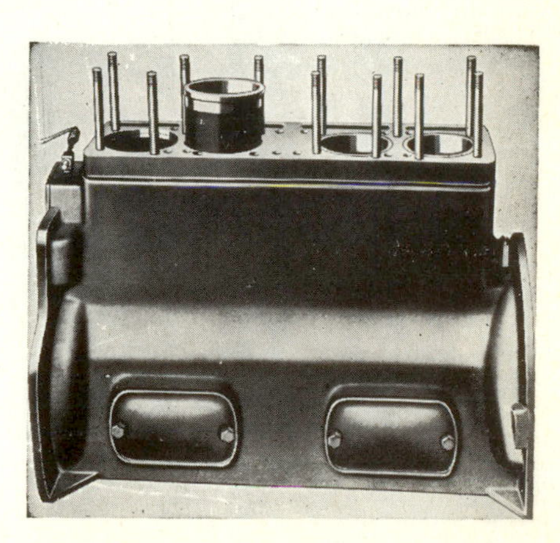

Figure 8-2 Cylinder block showing removable liners.

The grinding process consists in removing a thin layer of metal and producing an absolutely true bore by means of a somewhat coarse abrasive grinding wheel revolved at high speed. Honing is a finishing and polishing operation performed by a very fine-grained revolving stone. This operation produces a bore of exact dimensions and a mirror-smooth surface, which will provide the most efficient lubrication and the least possible wear.

Removable Cylinder Liners Some stationary engines, most tractor engines, and some truck engines are equipped with removable cylinder liners or sleeves (Figs. 8-3 and 8-4). The possible advantages are that (1) overhauling and rebuilding the engine are rendered relatively simple, easy, reliable, and satisfactory at a reasonable cost, since, by inserting new liners with new factory-fitted pistons, rings, and piston pins, the engine is practically restored to its original new condition so far as the cylinders and pistons are concerned; (2) in case one cylinder is scored or damaged, it can be easily repaired and the original bore and balance retained; and (3) a higher grade of cast iron or an alloy with better wear-resisting properties can be used for these liners, whereas to make the entire block of such material would be too expensive; special bore-finishing operations such as heat-treatment and honing to provide better oil control and longer wear are readily accomplished.

Cylinder liners may be classified as the "dry" type and the "wet" type. The latter serves as the entire cylinder wall and comes into direct contact with the coolant. Special sealing rings are placed at the top and

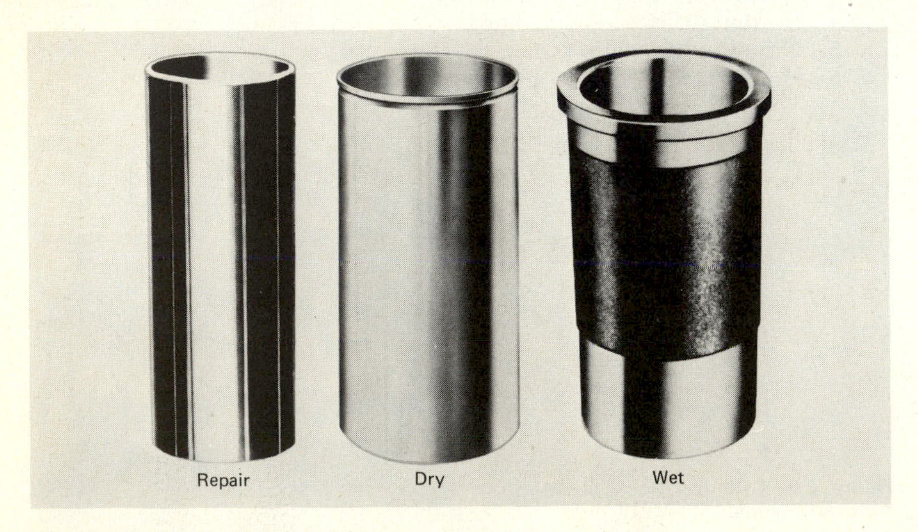

Figure 8-3 Types of cylinder liners.

Figure 8-4 Cylinder liner with piston and replacement parts.

bottom to prevent leakage of the coolant. ''Dry'' liners are usually much thinner and easier to replace and merely serve as the inner lining and cylinder wearing surface.

Crankcase and Oil Pan The crankcase is that part of the engine which supports and encloses the crankshaft and camshaft and provides a reservoir for the lubricating oil. It may also serve as a mounting for such accessories as the oil pump, oil filter, generator, starter motor, and ignition parts. Normally the upper portion of the crankcase is cast with or is integral with the cylinder block. The lower part of the crankcase is commonly called the *oil pan* and may be made of cast iron, cast aluminum, or pressed steel. This pan is securely bolted to the block and the joints sealed with suitable gaskets and oil seals.

Pistons The piston of an engine is the first part to begin movement and to transmit power to the crankshaft as a result of the pressure and energy released by the combustion of the fuel. Pistons are of the trunk type, that is, closed at one end and open at the other to permit direct attachment of the connecting rod and its free action. They are usually somewhat longer than their diameter, and are made as light as possible consistent with necessary strength, heat dissipation, and wearing qualities.

Materials used for pistons are gray cast iron, cast steel, and aluminum alloy. Cast iron or cast steel is ordinarily used in heavy-duty engines such as tractors, where high speeds and quick acceleration are not involved. Most automobile engines use aluminum-alloy pistons, because the lighter weight material permits higher operating speeds and greater speed flexibility. On the other hand, there are a few cases where heavy-duty engines use aluminum-alloy pistons because of this alloy's heat-conducting characteristic; that is, its use permits better control of the heat in the combustion chamber and therefore better control of the ignition and combustion of the fuel mixture.

Cast-iron pistons are more wear-resistant and require slightly less clearance in the cylinder than aluminum-alloy pistons. Cast-steel pistons are sometimes thinly coated with tin or a special metal to provide a smoother finish and a more wear-resistant surface. Some different types of pistons are shown by Fig. 8-5.

The approximate composition of a common piston alloy is:

Element	Percent
Aluminum	78.00–86.00
Silicon	11.25–15.00
Nickel	1.00– 3.00
Iron, max	1.30
Copper	0.50– 1.50
Magnesium	0.70– 1.30

The principal characteristics of this alloy are (1) low coefficient of expansion as compared with other aluminum alloys; (2) hardness and resistance to wear; and (3) good mechanical properties at high temperatures.

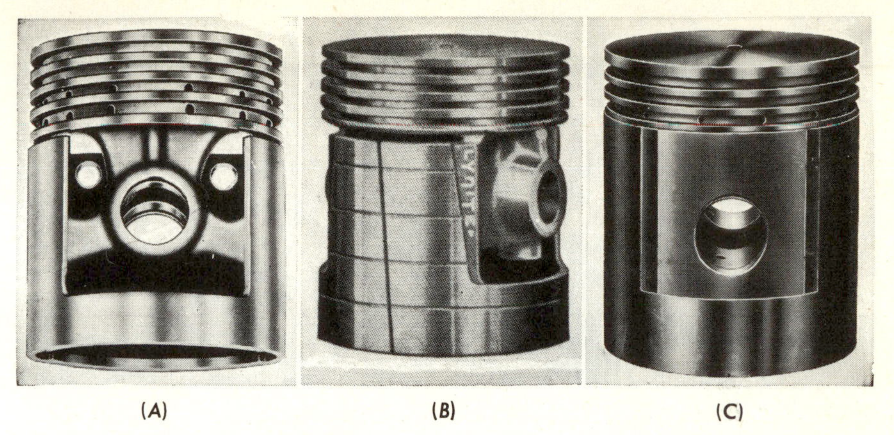

Figure 8-5 Common piston types. (A) Strut-type aluminum alloy, (B) split-skirt aluminum alloy, (C) plain cast iron.

Another piston alloy especially adapted to high-temperature conditions consists of:

Element	Percent
Aluminum	90.00–92.00
Copper	3.75– 4.25
Nickel	1.80– 2.30
Magnesium	1.20– 1.70
Iron, max	1.00
Silicon, max	0.70
Other impurities, max	0.20

The principal disadvantage of aluminum alloy for pistons is that it has a high coefficient of expansion. This means that the piston must be given slightly more clearance when fitted in the cylinder than is given one of cast iron. However, this problem has been partly overcome by embedding devices of steel or a special metal called *invar* at certain points in the piston when it is cast. These devices are called *struts*. They control the expansion by controlling the heat-flow path as well as serving to strengthen the piston structure. Reinforcing ribs on the inside of pistons also provide greater strength and better heat dissipation.

Figure 8-6 shows a special heavy-duty aluminum-alloy piston used in one make of diesel tractor. An iron band is built into the upper part and bears the top-ring groove. This is done to make the top ring and its groove more wear-resistant, because they are exposed to higher temperatures, carbon deposits, and other factors that induce more rapid wear. Note also the chrome-nickel steel plug in the top of the piston which provides heat resistance in the high-temperature area.

Pistons and piston rings are frequently given a special surface treat-

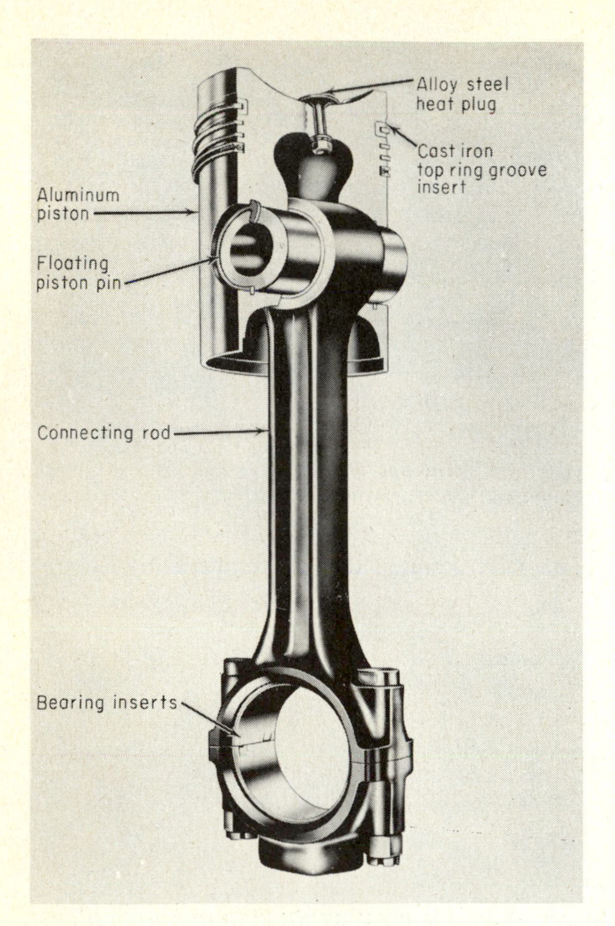

Figure 8-6 Piston and connecting-rod assembly for diesel tractor engine. (Courtesy of Caterpillar Tractor Company.)

ment to make them more wear-resistant. For example, some aluminum-alloy pistons are subjected to a process known as anodizing. This is an electrolytic process which converts a thin outer layer of the metal into aluminum oxide. This oxide is harder than the original metal and also more heat-resistant. Tin plating is another method of providing a smoother and better-wearing surface for both cast-iron and aluminum-alloy pistons.

The term *piston clearance* is applied to the space between the piston wall and cylinder wall. For cast-iron pistons, this clearance, at the closed end, when the engine is cold, should be about 0.0005 in for each inch of cylinder diameter. Most pistons are slightly tapered and larger in diameter at the open end and may have less clearance at this end, since this portion of the piston is not exposed to as much heat as the closed end and therefore expands less. Aluminum-alloy pistons require about twice as much clearance as cast-iron pistons; but, as previously explained, most aluminum-alloy pistons are now so constructed that they require no more clearance than other types. Some pistons are said to be *cam-ground;* that is, the skirt is not a true circle but has an elliptical shape. The variation is possibly 0.005 to 0.010 in, and the greatest diameter is across the surfaces of greatest wear, namely, at right angles to the piston pin. Too much piston clearance will cause loss of compression, oil pumping, and piston slap. Too little clearance will cause the piston to stick or seize in the cylinder as the engine gets hot. However, a piston may seize in a cylinder even though it has the proper clearance, as a result of improper action of the cooling system or lack of cylinder lubrication.

Piston Rings The primary function of the piston rings is to retain compression and, at the same time, reduce the cylinder-wall and piston-wall contact area to a minimum, thus preventing friction losses and excessive wear. Other important functions are the control of the oil and cylinder lubrication and the transmission of heat away from the piston and to the cylinder walls.

Piston rings are made of cast iron because it retains its wearing qualities and elasticity indefinitely. Furthermore, it is simpler and cheaper to replace the rings than the cylinders or liners. Hence, the rings should absorb the wear if possible, and this makes cast-iron rings more desirable. The top ring in some heavy-duty and automotive-type engines is chromium plated to improve its performance and wearing qualities. The number of rings per piston varies from three to five, depending upon the type of engine and the compression desired. The ordinary types of engines seldom have more than three or four rings, but diesel-type engines usually have five rings per piston.

Piston rings are classed as compression rings and oil rings (Figs. 8-7 and 8-8), depending upon their specific function and location on the piston. Compression rings are usually plain one-piece rings and are always placed in the grooves nearest the piston head. Oil rings are grooved or slotted and are located either in the lowest groove above the piston pin or in a groove near the piston skirt. Their function, obviously, is to control the distribution of the lubricating oil to the cylinder and piston surfaces and to prevent unnecessary or excessive oil consumption. The ring groove under an oil ring is provided with holes through which the excess oil returns to the crankcase.

When removed from the cylinder, piston rings are always slightly larger in diameter than the cylinder itself, and they must be compressed when inserted. Three kinds of joints are found (Fig. 8-7): the bevel joint, the plain butt joint, and step joint.

Piston-ring clearance is the distance or space at the joint of the ring

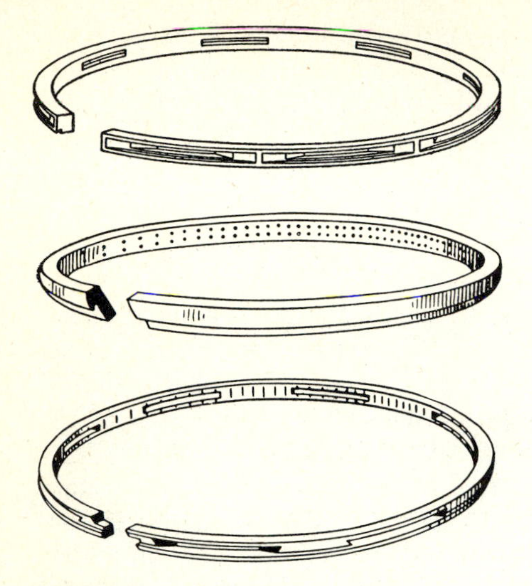

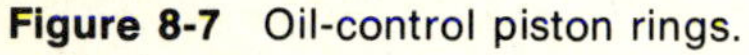

Figure 8-7 Oil-control piston rings.

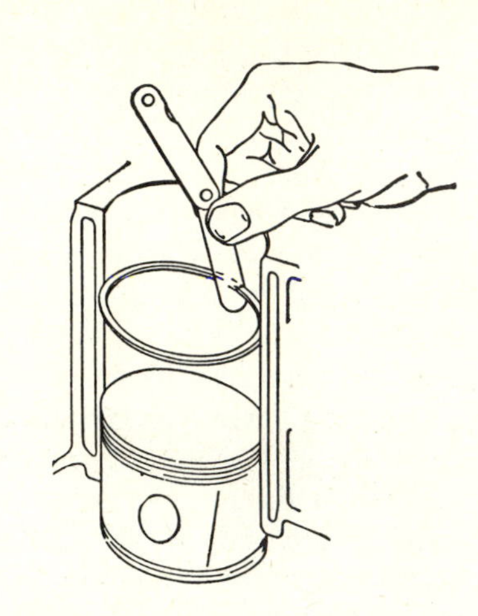

Figure 8-8 Checking piston-ring clearance.

when it is in the cylinder (Fig. 8-8). This clearance is necessary to allow for the expansion of the ring as it gets hot. Without such clearance the ring would buckle and break and consequently damage the cylinder and piston of the engine. Too much piston-ring clearance is apt to produce leakage of compression and possible waste of lubricating oil. Table 8-1 gives the SAE recommended end clearance for piston rings.

Special thin, flexible, spring-steel bands called inner rings (Fig. 8-9) or expanders are sometimes placed in the ring grooves under the regular rings in reconditioned engines for the purpose of creating better contact between ring and cylinder wall.

In fitting and replacing piston rings, care should be taken to see that (1) the grooves are free of carbon deposits; (2) each ring is free in its groove and has the correct end and side clearance; and (3) the joints of the rings are not in line when the piston is inserted in the cylinder.

Table 8-1 Piston-ring End Clearances

Cylinder diameter, in	End clearance, in
1–1$^{31}/_{32}$	0.005–0.013
2–2$^{31}/_{32}$	0.007–0.017
3–3$^{31}/_{32}$	0.010–0.020
4–4$^{31}/_{32}$	0.013–0.025
5–6$^{31}/_{32}$	0.017–0.032
7–8	0.023–0.040

large end of the connecting rod. The large end ordinarily has what is known as a split bearing held together by two bolts.

Connecting-Rod and Crankshaft Bearings One of the outstanding achievements in automotive engineering has been the remarkable improvement and progress made in the design of engine crankshaft and connecting-rod bearings whereby those bearings wear much longer even under excessively high speeds and loads. This exceptional bearing performance can be attributed to a number of factors, such as (1) the use of better bearing alloys; (2) greater precision in bearing and crankshaft manufacture; (3) the use of harder and more wear-resistant crankshaft steels; and (4) better engine lubrication and improved engine oils.

With the development of higher speeds and compression pressures, greater bearing pressures were produced, and it was found that older types of thick babbitt bearings would not stand up as satisfactorily as desired. Consequently, it became necessary to develop new bearing materials and designs, provide the crankshaft journals with a smoother and harder finish, and maintain precision construction throughout. The result was the introduction of thin-shelled precision or insert types of bearings and the use of new types of bearing alloys, such as lead-base babbitt, copper-lead, and aluminum.

The general construction of an insert bearing is shown in Fig. 8-11. The two-layer insert consists of a thin layer (0.010 to 0.030 in) of the bearing alloy bonded to a thicker bronze or steel back. A mild steel is used and is preferable to bronze from the standpoint of cost, strength, and general bonding characteristics. The three-layer insert is similar to the two-layer except that a very thin layer—about 0.001 in thick—of babbitt or other lead alloy is bonded to the usual bearing-alloy layer. This third layer, called an overlay or overplate, provides a cushion between the shaft

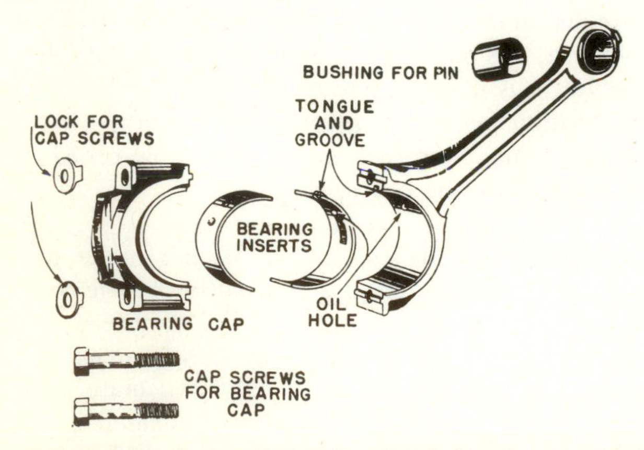

Figure 8-10 Connecting-rod assembly showing bearing inserts.

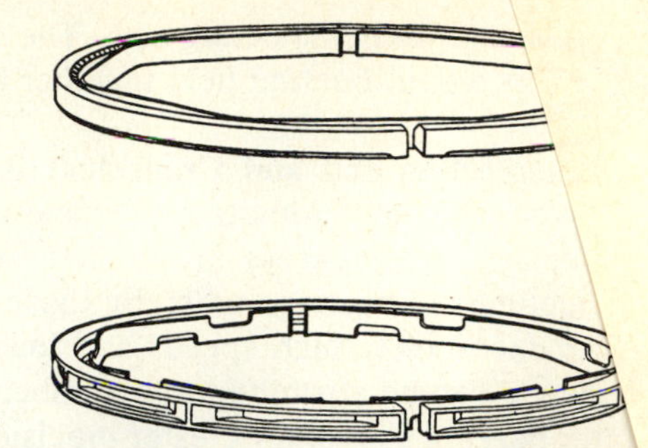

Figure 8-9 Piston-ring expanders.

Piston Pin The function of the piston pin is to join the connecting rod to the piston and, at the same time, provide a flexible or hinge-like connection between the two.

The pin passes through the piston-pin bosses and through the upper end of the connecting rod that rides within the piston on the middle of the pin. Piston pins are made of casehardened alloy steel with a precision finish. The hollow construction gives maximum strength with minimum weight. They are lubricated by splash from the crankcase or by pressure through passages bored in the connecting rods.

Three different methods are used to anchor piston pins so that they cannot work sideways and score the cylinder. The first is to clamp the pin to the connecting rod by means of a clamp screw or setscrew, so that the bearing will be at each end of the pin where it fits in the piston. In the second case, the pin is anchored to the piston by means of a setscrew and the bearing is in the connecting rod. In the third, or most common, method (Fig. 8-6), the piston pin is allowed to float, so to speak, or move in both the piston and connecting rod but is held in place by means of snap rings at each end of the pin.

Some kind of removable bushing, usually bronze, is placed either in the connecting rod or in the piston to receive the wear. When the piston-pin bearing becomes badly worn, this bushing can be removed readily and replaced. Frequently, however, it is necessary to replace both the piston pin and the bushing. The floating piston pin gives more uniform wear distribution. For this type, the connecting rod uses a removable bronze bushing, but bushings are not used in the piston bosses.

Connecting Rod Connecting rods (Figs. 8-6, 8-10) are made of *drop-forged* steel. They must be of some material that is neither brittle nor ductile; that is, they must stand a twisting strain but should not break or bend when subjected to such a strain. The I-beam type is the prevailing connecting-rod shape. It gives strength with less weight and material. That end of the rod fastened to the piston pin is known as the small end, and the other end, which is attached to the crankshaft, is spoken of as the

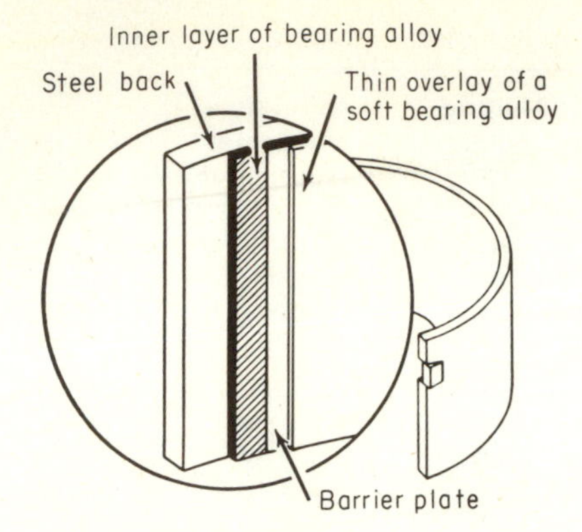

Figure 8-11 Three-layer bearing construction.

journal and the bearing during the "running-in" period as they conform to each other. The overplate usually wears away, but, in doing so, leaves a smooth fitting journal and bearing surface.

To give satisfactory results, a bearing alloy must have a number of specific characteristics such as (1) sufficient strength including load-carrying capacity, resistance to fatigue, high impact and compressive strength, and high-temperature strength; (2) good antifriction and surface characteristics so that it will adjust itself to the journal without any bonding action or surface failure and damage; (3) resistance to corrosion by combustion products or motor-oil contaminants; (4) high thermal conductivity to provide rapid frictional heat dissipation; (5) reasonable cost. Table 8-2 lists the properties of various bearing metals.

The principal bearing alloys and their approximate composition are:

Tin-base babbitt. Copper, 3.50 percent; antimony, 7.50 percent; tin, 89.00 percent.

Lead-base babbitt. Copper, 0.25 to 0.50 percent; tin, 5.00 to 10.00 percent; antimony, 9.00 to 16.00 percent; lead, 75.00 to 85.00 percent.

Copper-lead alloy. Lead, 25.00 to 40.00 percent; copper, 60.00 to 75.00 percent.

Aluminum alloy. Copper and nickel, 0.50 percent; silicon, 2.50 percent; tin, 7.50 percent; aluminum, 90.00 percent.

Tin-base babbitts have been used for many years for automotive engine bearings. However, as compression pressures and operating speeds and temperatures increased, it was found that tin-base babbitts lacked the necessary mechanical strength to meet reasonable service requirements. This was particularly true if the bearing temperatures were unusually high. Therefore other alloys were developed. At the present time the copper-lead alloys appear best adapted to heavy-duty engines and are widely used. Aluminum bearing alloys in the form of heavy solid bushings are also proving satisfactory for this heavy-duty service.

Table 8-2 Properties of Metals

Metal	Brinnell hardness number	Melting point, °F	Relative thermal conductivity
Tin	4.5–7.0	450	0.16
Lead	4.2	621	0.083
Antimony	. . . .	1,167	0.045
Aluminum	25–40	1,220	0.53
Silver	59	1,761	1.00
Copper (cast)	35	1,981	0.94
Nickel	90–110	2,651	0.22
Wrought iron	69–75	2,600	0.18
High-carbon steel	190–220	2,500	
Chromium	91	3,430	0.16

A most important consideration is the proper servicing and replacement of precision insert bearings. First of all, shims are never used with insert bearings, and bearing caps and inserts should never be filed under any conditions. The primary factor in insert-bearing replacement is correct oil clearance. This is the space allowed between the bearing surface and the journal surface for the oil film. If this space is too small, the film will be too thin to prevent metal-to-metal contact between the bearing and the journal. If the clearance is excessive or greater than necessary, oil will flow too freely and may result in a reduction in pressure and some increased oil consumption and wastage.

The oil-film clearance depends on the thickness of the bearing insert and the journal diameter; suggested clearances are indicated in Table 8-3.

Crankshaft The crankshaft, by means of its cranks and the connecting rods, converts the reciprocating movement of the piston into the necessary rotary motion. Most crankshafts are made of medium-carbon steel or a chrome-nickel alloy steel by the drop-forging process. A few high-speed, multiple-cylinder engines use a cast crankshaft, the material having a relatively high carbon and copper content. The casting process is simpler and less expensive because these crankshafts require heavy integral counterweights which make the forging method quite difficult.

The size of the crankshaft, the number of main bearings, and the number and arrangement of the cranks are dependent upon the type, size, and speed of the engine. Figures 8-12 to 8-17 inclusive show a number of typical crankshaft designs. The angle of the cranks with respect to each other for the crankshaft of a multiple-cylinder engine is determined by the number of cylinders and their arrangement. A symmetrical or balanced arrangement of cranks about the shaft provides better balance and reduces vibration. Crankshafts for high-speed multiple-cylinder engines are

Table 8-3 Recommended Oil Clearance for Various Types of Bearings for Pressure Lubrication

Diameter of crank-shaft journal or crankpin, in	Lead or tin-base babbitt, in		Copper-lead, in		Plated copper-lead, in		Aluminum, in	
	Connecting rod	Main journal	Connecting rod	Main journal	Connecting rod	Main journal	Connecting rod	Main journal
2–2.75	0.0008–0.0025	0.0010–0.0030	0.0015–0.0035	0.0020–0.0040	0.0008–0.0030	0.0010–0.0035	0.0015–0.0035	0.0020–0.0040
2.81–3.50	0.0010–0.0030	0.0015–0.0035	0.0020–0.0040	0.0025–0.0045	0.0010–0.0035	0.0015–0.0040	0.0010–0.0035	0.0015–0.0040
3.56–4.00	0.0020–0.0035		0.0030–0.0045		0.0020–0.0040		0.0035–0.0050	

equipped with counterweights throughout their length to provide complete balance and to reduce bearing stresses and wear at high speeds. These weights are usually made integral with the shaft.

The number of main crankshaft bearings often varies even for different engines having the same number of cylinders. For example, a crankshaft for a four-cylinder engine may have two, three, or five main bearings. A six-cylinder crankshaft may have three, four, five, or seven main bearings.

Great improvements have been made in crankshaft manufacture to reduce wear. One process involves the hardening of the surface layer of metal at the journals by an electrical treatment known as induction hardening or electrohardening. Also the entire shaft may be specially heat-treated and a method of precision grinding of the journal surfaces, known as superfinishing, gives this surface a mirror finish, thus reducing friction and wear.

Flywheel Flywheels are usually made of cast iron and their primary function is to maintain uniform engine speed by carrying the crankshaft through those intervals when it is not receiving energy from a piston. The size of the flywheel varies with the number of cylinders and the type and size of engine; that is, the greater the number of cylinders, the smaller and lighter the flywheel for the same total piston displacement because of the overlapping power strokes. The flywheel usually carries the ring gear that meshes with the starter pinion for self-cranking. The rear surface or side may be finished very smooth to serve as one of the pressure surfaces for the clutch plate. Engine-timing marks are frequently stamped on the flywheel.

CRANKSHAFT AND ENGINE BALANCE—VIBRATION CONTROL

The trend toward higher operating speeds for automotive engines has introduced new problems in design. One of these is proper mechanical balance and the control of vibration. An unbalanced mechanical condition in an automotive engine not only causes annoying and wear-producing vibration, but creates excessive stresses on the bearings and other supporting parts. These strains induce faster wear and mechanism failures.

Vibration in automotive engines may be caused by any or all of the following:

1. Unbalanced centrifugal forces of rotating members, particularly the crankshaft.
2. Torsional reaction of the crankshaft.
3. Unbalanced inertia forces created by reciprocating parts, particularly the pistons and connecting rods.
4. Torque reaction resulting from effect of the explosions on the cylinders.

Static and Dynamic Balance The crankshaft and flywheel are the principal engine parts having sufficient weight and rotative speed to create excessive centrifugal-force reactions. Ordinarily the flywheel is symmetrical in shape; hence, if it is uniform in structure and precision-made, it should have perfect balance at any speed. Therefore, the balance of the crankshaft alone is of major consideration.

A crankshaft should have both *static* and *dynamic* balance. For example, a single-cylinder crankshaft without counterweights is unbalanced statically because, if rotated on frictionless supports, it would stop only in one position—with the crank journal downward. If such a shaft were rotated at any appreciable speed, excessive strains and stresses would be set up within it midway between the supports as well as on the supports themselves, owing to centrifugal force created by the unbalanced crank journal. Now suppose suitable counterweights were added opposite the two crank throws (Fig. 8-12); the shaft if rotated would not come to rest in any one position each time because of static balance. Furthermore, the force reactions during rotation would be almost completely balanced and excessive bearing stresses reduced. The two-cylinder crankshaft (Fig. 8-13) has static balance even though it does not have counterweights; that is, the arrangement of the crank journals about the axis is such as to provide static balance.

However, neither one has *dynamic balance,* that is, balance when rotated at any appreciable speed. The centrifugal-force reactions would vary at different points along the shaft axis and thereby set up an unbalanced condition with respect to the shaft as a whole. For example, suppose the crankshaft (Fig. 8-13) is suspended in a vertical position by a strong wire attached on one end at the center point and then rotated at any speed. If there is perfect dynamic balance, it should rotate with its central or longitudinal axis remaining in a true vertical position. This shaft would not do this; the lower or free end would tend to swing outward and travel

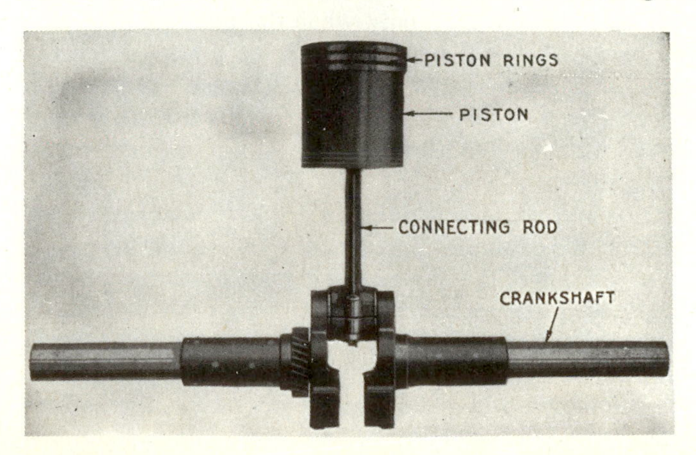

Figure 8-12 Single-cylinder counterbalanced crankshaft.

in a circular path, thus indicating a definite lack of complete balance at all points along the axis. To overcome this difficulty, the shaft must be equipped with counterweights (Figs. 8-14, 8-15). These must be placed as nearly opposite the crank journals and crank throws as possible so that it might be said that perfect static balance exists for any section of the shaft. This explains the purpose of the arrangement of weights found on nearly all multiple-cylinder crankshafts used in automotive engines, even though it may appear that the crank journals and throws, when symmetrically arranged about the shaft axis, should produce complete balance under all conditions. Figures 8-16 and 8-17 show four- and six-cylinder crankshafts, respectively, which are statically and dynamically balanced.

Torsional Vibration Torsional vibration means vibration resulting from the twisting reaction created in the crankshaft by the explosions in the cylinders and the inertia forces in the pistons and connecting rods. This will be more clearly understood by assuming that the crankshaft is made of very strong but slightly resilient rubber instead of steel and that the flywheel on one end is sufficiently heavy to resist any instantaneous changes in speed. When an explosion occurs, the impulse reacts on the crankpin through the piston and connecting rod, and deflects or twists the shaft between that particular crank and the flywheel. When the pressure decreases, the crankpin attains maximum deflection and begins a backward swing. However, it will not stop instantly at its normal position but will swing beyond this point and continue like a pendulum. Now suppose that, before this crankpin ceases to swing, the piston receives another power impulse which reacts on the pin just as it reaches the point of maximum backswing. This will increase the deflection or magnitude of the twisting action, and if this synchronized condition were maintained for any length of time, the shaft might fail. That speed at which such a condition exists is known as the *critical speed* and creates *torsional vibration*. The maximum vibration would be created only when the next explosion occurred on the first backswing of the crankpin. Some torsional reaction may exist in any crankshaft, even at low speed; but excessive reaction and vibration usually appear in the higher speed ranges. This trouble must be avoided or controlled, not only to eliminate the vibration itself but to prevent failure of the shaft.

Torsional vibration is controlled in different ways. If any engine, such as one used in a tractor, for example, operates at a reasonably low speed or within a limited speed range, the shaft can be made of such a size that its normal operating speed does not correspond with the critical speed. Or the shaft may be made very heavy so that any harmful critical speeds are higher than the maximum operating speed.

Most multiple-cylinder automotive engines operating at relatively high speeds have their crankshafts equipped with special devices for con-

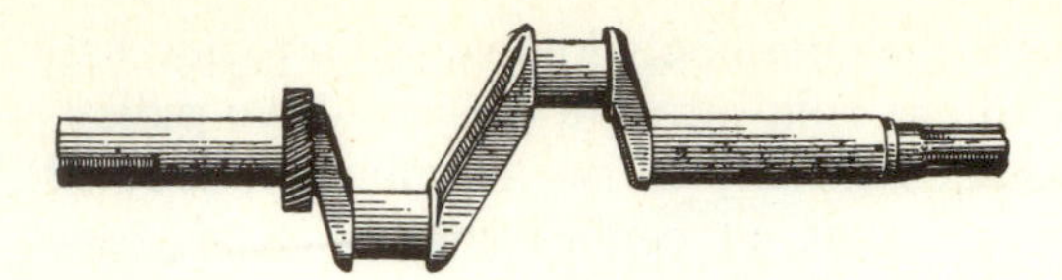

Figure 8-13 Plain two-cylinder opposed crankshaft.

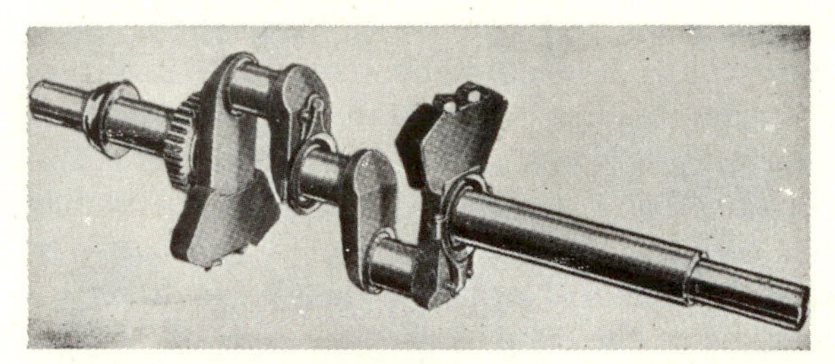

Figure 8-14 Two-cylinder three-bearing counterbalanced crankshaft.

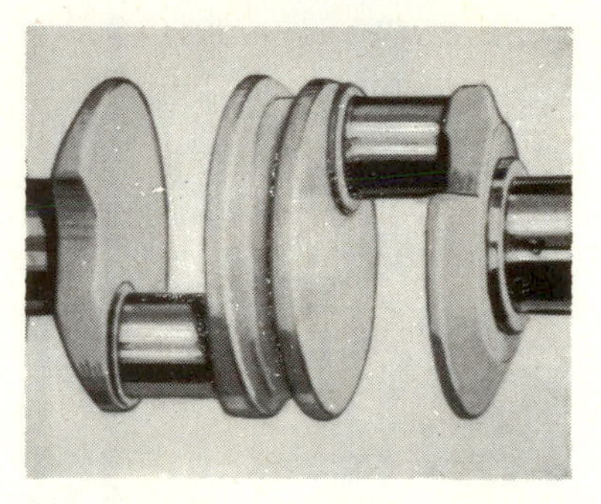

Figure 8-15 Statically and dynamically balanced two-cylinder crankshaft.

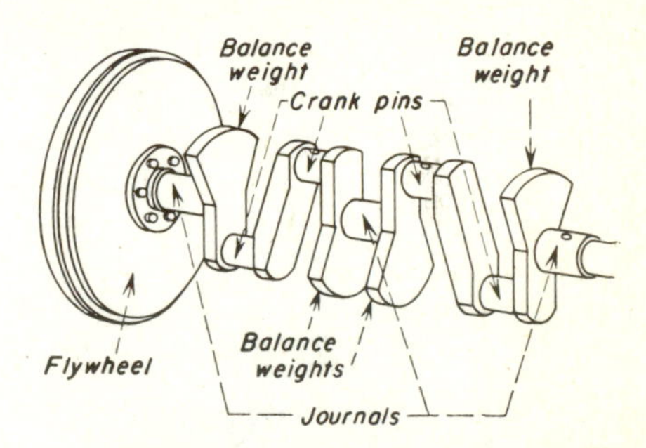

Figure 8-16 A statically and dynamically balanced four-cylinder three-bearing crankshaft.

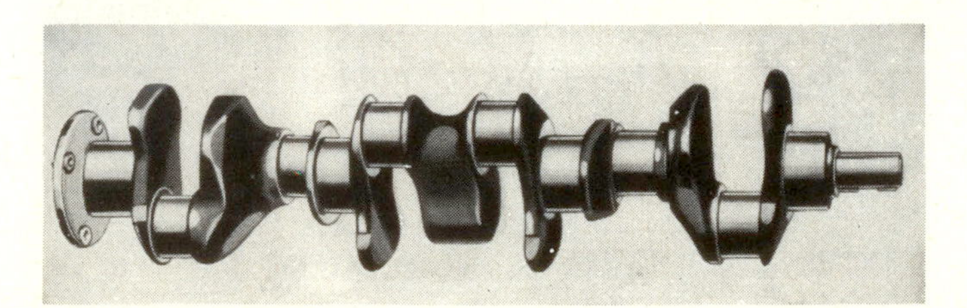

Figure 8-17 A statically and dynamically balanced six-cylinder crankshaft. (Courtesy Chevrolet Division, General Motors Corporation.)

trolling torsional vibration. These are commonly known as harmonic balancers or vibration dampers and are built into the fan-belt drive pulley. These devices (Fig. 8-18) consist essentially of two members connected with some degree of flexibility so that, at certain engine speeds, one member vibrates with respect to the other. If this vibration is out of phase with the torsional vibration of the crankshaft, it counteracts the latter and reduces it to some negligible value.

Balance of Inertia Forces Complete balance of those engine parts having a reciprocating movement is extremely important in eliminating or reducing vibration. These parts include principally the piston, piston pin, and connecting rod. An inertia force is one created when an object at rest is suddenly placed in motion or when a moving object is suddenly brought to a state of rest. In other words, inertia is that property of matter by virtue of which a body tends to remain at rest or to remain in motion in a straight line. Applying this to an engine piston and connecting rod, we find that these parts come to rest and change direction of travel twice in one crank revolution. Hence there is a pronounced, continuous, and rapid

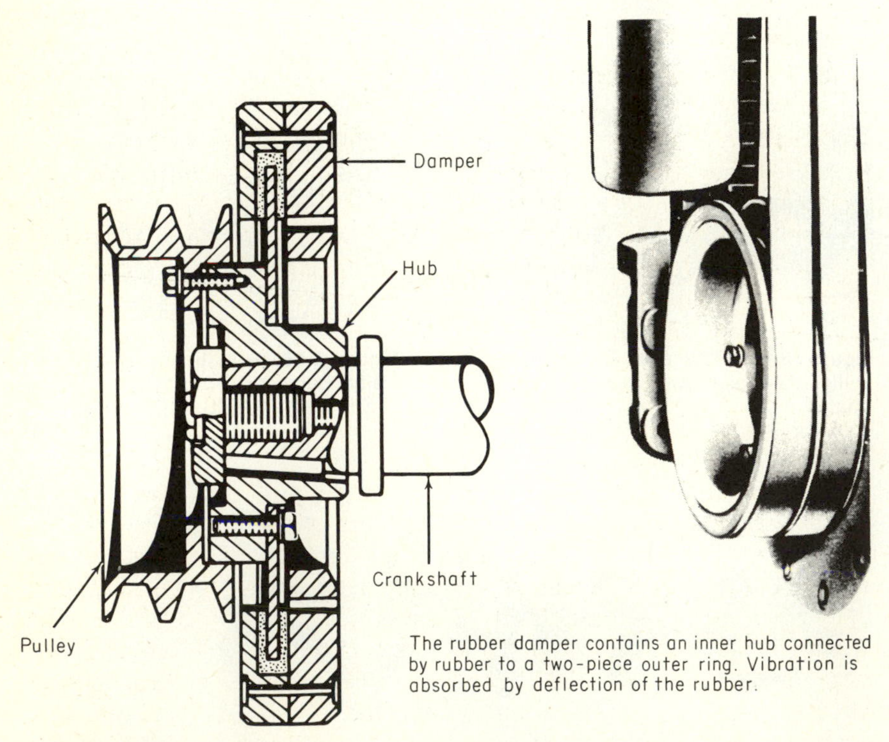

Figure 8-18 Cross-sectional view of a torsional vibration damper. (Courtesy of Caterpillar Tractor Company.)

change of velocity, which in turn sets up definite inertia forces. For example, if the crankshaft has a speed of 1,000 rpm, the piston and connecting rod come to rest, change direction of travel, and attain a definite and maximum velocity 2,000 times per minute. These changes in velocity and travel direction set up definite inertia forces which react on the cylinder and create vibration.

The effect and direction of the reaction from inertia forces is determined to a large extent by the cylinder arrangement. For a horizontal engine the forces would act in a horizontal plane, and for a vertical engine they would act vertically. If there were a definite unbalanced condition in both cases, the vibration would be likely to be more noticeable in the horizontal engine. In a vertical engine the reaction would be counteracted directly by the total mass or weight of the engine.

The balancing of these inertia forces in order to eliminate vibration can be accomplished in a number of ways. The problem is more difficult in large, heavy engines with one or two large cylinders than in light-weight, high-speed engines with several small cylinders. It is almost impossible to eliminate vibration in a large single-cylinder engine. Counterweighting the crankshaft, as previously explained, will help to some extent. Using counterweights and arranging the cranks and pistons for a two-cylinder engine as shown in Fig. 8-15 will very nearly eliminate the vibration. Certain inertia force reactions resulting from the connecting-rod movement are difficult to control completely. For a two-cylinder engine, the arrangement giving the least vibration is that of opposed cylinders with opposed cranks. Vertical, in-line engines with three, four, six, or eight cylinders and V-type engines with four, eight, or twelve cylinders give very little vibration from reciprocating forces if the cranks are properly arranged.

Some four-cylinder, high-speed engines develop vibration from what might be called secondary inertia forces. Even though the two end pistons are equal in weight and move in an opposite direction to the two inside pistons and all are attached to cranks in the same plane, the inertia forces are not completely balanced, because, in any engine, the pistons have a higher velocity during the upper half of the stroke than during the lower half. The resulting reaction, if excessive, acts along the vertical center line of the engine, midway between the axes of the two inner cylinders. It may be transmitted to the front and rear supports of the engine and create undesirable vibration. Lanchester, an English automotive engineer, developed a device for controlling this vibration. It consists of a helical gear, attached to one of the crank arms of the crankshaft, which meshes with another similar gear of one-half the size located in the bottom of the crankcase and at a right angle to the larger gear. The small gear meshes with a second one of like size lying parallel to it. The two small gears are attached to cylinders mounted on stationary studs and weighted heavier on one side, as shown in Fig. 8-19. These weights revolve in opposite

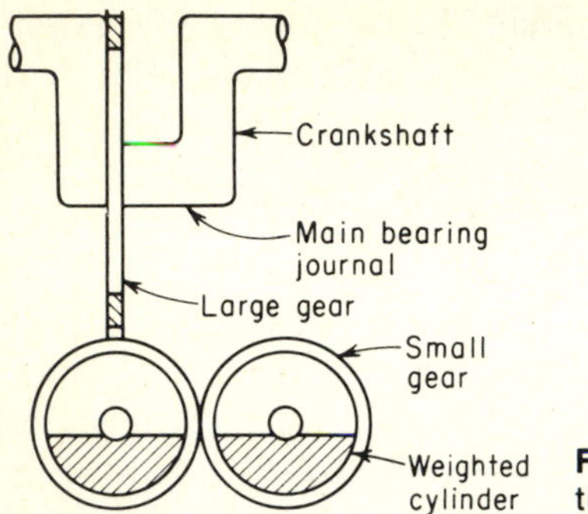

Figure 8-19 Lanchester-type vibration damper.

directions and at twice crankshaft speed and are so meshed with each other that the horizontal components of their respective centrifugal forces neutralize each other, while the vertical components are combined. Therefore, if the weights are properly proportioned, the centrifugal force created by their rotation reacts against the secondary unbalanced force created by the two pairs of pistons in their movement.

TRACTOR-ENGINE TYPES AND CONSTRUCTION

The first gas tractors built were equipped with large slow-speed horizontal engines having only one or two cylinders. Obviously, these engines required a strong frame, large wheels, and other supporting parts. Consequently the tractors themselves were very heavy, were difficult to start and handle, and had considerable vibration. Later on, when the possibilities of lighter weight tractors were observed, designers turned their attention toward the use of a higher speed, lighter weight power plant having at least two, and possibly four, cylinders, such as those used in trucks and automobiles. The result is that the trend in the tractor industry, as in automotive design, has been toward engines having two, four, or six cylinders, with lighter weight and higher operating speeds.

Tractor-Engine Characteristics Tractor power plants may be classified as two-cylinder twin horizontal engines and as four- and six-cylinder vertical engines. All tractor engines are either of the four-stroke-cycle, heavy-duty, carbureting type or of the diesel type. Gasoline, LP gas, and diesel fuels are the predominating fuels.

Horizontal Two-Cylinder Engines The horizontal two-cylinder-type engine (Fig. 8-20) has been used by some manufacturers. The advantages claimed for it are: (1) it has fewer and heavier parts with slower speed and, therefore, less wear and longer life; (2) it has greater accessibility; and (3) its belt pulley is driven direct and not through gears, because the

Figure 8-20 Cutaway view of a two-cylinder horizontal tractor engine.

crankshaft is placed crosswise on the frame. The use of two-cylinder engines is now confined largely to garden tractors and similar types of vehicles utilizing very limited power.

Vertical Three-Cylinder Engines Figure 8-21 shows a crankshaft for a three-cylinder vertical engine.

Four-Cylinder Engines The vertical four-cylinder type of engine, placed lengthwise of the frame, predominates in the tractor field at the present time. Some of the reasons for this are (1) good weight distribution

Figure 8-21 A crankshaft for a three-cylinder engine.

is secured; (2) manifolding and fuel-mixture distribution are simplified; (3) uniform and positive cylinder and bearing lubrication is facilitated; (4) vibratory effects caused either by moving parts or by the explosions act in a vertical plane and are thus less noticeable and nullified by the engine and tractor weight itself; (5) valve mechanisms, ignition devices, and other parts are made more accessible; (6) the clutch and transmission parts can be assembled in such a way as to give the entire machine a balanced construction and symmetrical appearance.

Six-Cylinder Engines A six-cylinder engine will produce somewhat smoother power and less vibration than an engine with fewer cylinders. The cranks are always arranged as shown in Fig. 7-14 and the usual firing order is 1-5-3-6-2-4.

Features of Tractor-Engine Design There is now a much greater similarity in tractor-engine construction and design than in former years. Such features as valve-in-head construction, removable cylinder sleeves, and cylinders cast en bloc are used in a majority of cases. Tractor engines are classed as heavy-duty engines; that is, castings, pistons, crankshafts, bearings, valves, and practically all important parts are made heavier and larger than would be necessary in an automobile engine, in order to withstand the unusually heavy strains and continuous heavy loads to which they are subjected.

Cylinders are cast en bloc, that is, all in one casting for the small and medium sizes of tractors. Larger engines may have the cylinders cast in pairs.

Tractor-engine cylinders are subjected to rather rapid wear, which eventually causes excessive power losses and oil pumping. If the block is not equipped with removable cylinder liners, the trouble can be remedied either by replacement with a new block or by reboring the old block and using oversize pistons. Most tractor engines use removable cylinder liners. The usual practice is to use new factory-fitted pistons with the new liners.

Cylinder heads for tractor engines are cast separately from the block. This permits ready access to the combustion chamber for cleaning out carbon deposits and for valve servicing. With the valves located in the cylinder head, valve reconditioning and adjustment are simplified.

Tractor pistons are, in most cases, made of cast iron with three to five rings. Piston pins are made hollow and usually are of the floating type (Fig. 8-6).

Crankshafts and Bearings All tractor crankshafts are one-piece drop forgings, which have been carefully machined, heat-treated, and precision finished. Two-cylinder crankshafts may have two main bearings (Fig. 8-12) or three main bearings (Fig. 8-14).

Four-cylinder crankshafts may have two or three main bearings, as shown in Fig. 8-16. The fewer the bearings, the larger the shaft diameter. The three-bearing crankshaft is most common. Six-cylinder crankshafts may have either three, four or seven main bearings. Both the main and the crank journals are very accurately machined and ground to within 0.001 or 0.002 in of the specified dimension.

Insert-type liners are used for both the main and crank journals on nearly all tractors. These may be standard tin-base babbitt inserts in the case of smaller engines or special heavy-duty bearing alloys in the case of larger engines.

Bore and Stroke The size of an engine is determined to a large extent by the cylinder diameter or bore, the length of stroke of the piston, and the total number of cylinders. There is no standard relationship between the cylinder bore and length of stroke but, in general, the stroke is 1.1 to 1.3 times the bore. An engine having a stroke greater than 1.3 times its bore would be considered a long-stroke engine. The trend in multiple-cylinder, high-speed engines is toward a shorter stroke compared with the bore. For example, in many automobile engines, the length of stroke is less than the cylinder bore. A bore to stroke ratio of 1.00 or less is particularly advantageous in very high-speed engines of six or more cylinders for the following reasons: (1) it makes possible some reduction in the amount of material needed in the engine; that is, the engine weight per horsepower can be reduced since the crank throws are shorter; (2) a shorter crank radius reduces the total moment of inertia of the crankshaft and thus reduces the probability of vibration at high speeds; (3) piston speeds and ring contact areas are less for any given crankshaft speed, which in turn means less cylinder wall and ring wear (see Table 8-4); (4) more cylinder head area is available to permit the use of larger valves, if needed. Assuming a given piston displacement, the effect of different bore-stroke ratios on the piston speed and ring contact area per stroke is shown in Table 8-4.

Table 8-4 Relationship of Bore-Stroke Ratio to Piston Speed and Ring-Contact Area

Bore and stroke, in.	Piston displacement, in^3	Bore to stroke ratio	Piston speed at 2,000 rpm, ft/min	Top ring contact area per stroke, in^2
$3^{13}/_{16} \times 3$	34	0.79	1,000	35.9
$3^1/_2 \times 3^1/_2$	34	1.00	1,167	38.5
$3^3/_8 \times 3^3/_4$	34	1.11	1,250	40.0
$3^9/_{32} \times 4$	34	1.22	1,333	41.3
$3^3/_{16} \times 4^1/_4$	34	1.38	1,417	42.6

Compression Ratio The compression ratio of an engine is the ratio of the cylinder volume existing when the piston is on bottom or crank dead center and the volume remaining above the piston when it reaches top or head dead center. For example, in Fig. 8-22, if the total cylinder volume with the piston on bottom dead center is 120 in^3 and this is reduced to 20 in^3 when the piston reaches top dead center, the compression ratio is the relationship of 120:20 or 6:1. As previously explained in Chap. 6, the power and over-all operating efficiency of an engine are dependent to a large extent upon its compression ratio; that is, in general, high compression ratios give better fuel utilization and thermal efficiency. However, as explained in Chap. 10, the fuel type and characteristics, the type of ignition system used, and other design factors determine the maximum practical compression ratio which can be used in an engine. In general, compression ratios as related to fuel type and quality are as shown in Table 8-5.

Piston Displacement and Engine Speeds The power output of an engine is largely determined by its piston displacement and crankshaft speed. Piston displacement is the space swept through by the piston in moving from one end of its stroke to the other and is expressed in cubic inches. For a single-cylinder engine, it is the cross-sectional area of the piston or cylinder multiplied by the length of stroke of the piston. For a multiple-cylinder engine, the total piston displacement is equal to the piston displacement of one cylinder, multiplied by the number of cylinders. For example, a 4½- by 6-in cylinder is one having a 4½-in bore or diameter and a 6-in piston stroke. If it had but one cylinder, the piston displacement would be calculated as follows:

$$\begin{aligned}
\text{Area of cylinder} &= (\tfrac{1}{2}\ \text{bore})^2 \times \pi \\
&= (2.25)^2 \times 3.1416 \\
&= 15.9\ \text{in}^2 \\
\text{Piston displacement} &= \text{area of cylinder} \times \text{stroke} \\
&= 15.9 \times 6 \\
&= 95.4\ \text{in}^3
\end{aligned}$$

A four-cylinder engine would have 4×95.4 in^3, or 381.6 in^3 displacement.

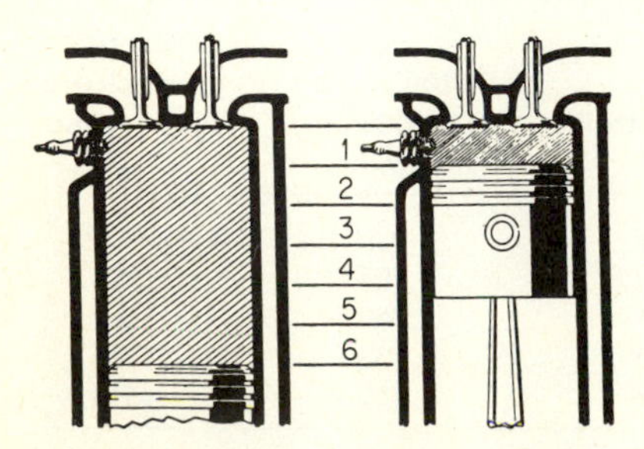

Figure 8-22 Sketch showing measurement of compression ratio.

Table 8-5 Engine Compression Ratios for Different Fuels

Fuel	Compression ratio	
	Minimum	Maximum
Gasoline, regular grade	7.0 to 1	8.5 to 1
Gasoline, premium grade	9.0 to 1	10.5 to 1
Gasoline, no lead	7.5 to 1	8.5 to 1
LP gas	8.0 to 1	9.0 to 1
Diesel fuel	15.0 to 1	18.0 to 1

Assuming that this engine has a crankshaft speed of 1,000 rpm, its total piston displacement per minute would be 2 × 1,000 × 381.6, or 763,200 in^3. If its power rating were 50 hp, then the piston displacement per minute per horsepower would be 15,264 in^3. A study of tractor-engine tests shows that the piston displacement per minute per horsepower ranges from 13,500 to 17,000 in^3, depending upon the operating efficiency. Hence the total piston displacement per minute of an engine is a reasonably good indicator of its maximum power output.

PROBLEMS AND QUESTIONS

1 Explain the meaning of piston clearance and give the amount recommended for a 4½-in plain cast-iron piston and for a similar aluminum-alloy piston.

2 Name the common types of bearing alloys and explain the specific qualities and applications of each.

3 Explain fully how crankshafts are designed to obtain proper balance under all conditions and give the effects of a poorly balanced crankshaft.

4 Explain the meaning of inertia forces and state how the reactions from them are controlled in high-speed engines.

5 What are the advantages of removable cylinder liners and why are they particularly desirable in a tractor engine?

6 To what extent and why is the compression ratio selected for an engine affected by the kind or type of fuel to be used?

7 An engine has a 3¾-in bore and 3⅝-in stroke. Compute the bore-stroke ratio, the piston displacement, and the top-ring contact area per stroke.

Chapter 9

Valve Systems and Operation

As stated in the preceding discussion, a single-cylinder engine has at least two ports or two valves, depending upon whether it is of the two- or the four-stroke-cycle type, respectively. The intake port or valve allows the fuel mixture to enter the combustion chamber on the intake stroke. The exhaust port or valve allows the products of combustion to escape following the explosion and expansion. As previously mentioned, some two-stroke-cycle engines have three ports, the extra port serving as a passage for the fuel from the fuel tank and carburetor to the crankcase.

Since the ports in a two-stroke-cycle engine are nothing more than openings in the cylinder wall that are opened and closed by the piston in its movement, no special valve-operating mechanism is required and the problem of valve timing is not present. The following information, therefore, applies to the four-stroke type of engine entirely.

Valve Construction Gas-engine valves are of the poppet or mushroom type (Fig. 9-1). The valve itself consists of a flat head with a beveled edge called the face, and the stem. The valve opening in the cylinder block or head has a similar beveled edge called the seat. A strong

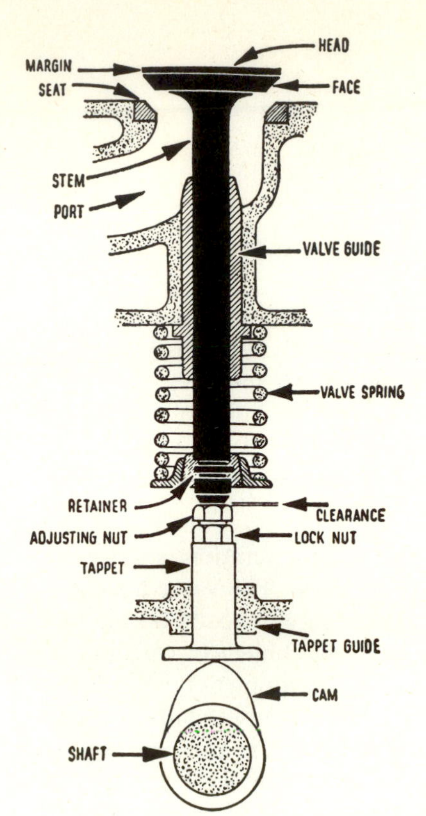

Figure 9-1 L-head valve construction and operation.

spring, held in place by a retainer and key or wedge, holds the valve face tightly against the seat and thus prevents leakage on the compression and power strokes. The usual face and seat angle is 45°. However, a 30° angle is frequently used for intake valves.

Valves are subject to considerable strain and wear owing to the high temperature to which they are exposed and the speed at which they must operate. Obviously the exhaust valve becomes much hotter than the intake because it is exposed to an almost continuous flame. In fact, it probably attains a dull red heat under load conditions. In order to resist breakage, corrosion, warping, and rapid wear, exhaust valves are made of special alloy steels containing relatively high percentages of chromium, nickel, and silicon as well as smaller amounts of other metals. Inlet valves operate much cooler and therefore are less subject to burning, corrosion, and wear. The composition of some typical steel alloys for valves is shown in Table 9-1. The stem of the valve need not be as hard as the head but must be hard enough to resist rapid wear. Sometimes the end is hardened to reduce tappet or rocker-arm wear.

The valve stem operates in a sleeve or guide. In a few cases, the guide

Table 9-1 Steel Alloys for Values

Element	Intake, percent	Exhaust, percent
Carbon	0.45– 0.80	0.20– 0.70
Chromium	0.50–20.00	15.00–21.00
Manganese	0.40– 0.90	0.60– 9.00
Nickel	0.55– 1.30	1.90–11.50
Silicon	0.30– 3.30	0.15– 3.00

is merely a hole bored in the block or head, but in most automotive engines the guide is removable. It may be either one-piece or split and is usually made of cast iron. The valve stem should work freely in the guide but a precise fit is extremely important to control lubrication and prevent compression loss. Some engines are provided with valve stem seals to aid in such controls.

Valve Rotators Effective and positive valve operation is extremely important in obtaining efficient engine performance. The valve must have complete freedom of action and remain free of any deposits on the face, seat, and stem. Exhaust valves, particularly, are subject to certain troubles arising from the high temperatures and the resulting deposits of combustion products. The normal operation of a simple poppet valve involves only reciprocating movement of the stem and head. It is quite evident that if the valve can be given some slight rotative movement with respect to its seat, more uniform wear, longer life, and cooler operation will be attained. The first successful valve rotators were developed about 1945 and are now used extensively in tractor, diesel, truck, and other heavy-duty engines. In general, rotators are used only on exhaust valves but a few engines use them on the intake valves.

Two principal types of mechanisms have been developed to provide valve rotation. The first is the release-type rotator (Fig. 9-2). The mechanism is composed of a special valve-spring retainer *A*, a cup *B*, and a pair of half-round keys or locks *C*. A special key groove in the valve stem *D* is required with this type of rotator. By means of the clearance *E* between the valve-spring tip and rotator cap, the cap, keys, retainer, and valve spring are lifted by the rocker arm or cam follower before the valve is moved. This arrangement momentarily releases the valve from its spring load and thus allows it to turn as it is lifted off its seat. Turning or rotation is induced by forces arising from vibrations of the valve train.

The other type of rotator in general use is a positive type known as the *Rotocap* (Fig. 9-3) which turns the valve by the positive action of the mechanism. The Rotocap replaces the standard valve-spring retainer in most instances, although it is sometimes placed between the valve spring and its seat in the cylinder head or block. In the valve-closed position, the

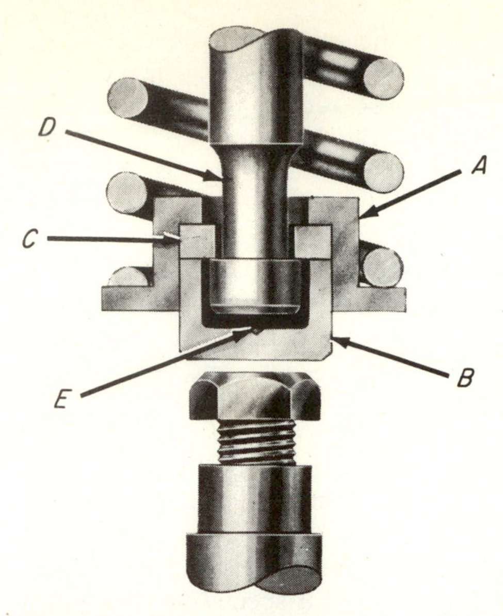

Figure 9-2 Release-type valve rotator. (Courtesy of Ethyl Corporation.)

spring washer *A* rests lightly under the steel balls *B* which are held at the top of their inclined races by small coil springs *D*. As the valve opens, the Rotocap is compressed, thereby flattening the washer and forcing the balls down the inclined races. As the balls roll down the races, they cause the spring washer and retainer cap *C* to turn a slight amount with respect to each other, thus rotating the valve.

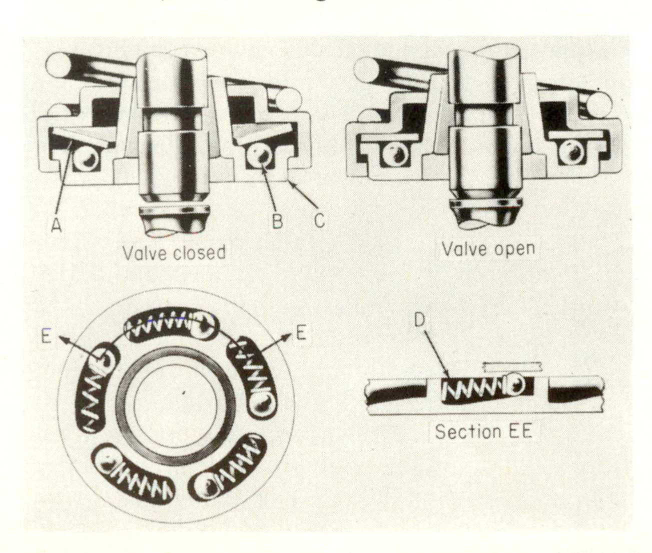

Figure 9-3 Positive-type valve rotator. (Courtesy of Ethyl Corporation.)

Valve Springs The basic function of the valve spring is to close the valve after it has been opened and to hold it tightly closed against its seat. Furthermore, modern, high-speed engines require that each valve complete its open-and-close cycle in a fraction of a second. For example, each valve in a single-cylinder engine having a crankshaft speed of 1,000 rpm must operate 500 times per minute or 8 times per second. This means that the spring must be of such design and strength that it will be positive in action and does not weaken during prolonged use. Also it should not be affected by increase in temperature. Valve springs are made of a special alloy steel wire having the desired qualities to give good valve performance, depending upon the engine type, design, and speed.

Valve-Seat Inserts Normally the valve seat is a part of the cylinder block or head, the material being cast iron. Obviously, if a hard-steel valve is constantly hitting against a cast-iron seat under high-temperature conditions, there will be rapid wear, burning, and corrosion of the seat. Hence most engines are now equipped with seat inserts (Fig. 9-4), particularly for the exhaust valves. They are made of a special wear-resistant steel alloy and are pressed into place as shown. These inserts wear very little but can be replaced if necessary.

Valve Location and Arrangement The two common valve arrangements are (1) the L-head (Fig. 9-5) and (2) the valve-in-head (Fig. 9-6). A third arrangement (Fig. 9-7) consists of horizontal valves in the head operated by a long rocker arm directly from the camshaft. The valve-in-head arrangement is found in many types of engines and is used almost entirely in farm-type and other heavy-duty multiple-cylinder engines. Nearly all tractor engines are of this type. Some advantages given are: (1) the valves are removable with the cylinder head and, therefore, are accessible and easily serviced and replaced; (2) in case of a damaged valve seat, it is necessary to purchase only the cylinder head, while in other types it might

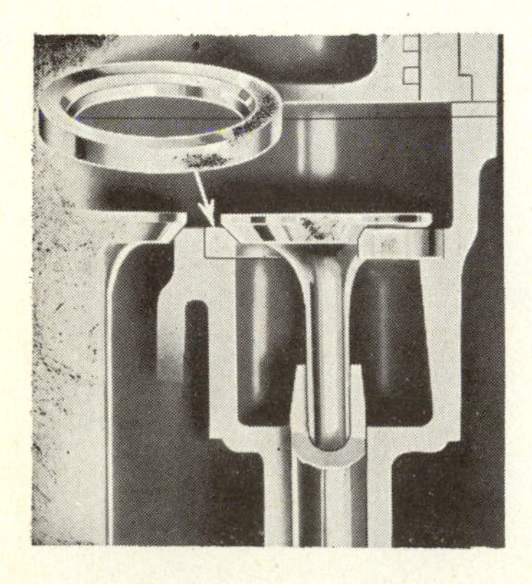

Figure 9-4 Valve-seat insert.

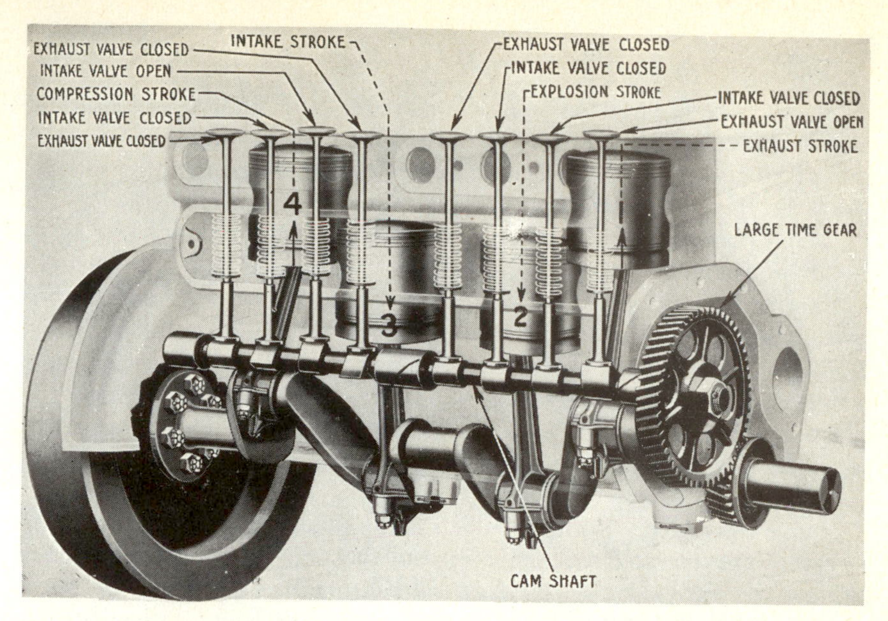

Figure 9-5 Valve arrangement and operation for typical multiple-cylinder L-head engine.

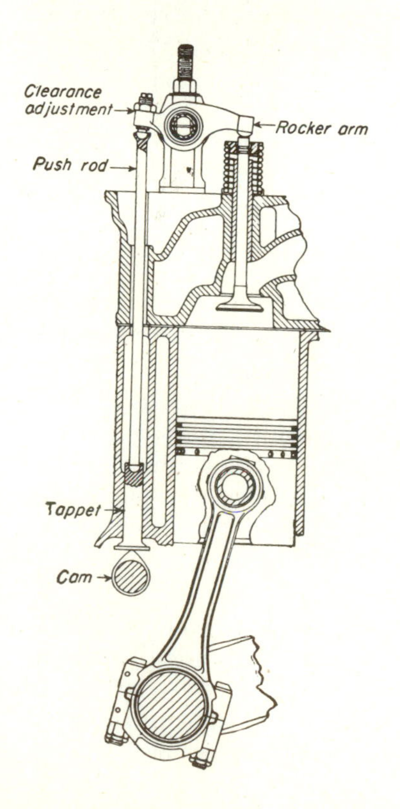

Figure 9-6 Valve-in-head construction and operation.

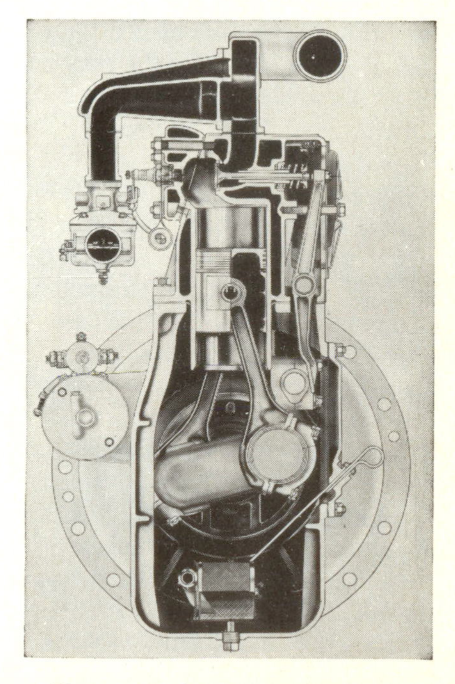

Figure 9-7 Valve-in-head construction with long rocker arm and horizontal valve.

be necessary to replace the complete cylinder or cylinder block; and (3) uniform and effective cooling of the valves is greatly facilitated. The principal disadvantages are (1) a more complicated valve-operating mechanism and (2) more places to wear, hence more frequent adjustment.

The L-head arrangement is limited in use and confined largely to smaller power units. The valve-operating mechanism is simple because the cam and tappet act directly against the valve stem. However, the valve mechanism is somewhat inaccessible, and the intake and exhaust manifolds must usually be removed to permit adjustment of the valve clearance. Since the valves and openings are on the side of the block, the combustion space in the head is shallower and spread out with a part of the space to one side of the piston and over the valves.

Valve Operation Both the intake and exhaust valves on most types of engines are operated mechanically, but on some small, single-cylinder engines the intake valve is said to operate automatically. When thus operated, it is equipped with a very light spring. On the suction stroke of the piston, the valve is drawn open by the suction or by decreased pressure in the cylinder; the fuel charge is drawn in through the open valve. Near the end of the stroke, as the suction decreases, the light spring causes the valve to return to its seat. This method of intake-valve operation is very simple, but unless the valve is in perfect condition it may not seat properly, creating loss of compression. It is also noisy and not well adapted to high-speed engines, for the reason that there is no control of the time of opening and closing of the valve. Therefore the engine may not receive a full charge of fuel mixture.

Mechanically Operated Valves In many single-cylinder engines and in all multiple-cylinder types, the intake as well as the exhaust valves are operated mechanically. For an L-head engine (Figs. 9-1 and 9-5), the mechanism consists of (1) a small gear on the crankshaft; (2) a half-time or cam gear; (3) a cam or, in the case of multiple-cylinder engines, a camshaft with a number of cams; (4) a tappet or short push rod. As the cam gear rotates the cam, the latter moves the tappet, and it in turn pushes the valve open. In addition to the parts enumerated above, the valve-in-head arrangement (Fig. 9-6) includes a push rod and rocker arm.

Camshafts and Driving Mechanisms It has already been stated that, in any four-stroke-cycle engine, there are four strokes of the piston per cycle, the first stroke of the cycle being known as the intake stroke and the last stroke as the exhaust stroke. Therefore, the intake valve must open and close on the intake stroke, and the exhaust valve must open and close on the exhaust stroke; that is, each valve must operate one time per cycle. Consequently, the cam gear and cams that open and close these valves must make one revolution per cycle. The cam gear, however, is driven by a gear on the crankshaft that makes two revolutions per cycle. Therefore, the speed of the cam gear will be one-half that of the crankshaft, and it will have twice the number of teeth found on the crankshaft gear. Likewise, in multiple-cylinder engines having more than two valves, the speed of the cam gear will be one-half the speed of the crankshaft, because, as previously explained, all cylinders complete a cycle during the same two revolutions of the crankshaft. Therefore, all valves must open and close once during any two revolutions, and the camshaft, in order to open and close all these valves one time, must make one revolution to two of the crankshaft.

Camshafts are driven either by gears (Fig. 9-5) or by the tooth-type silent chain (Fig. 9-8). Freedom from noise is most important, and, for this reason, the chain drive is preferable in very high-speed automotive engines. Gears are predominant in the slower-speed heavy-duty engines such as tractor engines and similar power units. Quiet operation in gear drives is obtained by the use of nonmetallic materials such as bakelite instead of iron for the larger camshaft gear and also by the use of helical teeth.

Hydraulic Valve Lifters or Adjusters The hydraulic-type valve lifter or lash adjuster has been designed and adopted by some automotive-type engine manufacturers to eliminate certain objectionable factors involved in valve clearance and its adjustment. Basically, the hydraulic valve lifter gives better engine performance by keeping the valve operating parts in

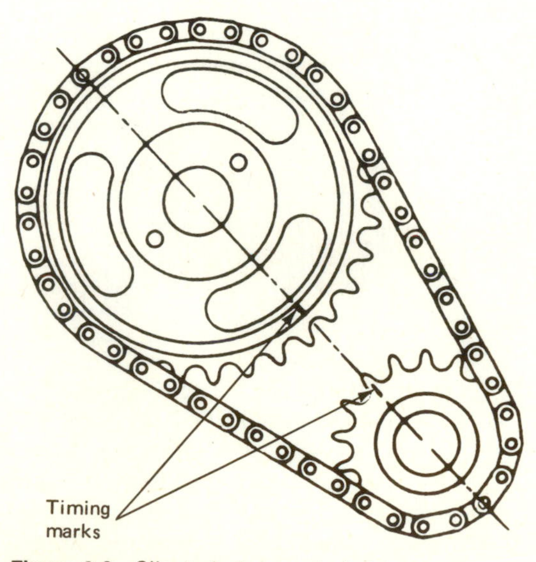

Figure 9-8 Silent-chain camshaft drive.

full contact with each other at all times, as well as eliminating periodic clearance adjustment. Specific advantages are (1) elimination of tappet clearance noise, (2) elimination of periodic valve clearance adjustments, (3) automatic compensation for expansion and contraction of the valve mechanism and motor block due to temperature changes and normal wear, (4) longer valve life owing to elimination of pounding, and (5) better engine performance because of precise control of valve timing.

Figure 9-9 shows one type of hydraulic valve lifter, which operates as follows: The lifter body-guide cavity is connected by small drilled holes to the engine lubricating system. Referring to position 1, the oil enters the lifter body through the body-feed holes, flows into the inside of the plunger through the holes in its side, and continues to flow through the hole in the bottom of the plunger, around the ball, and down through holes in the ball retainer, completely filling the ball retainer cavity. As the lifter is raised, position 2, the oil below the plunger tries to escape past the ball check. This rush of oil around the ball check forces the ball check to seat on the plunger, which, in turn, seals the hole at the bottom of the plunger. The lifter then follows the cam as a relatively solid unit. As the lifter rises on the cam and the full valve assembly load is applied, a predetermined and closely held clearance between the plunger and the body permits a controlled amount of oil to escape from below and past the plunger. This condition—the relative movement of the plunger with respect to the body after the ball check is seated—is termed leakdown. As the lifter plunger

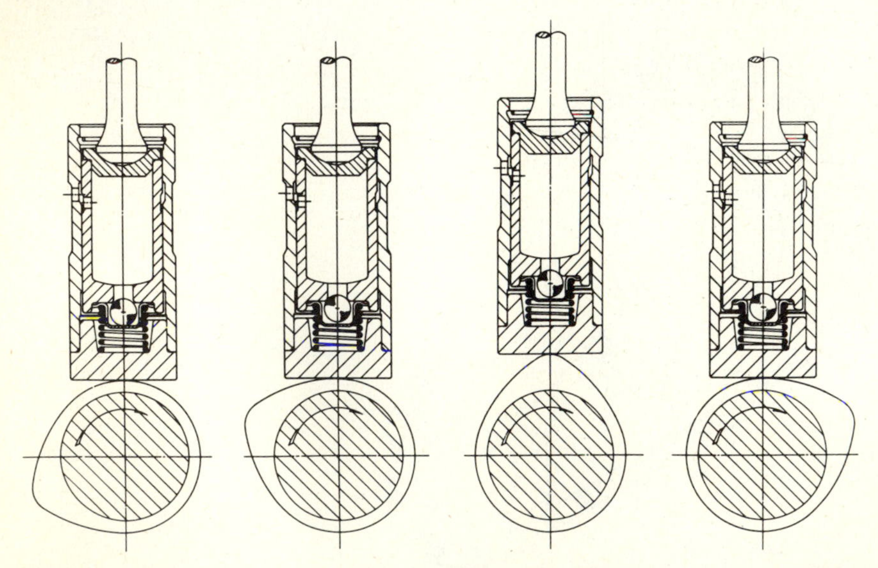

Figure 9-9 Hydraulic-type valve lifter. (Courtesy of Diesel Equipment Division, General Motors Corporation.)

reaches the cam nose, position 3, the plunger has leaked down a very minute distance relative to the body as compared with its location in position 2. As the lifter reaches the closing ramp, position 4, the plunger is lower in relation to the body than in position 2, where the ball had just seated. As the lifter continues to ride down the ramp, the ball check opens if the valve gear has remained the same or contracted. The spring under the plunger compensated by taking up the clearance and the cycle is repeated. However, if the valve gear has expanded more than the relative amount of movement of the plunger in the body during leakdown, then the ball check will not unseat until enough cycles have taken place to account for the expansion of the valve gear. Therefore, when the engine structure and valve gear expand and contract with changes in engine temperatures and other differentials, the lifter automatically adjusts its own length to compensate for these changes. When temperature changes require shortening of the lifter length, the engine valve spring forces the plunger down because of the leakdown characteristics, thus constantly correcting for this condition. When lengthening of the lifter length is required, the lifter spring raises the plunger, causing oil to flow into the spring chamber.

Figure 9-10 shows a second type of hydraulic valve lifter which is located in the valve train next to the rocker arm. Its general construction and operation are essentially the same as Fig. 9-9; however, the oil is fed to it from the rocker arm through a drilled hole in the valve adjusting screw and a mating hole in the plunger cap into the supply chamber of the adjuster.

Hydraulic valve lifters give little trouble and seldom require attention. After an engine has been standing for some time, a certain amount of valve-lifter noise will occur when the engine is first started. This is due to leakdown on those lifters which were holding valves open against spring pressure when the engine stopped. Oil pressure will refill these lifters after a few seconds of engine operation and the noise will disappear. Cleanliness and the use of a high-quality engine oil of correct viscosity are most important in an engine equipped with hydraulic valve lifters.

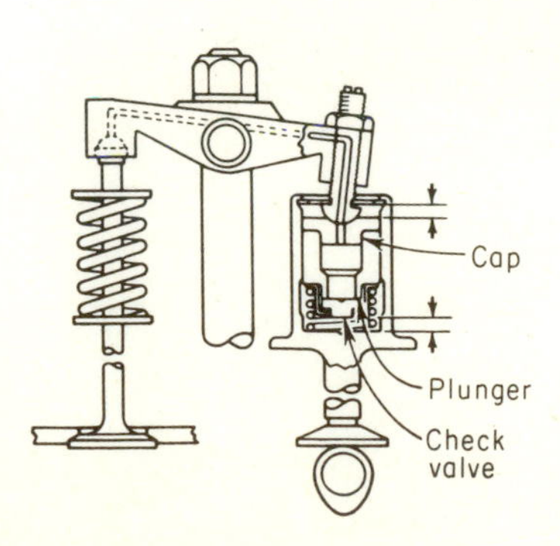

Figure 9-10 Hydraulic-type lash adjuster.

Valve Timing The timing of any valve is specified with respect to the instant it begins to open and the instant it closes in relation to the crankshaft rotation and the piston position; that is, if a valve starts to open and then closes at the specified time with respect to the crankshaft and piston position for that cylinder, it is said to be correctly timed. These opening and closing points are determined by (1) the timing of the crankshaft and camshaft gears; (2) the shape of the cam; and (3) the tappet or rocker-arm clearance. A study of some typical cam shapes (Figs. 9-1 and 9-6) shows that the valve action is a gradual one; that is, the valve is pushed open gradually rather than abruptly, reaches its maximum lift and remains so only for an instant, and then closes gradually.

It would be assumed, ordinarily, that a valve would start to open at the beginning of its stroke and be completely closed at the end. Such, however, is not necessarily the case. It has been found that most engines operate efficiently only when the valves open and close at certain points in the cycle and remain open a certain length of time. Therefore, the timing of the valves in any engine is very important.

Figure 9-11, commonly known as a valve-timing spiral or diagram, illustrates the timing of the average engine. The vertical line might be called the dead-center line with crank dead center (CDC) at the bottom and head dead center (HDC) at the top. The crankshaft and piston may be said to be on dead center when the centers of the piston pin, crankshaft, and crankpin all fall in this line. The spiral represents the travel of the crankpin or connecting-rod journal. Starting at HDC, the crank rotates in the direction of the arrow toward CDC, moving the piston through the intake or suction stroke, followed by compression from CDC to HDC, expansion from HDC to CDC, and exhaust from CDC to HDC, thus completing the cycle.

Referring to the figure, it is noted that the intake valve should begin to

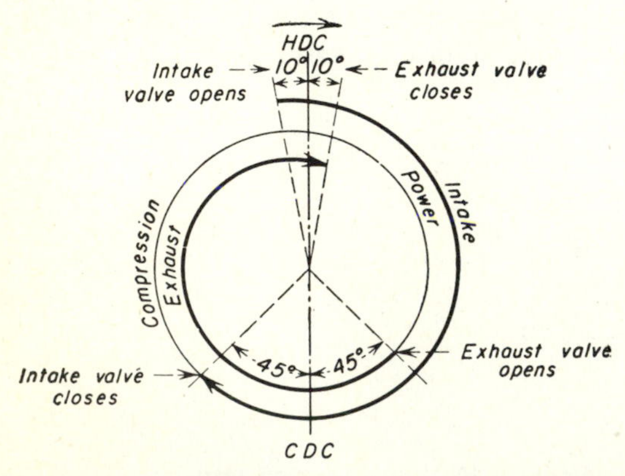

Figure 9-11 Diagram showing usual timing of valves.

open when the crankshaft lacks about 10° of reaching HDC, and that it remains open throughout the stroke and does not close until the crankshaft has reached a point about 45° after CDC. The exhaust valve begins to open before the piston reaches the end of the power stroke, that is, when the crankshaft is making an angle of about 45° before CDC with the piston on the power stroke. It remains open through the exhaust stroke and closes when the crankshaft has rotated about 10° after the piston has reached HDC. The figures given are not correct for all engines but are very near the average.

Range of Operation and Cam Shape The number of degrees through which the crank rotates, from the point where the valve starts to open to the point where it is just completely closed, is known as the range of operation of that valve. The average range of operation for both valves, according to the figure, is 235°.

The range of operation is determined by the cam shape and the valve clearance (Figs. 9-1 and 9-6). Too much valve clearance decreases the range of operation, and too little clearance increases it. Figure 9-12 shows some typical cam shapes and their effect on the timing of the valve. The camshaft diameter is M and N is the cam lobe height or valve lift. Both cam A and cam B have the same lift, but B has a greater angle of action and would give a range of operation of 260°. Likewise, cam B has a longer "dwell" interval which would be desirable for the intake valve on a very high-speed engine in order to permit a maximum charge of fuel mixture to enter the cylinder. The cam shape varies with respect to the intake and exhaust valves on any engine and, particularly, with respect to different types of engines, depending upon engine size and speed and other factors.

Even though the range of operation is correct, the valve may open too early and therefore close too early by the same amount, or it may open too late and therefore close too late by the same amount. This is due to the fact that the cam gear is out of time; that is, the cam is not coming around at the correct time to open and close the valve at the proper time with

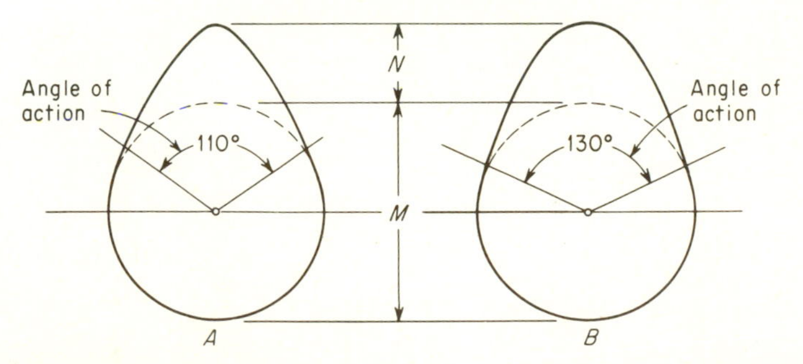

Figure 9-12 Valve cam profiles.

respect to the travel of the piston. To time the valve, the cam gear must be removed and rotated the correct amount in the right direction and remeshed with the gear on the crankshaft. This shifting of the gear will not change the range of operation of the valve.

Most manufacturers place marks on their engines (Fig. 9-13) by which the correct valve and ignition timing of the engine may be determined and checked. However, the system of marking used is not standard; that is, every manufacturer has a somewhat different system. Consequently, the instruction book supplied with an engine should be studied carefully so that the method used will be well understood.

Effects of Incorrect Intake-Valve Timing As previously discussed, the intake valve starts opening at the beginning of the intake stroke and remains open until the crankshaft has rotated considerably past crank dead center. At first thought, it would seem more reasonable to close the intake valve at the end of the intake stroke, but experience has proved that allowing this valve to remain open, after the piston has started back on the compression stroke, enables the engine to generate more power and operate more efficiently. The explanation is that the charge rushing into the cylinder has not completely filled the space when the piston reaches the end of the intake stroke, but if the valve is allowed to remain open somewhat longer, the fuel mixture will continue to pour in because of its inertia; thus a more complete charge will enter the cylinder and greater power will be generated. There is some variation in the closing of this valve in different types of engines, but in general the point is determined by the speed of the engine; that is, a high-speed engine would have the intake closing possibly as much as 45 to 60° after CDC.

Effects of Incorrect Exhaust-Valve Timing In practically all types of engines, regardless of size, number of cylinders, speed, or other factors,

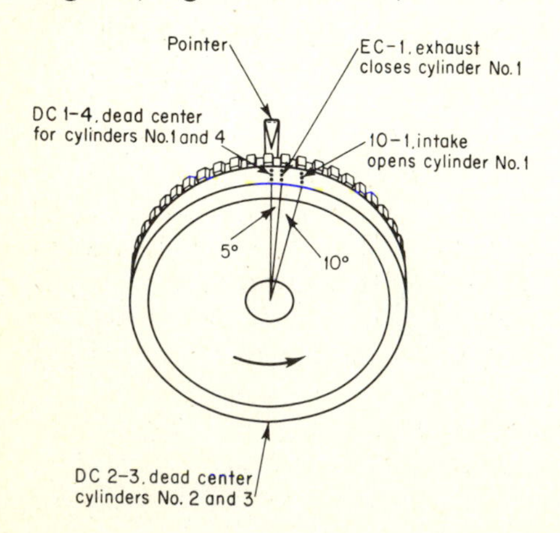

Figure 9-13 Flywheel timing marks.

the exhaust valve starts opening before the end of the power stroke of the piston, that is, when the crankshaft lacks about 45° of having completed the stroke. Experience has proved that this early opening gives more satisfactory results in that the engine generates more power, runs cooler, and uses less fuel than when the valve is opened later, or at a point near the end of the stroke. It would seem that, in opening the exhaust valve before the completion of the expansion stroke, this power would be lost or wasted. But what actually happens is that this unspent pressure is utilized to better advantage to force the exhaust gases out, thus producing better cleaning of the cylinder.

Overheating of the engine is one of the most pronounced effects of too late exhaust-valve opening. This overheating develops because the hot gases are allowed to remain in the cylinder longer and to follow the piston through the entire stroke, thus coming in contact with a greater cylinder area. The closing of the exhaust valve occurs shortly after the completion of the exhaust stroke. If the valve were closed before or even at the end of the stroke, a small amount of burned residue might remain in the cylinder.

Multiple-Cylinder Valve Timing—Valve Clearance The correct timing of the valves in any internal-combustion engine is very important. It is essential to proper operation of the engine if efficient fuel consumption, maximum power, and smooth running are desired. In the design of an engine, the manufacturer determines accurately the correct timing for his engine and sees that the valves open and close so as to give the best possible results. In many engines, certain teeth on the cam gear and crankshaft are marked or centerpunched so that, when either is removed and replaced (if the marks are placed together), the valves will be in time. Marking of the timing gears in multiple-cylinder engines is important and almost universally practiced.

The timing of the valves in multiple-cylinder engines is a comparatively simple matter, consisting in (1) meshing the timing or cam gears so that the marked teeth are together and (2) providing the correct valve clearance or lash. Valve clearance is defined as the space allowed between the end of the valve stem and the rocker arm (Fig. 9-14), or, in the case of the L-head type, between the valve stem and the tappet (Fig. 9-1).

The valve-in-head construction requires more mechanism than the L-head construction. Therefore, slightly more wear is possible, and frequent adjustment is usually necessary. Figure 9-13 shows the usual mechanism involved and the method of adjustment. Figure 9-1 shows the L-head arrangement and the usual provisions for adjusting valve clearance.

A certain amount of valve clearance is necessary to allow for the expansion of the valve stem and engine parts when they get hot and thus permit the valves to seat properly. At the same time, too much clearance

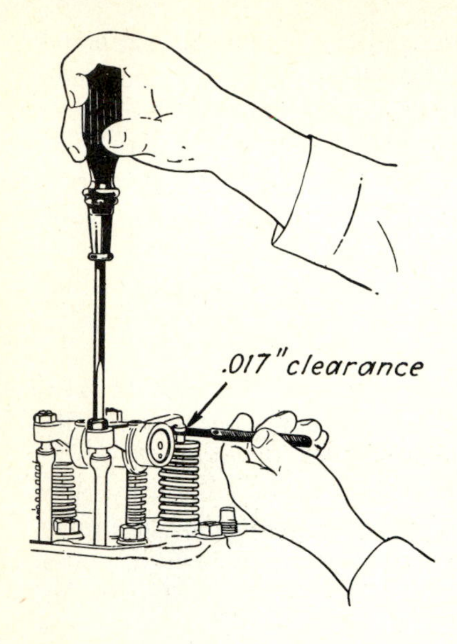

Figure 9-14 Valve-in-head clearance adjustment.

affects the timing of the valves and the smooth operation of the engine and also produces noise. In some cases the adjustment is made when the engine is cold, while in others the engine should be warmed up or at its normal running temperature. In many cases, the exhaust valve is given more clearance than the intake because its operating temperature is much higher. This clearance varies as follows, depending upon the size and type of engine: intake valve, 0.006 to 0.012 in for auto engines and 0.010 to 0.020 in for tractor engines; exhaust valve, 0.010 to 0.020 in for auto engines and 0.012 to 0.030 in for tractor engines.

The importance of correct valve-clearance adjustment cannot be overstressed, because if it is neglected, more or less faulty operation may result. The following rules for adjusting valve clearance apply to either the L-head or the valve-in-head type and should be observed carefully:

1 Read the instruction book supplied with the tractor and see whether the manufacturer advises making the adjustment when the engine is hot or when cold.

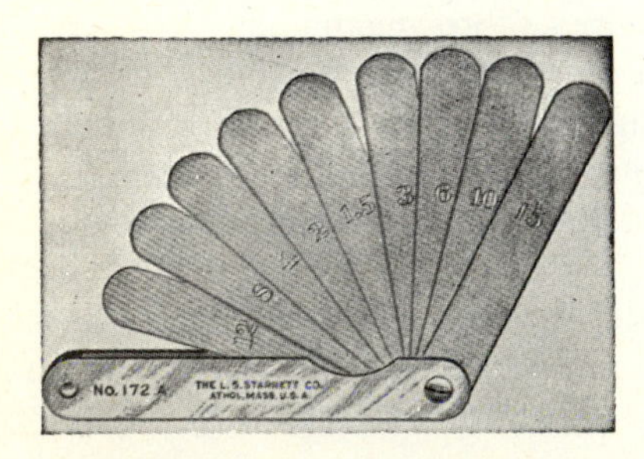

Figure 9-15 Thickness gauge.

2 When making the adjustment, always be sure that the push rod and tappet are resting on the low part of the cam.

3 Some manufacturers mark the flywheel as shown by Fig. 9-13. If this is done, check the valve adjustment and timing by referring to these flywheel marks. The instruction book will explain just how to do this.

4 Valve clearance should always be adjusted whenever the valves are ground or reconditioned. In valve-in-head engines always readjust the clearance whenever the cylinder head bolts are tightened.

5 For accurate adjustment, use a standard thickness gauge (Fig. 9-15).

6 Always see that the lock nuts are tight.

PROBLEMS AND QUESTIONS

1 Compare the relative composition of intake and exhaust valves and give reasons for any difference.
2 What factors may affect the proper clearance of valves? State the effects of (a) excessive clearance and (b) insufficient clearance for each valve?
3 What are the advantages of hydraulic valve lifters?
4 Why is valve clearance frequently adjusted with the engine running?
5 Explain what differences there would be in the timing of the valves in a four-cylinder slow-speed tractor engine compared with a six-cylinder high-speed automobile engine. Give reasons.

Chapter 10

Fuels and Combustion—Exhaust Emission and Control

One disadvantage of the internal-combustion engine is that the fuels required for its successful operation must be of a specific nature and, therefore, are somewhat limited with respect to source and supply. Under certain conditions these fuels may be difficult to secure, and their use becomes economically prohibitive. Some of the essential characteristics of such fuels are:

1 They must have a reasonably high energy value.
2 They must vaporize at least partially at comparatively low temperatures.
3 The fuel vapors must ignite and burn readily when mixed in the proper proportions with oxygen.
4 Such fuels and their combustion products should not be unduly harmful or dangerous to human health or life.
5 They must be of such a nature that they can be handled and transported with comparative ease and safety.

CLASSIFICATION AND COMPOSITION OF FUELS

Gaseous Fuels Fuels for internal-combustion engines may be classified in two ways: (1) as either gaseous or liquid, according to the physical state before going into the engine cylinder; and (2) as artificial or natural, according to whether the fuel is obtained from natural sources or is a manufactured product.

The common gaseous fuels derived from nature are (1) natural gas and (2) propane and butane or LP (liquefied-petroleum) gases. Natural gas is used directly as it comes from the well, without being subjected to any complex refining or purifying action. It is colorless and odorless and consists largely of methane (CH_4) and other simple hydrocarbon gases and small amounts of carbon dioxide and nitrogen. It is distributed under pressure through pipe lines and used largely as a fuel for cooking, heating, lighting, and refrigeration, and for industrial purposes.

Natural gas is a very satisfactory fuel for internal-combustion engines, but its use must be confined to stationary power units, which can be permanently connected to a supply line. The only special equipment needed is a suitable pressure regulator or control and a mixing device similar to a carburetor.

In the manufacture of iron and steel, certain so-called by-product gases are produced such as blast-furnace gas, coke-oven gas, and producer gas. These artificial gaseous fuels are limited in use to the production of power in or near these plants and are not available for general use.

In some countries in which petroleum fuels are expensive and difficult to obtain, equipment has been devised for use on automobiles, trucks, and tractors which generates a gaseous fuel from coal, wood, charcoal, sawdust, or similar materials. This gas is called *producer gas* and consists of a mixture of carbon monoxide, hydrogen, methane, nitrogen, and carbon dioxide. The equipment is quite complex and bulky and includes devices to clean and cool the gas and separate out tarry and other undesirable matter, as well as the mixer for supplying the engine with the correct quantity of air and gas mixture.

Butane and propane or a mixture of the two are used to a considerable extent as engine fuels. Referring to Table 10-1, both are simple hydrocarbons of the paraffin series and become gases at normal atmospheric pressure and temperature. For this reason they are stored and transported under pressure in liquid form in heavy tanks. This explains why they are called *liquefied-petroleum gases*. Usually they are more or less by-products of the petroleum-refining process. They may be recovered (1) from the dry gas and gasoline as it is removed from crude oil; (2) from recycling plants from the wet gas drawn from natural-gas wells; or (3) from the normal processing of crude oil into commercial gasoline and distillates (see Figs. 10-1 and 10-3).

Liquid Fuels The most common liquid fuels are gasoline and diesel fuel. Liquid fuels have the advantage over gaseous materials of being

Table 10-1 Characteristics of Paraffin Hydrocarbons

Formula	Name	Weight per gallon, lb	Gravity API	Gravity Specific	Boiling temp., °F	Self-ignition temp., °F	Heating values, Btu/lb High	Heating values, Btu/lb Low
CH_4	Methane	—	—	—	−258	1346	23,850	21,597
C_2H_6	Ethane	3.11	247.0	0.374	−128	1050	22,400	20,597
C_3H_8	Propane	4.23	147.0	0.508	−44	995	21,650	20,015
C_4H_{10}	Butane	4.86	110.8	0.584	31	961	21,300	19,795
C_5H_{12}	Pentane	5.25	92.8	0.631	97	933	21,100	19,597
C_6H_{14}	Hexane	5.53	81.6	0.664	155	909	20,900	19,432
C_7H_{16}	Heptane	5.73	74.2	0.688	209	893	20,850	19,408
C_8H_{18}	Octane	5.89	68.6	0.707	258	880	20,750	19,329
C_9H_{20}	Nonane	6.01	64.5	0.722	303	871	20,700	19,292
$C_{10}H_{22}$	Decane	6.11	61.3	0.734	345	866	20,600	19,205
$C_{15}H_{32}$	Pentadecane	6.41	52.3	0.770	518	—	20,450	19,091
$C_{18}H_{38}$	Octadecane	6.49	49.9	0.780	603	—	20,400	19,047

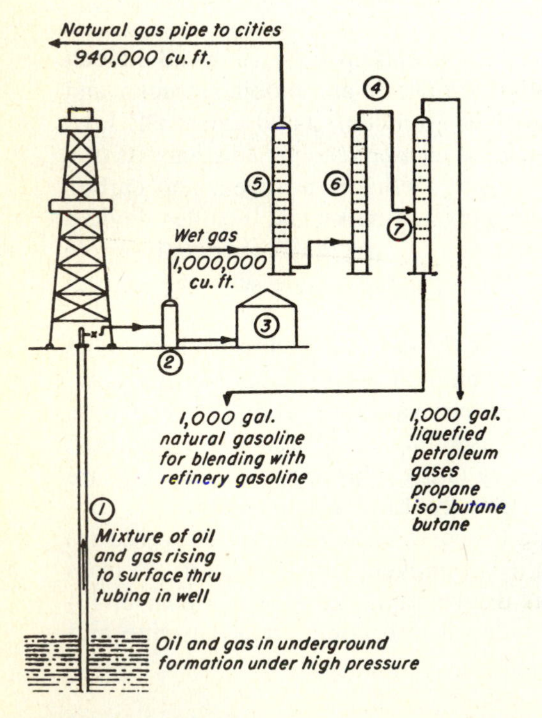

Figure 10-1 Natural gas well with storage and distribution equipment. (Handbook of Butane-Propane Gases.)

more concentrated and readily portable. They are obtained almost entirely from petroleum, a product of nature. The only artificial liquid fuel of any consequence that might be used for internal-combustion engines is alcohol. It has never been used extensively, however, and will be discussed later. Gasoline and diesel fuels are heavier products resulting from the distillation and refining of crude oil.

Origin of Petroleum Crude oil is obtained from natural deposits existing usually at depths varying from several hundred to thousands of feet below the surface of the earth. For this reason petroleum and its products are said to be of mineral origin or derived from nature. The first crude oil of any consequence was discovered in Pennsylvania in 1859 and came from a well having a depth of about 70 ft. This state continued to be the only producer for a number of years, but eventually oil was discovered in Ohio, Indiana, and Illinois, and later in California, Texas, Oklahoma, Kansas, Louisiana, Mississippi, and other states and in Canada, Mexico, South America, and other foreign countries.

There is considerable variation in the character of crude oil obtained in these different regions, but they are generally placed in one or the other of three classes, namely, asphalt-base crudes, paraffin-base crudes, and mixed-base crudes. An asphalt-base crude is usually a heavy dark liquid and, when distilled, produces an end product of a black, tarry nature. Paraffin-base crudes are lighter in weight and color and upon distillation produce a residue that is light in color and resembles paraffin. The ordinary paraffin wax is obtained from such crudes. Some crude oils have been found which contain both asphaltic and paraffin materials and are therefore known as mixed-base crudes.

Crude oils, as obtained from these various regions, differ considerably from the standpoint of the final products of distillation and vary greatly in the amount of the lighter products of distillation that they yield, such as gasoline and kerosene. The so-called light crudes, that is, those which are comparatively thin and lighter in weight per unit volume, produce a higher percentage of gasoline and kerosene and a lower percentage of lubricating oil and heavy products, whereas the heavier crude oils contain less gasoline and kerosene and produce a greater quantity of lubricating oil and heavy products. It was thought originally that the gasoline and lubricants from certain crude oils were superior to those obtained from others, but with the modern refining processes the products produced from these different crudes do not vary greatly in character or quality.

Chemical and Physical Characteristics of Petroleum Products A brief and elementary consideration of the chemical and physical characteristics of these petroleum products is important and will assist in explaining their action as fuels and lubricants. Crude oil and the products obtained from it are known as *hydrocarbons;* that is, they are products made up almost entirely of carbon and hydrogen, having such general chemical formulas as C_nH_{2n}, C_nH_{2n+2}, etc. A simple hydrocarbon, therefore, would be CH_2; another, CH_4; a third, C_2H_6; and so on. A careful analysis of crude oils and the common fuels obtained from them indicates that they contain about 84 percent carbon and 15 percent hydrogen by weight, with the remaining 1 percent made up of nitrogen, oxygen, and sulfur. Table 10-1 shows some typical simple petroleum products and their physical characteristics. It will be noted that as the molecular structure becomes more complex the material changes from a gas to a light liquid and then to a heavy liquid at ordinary temperatures. For example, butane (C_4H_{10}) is a gas at 32°F, while hexane (C_6H_{14}) and octadecane ($C_{18}H_{38}$) are normally liquids and become gases at temperatures of 155 and 603°F respectively. Most liquid fuels including gasoline and distillates are thus made up of a mixture of these somewhat less complex hydrocarbons, while heavier products such as diesel fuels, fuel oils, lubricants, and greases are composed of the heavier hydrocarbons having a very complex molecular structure. Table 10-2 gives the principal characteristics of the common liquid fuels. It should be noted that there is very little difference in the heating value of crude oil and its products, even though there is a marked variation in the weight of a given volume of each. As shown, alcohol has about the same gravity as kerosene but a much lower heating value.

Heating Value of Fuels The heating value or the amount of heat energy contained in any fuel is measured by means of the heat unit known as the *British thermal unit* (Btu). This unit is the quantity of heat required to raise the temperature of 1 lb of water 1°F. The heating values of the common liquid fuels are given in Table 10-2.

Table 10-2 Characteristics of Liquid Fuels

Fuel	API[1] test, deg.	Specific gravity	Weight per gallon, lb	Initial boiling point, °F	End point, °F	Heating value, Btu[2] per lb	
						Low	High
Gasoline	65–56	0.720–0.755	5.99–6.28	85–105	300–435	19,000	21,000
Kerosene	45–40	0.802–0.825	6.68–6.87	340–360	500–550	18,000	20,000
Diesel tractor fuel	40–30	0.825–0.876	6.87–7.30	325–460	600–725	18,000	20,000
Crude oil	57–10	0.751–1.000	6.25–8.30	—	—	18,000	20,000
Denatured alcohol	45–40	0.802–0.825	6.68–6.87	150	—	10,000	13,000
Propane	147	0.509	4.24	−44	—	20,015	21,600
Butane	111	0.584	4.86	32	—	19,795	21,300

[1] American Petroleum Institute.
[2] British thermal unit.

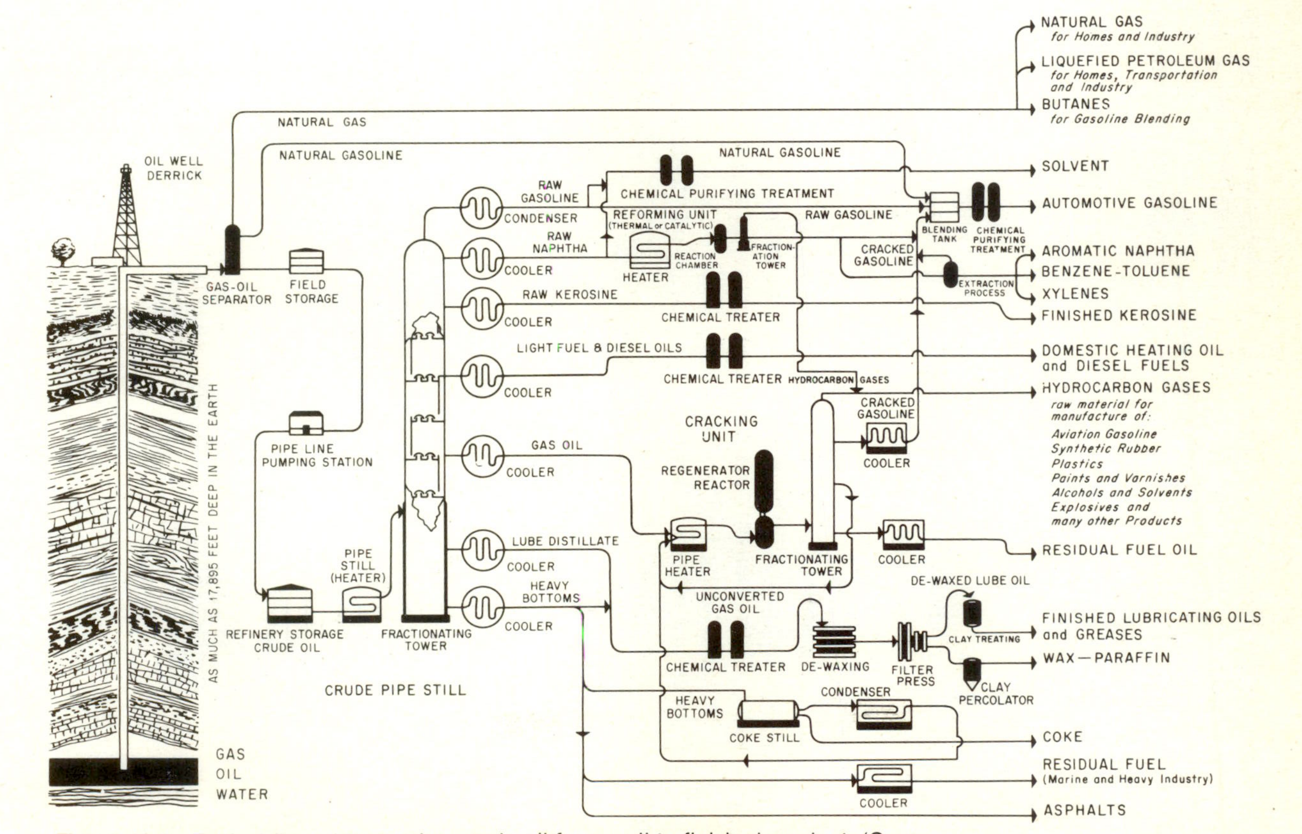

Figure 10-2 Typical flow chart tracing crude oil from well to finished product. (Courtesy American Petroleum Institute.)

Petroleum Refining The production of crude oil and its conversion into hundreds of commercial products by the various distillation and refining processes has become one of the greatest industries of the age. As a result of economic and other factors, changes and improvements in refining methods are being constantly introduced. The manufacture of the numerous crude-oil products has developed into a highly complex industry, and it is difficult to convey to the average layman, by any simple or limited explanation, just how the various products are made. A study of a typical flow chart (Fig. 10-2) shows that the first stage in the refining process involves the breakdown of the crude, by straight distillation, into about seven groups, the lightest of which is entirely gaseous in nature, the heaviest being a semisolid or solid such as asphalt and coke. The second stage consists in converting the fractions into their respective end products by means of distillation, chemical action, and special treatment. It will be noted that gasoline is obtained from the raw product in two ways: (1) by direct distillation and (2) by a special process called *cracking,* which converts a part of the heavy "gas oil" fraction into fuel. The refining process for manufacturing fuels is shown by Fig. 10-3. Lubricants for various purposes are obtained from the lubricating-oil "cut," which is heavier than that of gas oil.

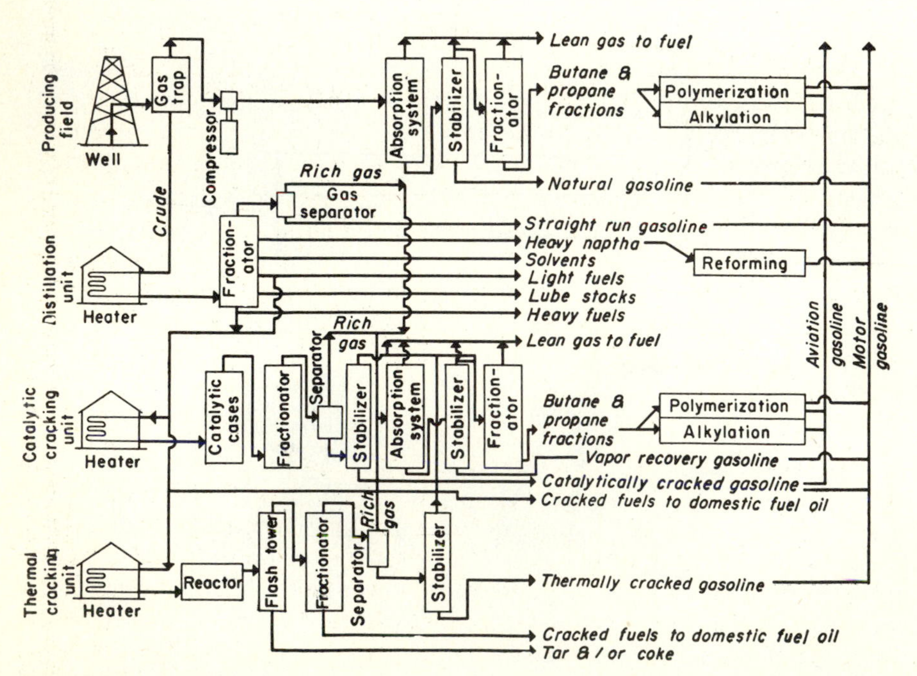

Figure 10-3 Refinery processes for manufacturing fuels. (Courtesy of Society of Automotive Engineers.)

GASOLINE

Gasoline Manufacture Gasoline originally was a product obtained only by the distillation of the crude oil. It was formerly somewhat lighter and more volatile than the average gasoline now found on the market. As indicated in Fig. 10-2, it can be produced in three distinct ways and classified as natural, raw, and cracked, depending upon the manner of production. So-called raw or straight-run gasoline is produced by the ordinary distillation of crude oil. The crude is heated to a given temperature for a certain length of time, and the resulting product that distills over is gathered in a condenser. This distillation product is not ready for use, however, but is known by the refiner as crude benzene or crude naphtha. Before being placed on the market, it must be subjected to a refining process, which consists in heating it with chemicals and redistilling it to separate out any heavy materials that may have passed over during the first distillation and to remove any impurities.

Cracked gasoline is manufactured from heavier distillation fractions, particularly gas oil. As the term implies, the process consists in breaking up the so-called heavy molecules into the lighter ones that make up gasoline. The process consists essentially in placing these heavier distillation products in receptacles and heating them under pressure. This results in certain lighter products that pass off, are condensed and further treated and refined, and eventually resemble straight-run gasoline.

Natural gasoline is manufactured from the gas that issues from oil wells or is obtained in the distillation of the crude oil. The process consists in compressing and liquefying the gases and then distilling these liquids under pressure, thus producing a very light and volatile product. This gasoline makes up only a very small percentage of the total amount manufactured. Ordinarily it is not used in its original condition, owing to its great volatility, but it is mixed with heavier, lower quality gasoline to form what is known as a *blended* product. Much of the gasoline now found on the market is known as blended gasoline and consists in many cases of all three types—natural, raw, and cracked—as mixed together in the refining process.

Grades of Gasoline Gasoline for automobiles and stantionary power units is available in four grades: (1) regular, (2) premium, (3) unleaded, and (4) super-unleaded. Regular gasoline contains tetraethyl lead to increase its antiknock qualities. It will give satisfactory results at the compression ratio found in most tractors and power units. Premium gasoline, which contains a larger quantity of tetraethyl lead, is a higher quality fuel from the standpoint of distillation range, performance, antiknock rating and should be used in engines with unusually high compression ratio. Unleaded gasoline contains no tetraethyl lead but is made up primarily of aromatics which are produced by the cracking units during refining and has a high antiknock quality. Methyloyclopentadienyl manganese tricarbonyl has been used as an antiknock agent in unleaded gasoline but was replaced by methyl tertiary butyl ether. Super-unleaded gasoline has a higher octane rating than unleaded because of the refining method and the use of a larger quantity of antiknock material.

Diesel Tractor Fuel Fuels for compression ignition engines resemble low-grade distillate, particularly with respect to gravity and distillation range. However, they are largely straight-run rather than cracked products. Present-day high-speed, multiple-cylinder diesel engines require a fuel having certain specifications.

To meet the needs of the various types and sizes of modern diesel engines, the American Society for Testing Materials (ASTM) has established three grades, 1D, 2D, and 4D. The specifications for the two best grades are shown in Table 10-3. Either grade 1D or 2D may be used. The grade 2D is heavier and will produce more work per gallon. However, grade 1D will give better performance under certain conditions. Table 10-4 shows the general conditions to which each grade is best adapted. Under some conditions, a mixture of these two grades might prove more satisfactory.

Two important qualifications of diesel fuels, particularly those used in high-speed engines, are (1) freedom from solid matter, sediment, and moisture; and (2) viscosity. An absolutely clean fuel is highly essential to the satisfactory operation of the injection mechanism, including the injection pumps and nozzles. Even slight traces of fine sediment or moisture may cause trouble.

A fuel having a certain viscosity range is likewise necessary to permit the injection mechanism to handle it properly and, at the same time, to provide some lubrication. A fuel having free-flowing characteristics at all temperatures at which the engine may be operated is also important.

Table 10-3 ASTM Diesel Fuel Specifications

	Fuel grade	
Characteristics	**1D**	**2D**
API gravity	35–40	26–34
Lb/gal	6.95	7.31
Btu/gal, average	137,000	141,800
Viscosity, 100°F	45 s	32 s
Volatility, 10% point	350–475°F	400–500°F
End point	500–625°F	625–700°F
Sulfur, percent by wt. max.	0.5	0.1
Flash point	100°F	125°F
Cetane number	45	40

Table 10-4 Physical Properties of Gasoline-Alcohol Blends[1]

Fuel	Percent alcohol by weight	API gravity	Weight per gallon lb	Heat per lb BTU (high)	Heat per gal BTU (high)	Increase in octane rating
Alcohol[2]	100	46.0	6.637	13,082	86,825	—
Gasoline (third grade)	0	62.8	6.062	20,463	124,047	—
Gasoline-Alcohol	5	62.3	6.078	20,094	122,133	2.5
Gasoline-Alcohol	10	61.7	6.097	19,725	120,262	4.9
Gasoline-Alcohol	15	60.9	6.122	19,356	118,497	7.7
Gasoline-Alcohol	25	59.5	6.167	18,618	114,817	10.1

[1] Barger, E. L. 1941. Power Alcohol in Tractors and Farm Engines. ASAE 22(2):65–67.
[2] Ethanol alcohol, anhydrous, produced and denatured for use as a motor fuel.

Cetane Rating of Diesel Fuels The marked increase in the application of high-compression ignition to high-speed engines of the automotive type has disclosed the necessity of utilizing, in these engines, fuels having certain specific physical and chemical qualities that will provide the necessary combustion characteristics when injection occurs.

For the purpose of comparing and designating the ignition qualities of such fuels, the so-called cetane rating method has been developed. The cetane rating of a diesel fuel is its designated ignition quality as determined by comparing it with a standard reference fuel consisting of a given blend of cetane ($C_{16}H_{34}$) and alpha-methylnaphthalene ($C_{11}H_{10}$). For example, a 40 cetane fuel is one having the same ignition qualities as a blend containing 40 percent cetane and 60 percent alpha-methylnaphthalene. The actual rating of a given fuel must be determined by means of a standard test engine and standard conditions.

Alcohol and Alcohol Blends Owing to the constantly growing demand for petroleum fuels such as diesel and gasoline for use in internal combustion engines, the possibility of a reduction in their supply along with an increase in their cost has stimulated investigations concerning possible sources of suitable substitutes. Thus far alcohol is receiving much attention either as a straight product or as a blend with a petroleum fuel such as gasoline. A blend of 90 percent unleaded gasoline and 10 percent alcohol is used to some extent as fuel for spark-ignition tractors and automobile engines and is known as gasohol.

Two kinds of alcohol might be used as a fuel for spark-ignition engines namely, ethanol (C_2H_6O) and methanol (CH_4O). Ethanol can be produced from (1) grain crops, such as corn, wheat and sorghum, (2) sugar crops, such as sugarcane, sugar beets and sweet sorghum, (3) starches, such as potatoes, or (4) crop residual such as corn stalks; this is known as grain alcohol. Methanol alcohol can be produced from coal, natural gas, petroleum, or wood. It is commonly known as "wood alcohol". Neither of these is available in a pure form for use as a fuel for internal combustion engines but either can be used in another form known as denatured alcohol. Denatured alcohol consists largely of ethanol (grain) alcohol with some methanol alcohol mixed with it, together with pyridine, which gives it a distinct odor and color.

The principal disadvantages given for the use of alcohol as an engine fuel are (1) it has only about 67.5 percent of the heating value of gasoline (Table 10-4); (2) it does not vaporize as readily as gasoline; (3) the cost of manufacture and production is relatively high for anhydrous alcohol. Methanol and ethanol require about three times more heat to make them evaporate than the same volume of gasoline. This is because alcohol, being chemically pure, should have one boiling temperature, 173.3F (78.5C) and a straight line distillation curve while gasoline will not (Fig. 10-6).

The advantages of adding alcohol to gasoline include (1) it raises the octane rating of the fuel (Table 10-4); (2) it forms no soot or carbon in the engine; (3) it burns without producing smoke or disagreeable odors; (4) the exhaust is free of nitrogen oxides without special cleanup devices because alcohol burns cooler; (5) cool burning reduces the possibility of burning valves; (6) a higher compression ratio can be utilized thereby increasing engine efficiency.

Any gasoline engine can operate on straight alcohol by making changes in the carburetor to eliminate plastic parts and by adding a heat source to aid combustion for cold weather starting. The carburetor jet should be modified to provide about 1.5 times more fuel, either by using a larger jet or by enlarging the old one to carry more fuel.

Anhydrous or absolute ethanol alcohol will mix with gasoline in all proportions. Under certain conditions commercial alcohol of 95 percent purity or less when mixed with gasoline will have phase separation. It can be made to blend with gasoline by adding blending agents, such as propyl, butyl and amyl alcohols, and benzol or acetane.

Gravity Method of Testing Fuels Formerly it was the practice to express the comparative quality of fuels, especially gasolines, in terms of gravity; that is, the owner or operator of an automotive vehicle such as an automobile or a tractor was led to believe that a so-called high-test gasoline was a higher quality product and, therefore, more desirable than a low-test gasoline. As the demand for gasoline increased it became necessary for the oil refiner to meet the situation by converting more of the heavier fractions of crude oil into gasoline. Thus the commercial gasoline now being produced is a slightly heavier product than formerly and possibly somewhat less volatile. However, this apparent lowering of the quality of the fuel has had very little effect upon satisfactory engine operation, because the engine manufacturers have promptly taken steps to adapt their products to these fuels. In fact, the present-day automotive engines operate more smoothly and with less trouble on these lower gravity fuels than did those of a few years ago on fuels that were considered as more desirable in quality.

The gravity of a fuel or an oil may be expressed as *specific gravity* or as API (American Petroleum Institute) gravity, formerly known as Baume gravity. The latter scale is preferred in the United States in connection with petroleum products. The *specific gravity* of a liquid is the ratio of its weight to the weight of an equal volume of water at 60°F. The relationship of the API gravity scale to specific gravity is expressed by the following formula:

$$\text{Degrees API} = \frac{141.5}{\text{sp. gr. } 60°/60°\text{F}} - 131.5$$

The device used for testing the gravity of a liquid is known as a *hydrometer*. Usually two hydrometers are necessary, one for liquids lighter than water, and another for liquids heavier than water. Figure 10-4 illustrates how the hydrometer is placed in the liquid and read.

Table 10-2 gives the API gravities of the more common liquid fuels. The values vary from about 65° for a high-test, very volatile gasoline to 30° for diesel fuel. It is thus observed that the lighter the fuel and the lower its specific gravity, the higher the API gravity value in degrees. The API gravity of pure water is 10°.

Distillation Test of Liquid Fuels A more reliable and widely used method of determining the value of liquid fuels is known as the ASTM distillation test. In making such a test a given quantity (100 cc) of the fuel is placed in a flask (Fig. 10-5), heated, and its so-called initial boiling point noted from the thermometer inserted in the top of the flask. Then, as the heating is continued, a certain amount vaporizes, passes off, and is condensed. The temperatures at which certain percentages of the fuel pass off are noted, and finally the end point is observed, that is, the temperature at which the last drop evaporates. The usual practice is to plot a curve of the distillation values for a given sample and compare it with that of a similar fuel of recommended or standard requirements and specifications. Figure 10-6 shows distillation curves for the more common liquid fuels.

Significant information indicated by a fuel-distillation curve in relating volatility characteristics of gasoline to engine performance is obtained from the temperatures at which 10, 50, and 90 percent of the fuel is evaporated. The initial boiling point is the temperature at which the first drop of liquid leaves the condenser. The lower the initial point, the greater is the

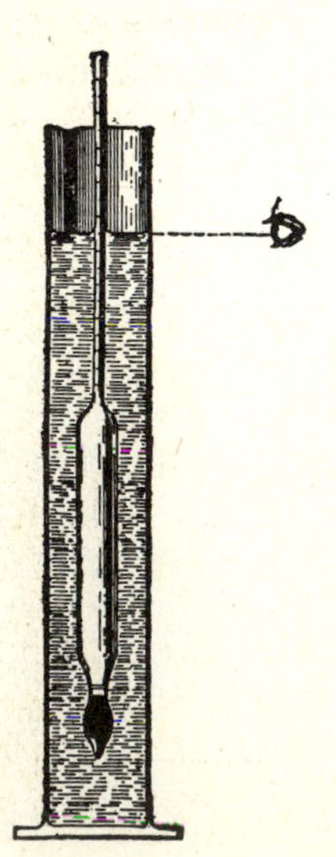

Figure 10-4 Use of hydrometer in determining gravity of fuels.

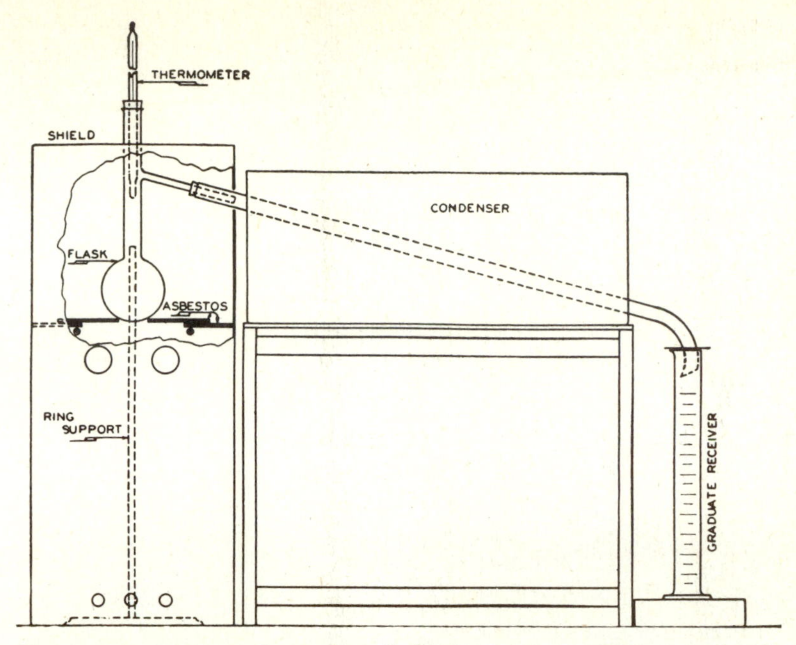

Figure 10-5 Apparatus for making distillation tests of gasoline and other liquid fuels.

"front-end" volatility of the gasoline, and, in general, the higher the vapor pressure. Although the initial point is of some importance in cold engine starting, it is difficult to measure accurately. The 10 percent point can be measured more accurately and therefore is more useful. For this reason the initial boiling point is of less importance as a specification for gasoline.

The "10 percent point," as it is commonly called, is of primary importance as a specification related to engine starting. The lower this temperature, the better the starting characteristics of the gasoline, other factors being equal. The 10 percent point is also used to indicate the relative vapor-locking tendencies of fuels having equal vapor pressures.

The "50 percent point" is important as an index of the engine warm-up characteristics of gasoline; that is, the lower the 50 percent point temperature, the faster the warm-up. The presence of high-boiling components in the gasoline is likely to cause poor mixture distribution in the intake manifold and therefore may affect engine performance during acceleration. Also the presence of a large portion of such materials may lead to crankcase dilution. This results from the failure of the fuel to evaporate and burn. The unburned portions may enter the crankcase past the piston rings and dilute the oil. The 90 percent temperature provides a good indication of the fuel performance in this respect, and it has become an important part of gasoline specifications. The end point has little significance and is not usually included in gasoline specifications. The 90 percent point is used in preference to the end point because the latter is difficult to determine accurately.

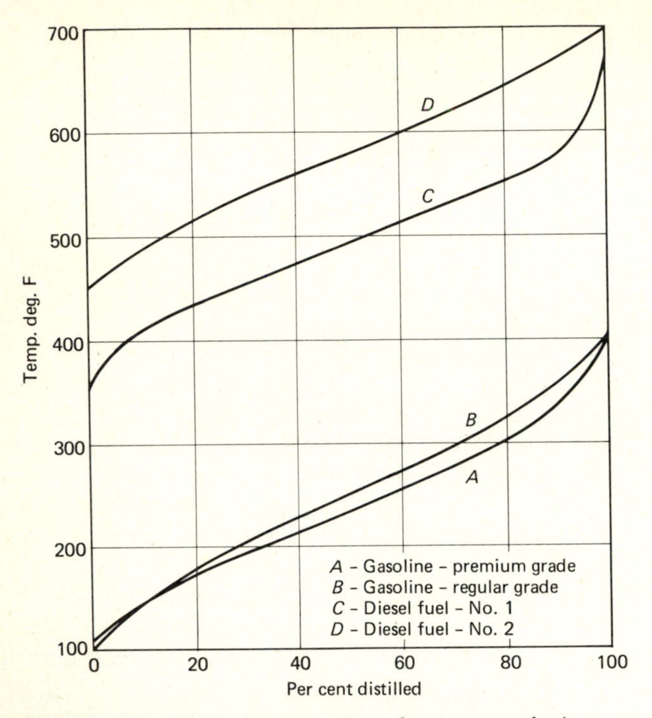

Figure 10-6 Distillation-test curves for common fuels.

Detonation—Preignition—Knock Rating The tendency of designers of automotive and tractor-type engines to resort to higher compression pressures, in order to obtain more power and speed and increase efficiency, has resulted in pronounced fuel-knocking effects in many of these engines. This knocking, properly termed *detonation,* causes an unpleasant, sharp, clicking sound that is most noticeable when the motor is operating at low speed with wide-open throttle. Detonation is often confused with preignition, but there is a distinction. Preignition occurs when the charge is fired too far ahead of the compression-dead-center position of the piston, owing either to excessive spark advance or to premature, spontaneous ignition resulting from excessive heat in the cylinder. The resulting noise is that of the pistons and bearings as the piston completes its stroke against abnormally high pressure. Ordinarily, retarding the spark will eliminate this type of knock.

Detonation occurs during the process of combustion of the mixture within the cylinder after ignition has taken place. It is often referred to as "knock," "pink," or "ping."

The problem of control of detonation is handled in a number of ways, as follows:

1 Using specially designed cylinder heads and pistons to so shape the combustion space that the last-to-burn portion of the charge will be spread out into a thin sheet and its temperature held down more effectively.

2 Providing more effective water circulation and cooling around the cylinder head and exhaust valves.

3 Using cylinder head and piston materials—particularly aluminum alloys—that provide more rapid heat dissipation.

4 Using a properly designed spark plug and locating it in the "hot region" and preferably near the exhaust valve.

5 Eliminating or reducing carbon deposits. Carbon is an excellent heat insulator and thereby induces detonation by creating ineffective cooling of the combustion space.

6 Maintaining the correct mixture of fuel and air and the proper spark setting and valve adjustment.

7 Using specially treated or so-called "doped" or antiknock fuels.

Certain chemicals may be added to gasoline to reduce its tendency to detonate. The effect of these so-called "dopes" upon combustion is to raise the autoignition temperature and also to slow down the combustion process. This affords time for the piston movement to provide some increase in volume for the hot gases. There are many substances that will successfully slow down combustion but that are unsatisfactory for one reason or another.

Tetraethyllead compound is the most popular gasoline knock suppressor in use today. Gasoline treated with this compound is marketed under the trade name Ethyl. Ethylene dibromide is added to prevent the formation of lead oxide, which would otherwise deposit on spark plugs, valve seats, and valve stems. The red aniline dye serves only for identification. The amount of tetraethyllead added to gasoline varies from about 1.4 to 2.7 cm^3/gal, depending upon the desired octane requirement. Although it is very effective in slowing the combustion process and preventing detonation, the addition of ethyl increases the price of the fuel which is only partly offset by the increased fuel economy resulting from the use of higher compression ratios.

Antiknock (Octane) Rating Since fuel detonation has become such an important factor in fuel selection and its relation to proper engine performance, a means of designating the antiknock quality of certain fuels and particularly the different grades of gasoline seemed desirable. Such a rating system was developed by automotive and petroleum engineers. It is based upon the fact that certain pure hydrocarbons have a very high antiknock quality while others are very poor in this respect. For example, isooctane (C_8H_{18}) has excellent antiknock qualities and is given a rating of 100. Normal heptane (C_7H_{16}), on the other hand, would knock excessively even under low-compression conditions and was assigned a value of zero. Therefore the antiknock value of a fuel is determined by comparing it with a mixture of isooctane and heptane, and the fuel is given an octane rating value based upon the percentage by volume of isooctane in an isooctane-heptane mixture; that is, a fuel having the same knock characteristic as a 70-30 percent isooctane-heptane mixture is called a 70 octane fuel.

Numerous attempts have been made to find some physical or chemical property of motor fuels that would predict the knocking tendency that the fuel would have when used in a motor. These attempts have been only partially successful, and hence standard methods of determining the octane ratings of fuels have been developed. These methods involve the use of a standard test engine (Fig. 10-7) specifically designed and equipped for testing the knock rating of fuels. The first engine was introduced in 1932 and was known as the *CFR* (*Cooperative Fuel Research*) *engine*. This method was later approved by the American Society for Testing Materials and is now known as the *ASTM* (*CFR*) *motor method*. About 1939, as a result of changes in automotive engine design and fuel characteristics, it was decided that the motor method was no longer fully adequate and

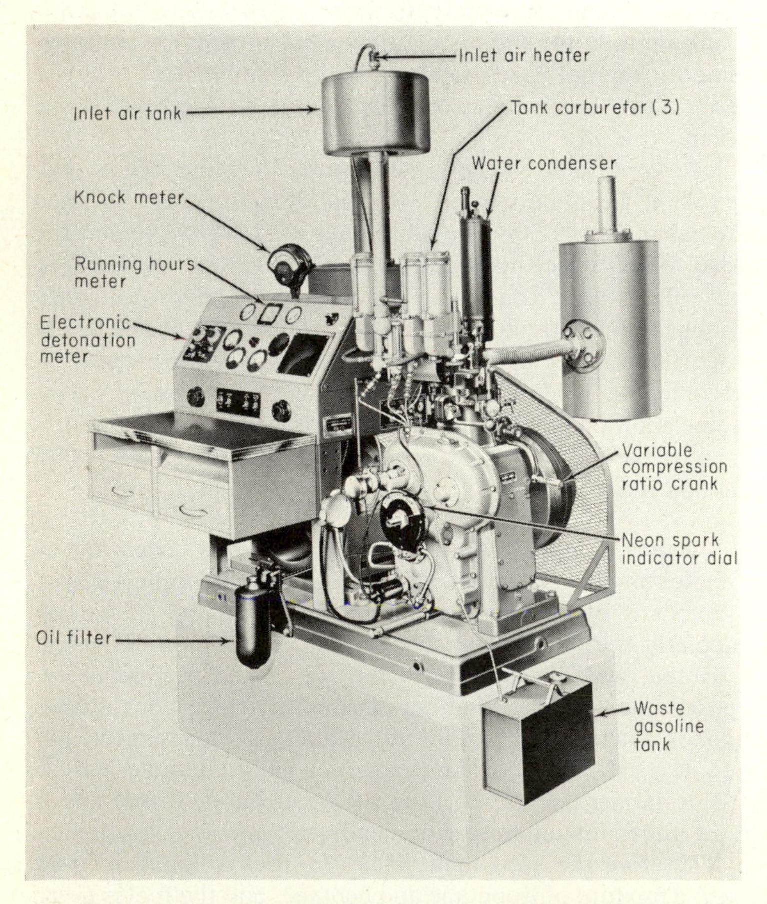

Figure 10-7 Waukesha TSTM-CFR Octane rating test engine. (Courtesy of Waukesha Motor Company.)

reliable as an index of the road performance of fuels. Hence certain changes were made in the original CFR test engine and test specifications, particularly with respect to engine speed, spark timing, and inlet air temperature. The outcome was a modified knock test method known as the *CFR research method,* which the ASTM adopted as a tentative standard in 1948. In both the motor method and the research method, the octane number of a fuel is determined by comparing its knocking tendency with those for blends of the reference fuels of known octane number under standard operating conditions. This is done by varying the compression ratio for the sample to obtain the standard knock intensity as defined by a guide curve and as measured by a bouncing pin and knock meter. When the knock meter reading for the sample is bracketed between those for two reference blends differing by not more than two octane numbers, the rating of the sample is calculated by interpolation.

In general, there is a difference in the octane value for a given fuel when tested by each method. With few exceptions, the research method gives higher readings than the motor method. Figure 10-8 shows the trends in the antiknock quality of regular and premium-grade gasoline over a 30-year period.

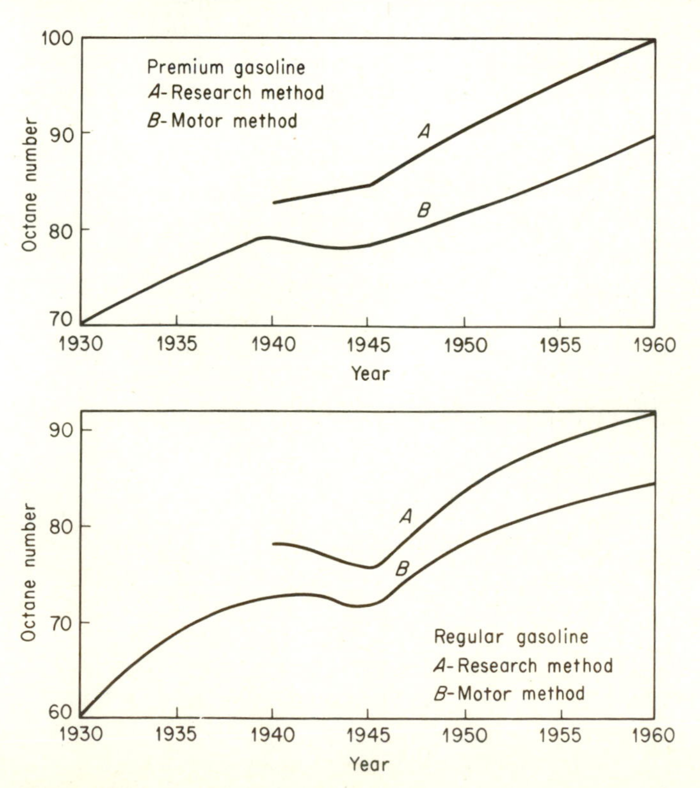

Figure 10-8 Trends in antiknock quality of gasoline (since 1960 trend curves have changed only minimally).

Vapor Lock Vapor lock, gas lock, or air lock is the partial or complete interruption of the fuel flow in the fuel-feed system as a result of vaporization of the fuel and the formation of vapor or gas bubbles at some point. It may be due to the use of a fuel having a too high percentage of light or volatile material or to the location of the fuel tank, lines, pump, and so on, with respect to the hotter parts of the engine. Vapor lock often occurs with engines operated at high altitudes as a result of the lower boiling point.

Difficulties from vapor lock show themselves most commonly in failure of the engine to idle after a fast, hot run or in traffic; sometimes in intermittent or uneven acceleration after idling; and sometimes by irregular operation during a sustained high-speed run. Complete stoppage of the engine rarely occurs; when it does, the cause may be traced either to vapor lock in the line leading to the carburetor or to boiling in the filter bowl.

Combustion of Hydrocarbon Fuels The term *combustion* is applied to the process by which a fuel unites chemically with oxygen, producing what is known as an oxide and often generating heat of considerable intensity, and sometimes light. It may be a very slow or a very rapid action. For example, the rusting of a piece of iron is a comparatively slow process, resulting in the union of the iron with oxygen, forming what is known as red iron oxide or rust. In a common wood or coal stove there is a more rapid union of oxygen with the carbon of the wood or coal, resulting in a high temperature and a heating effect.

In the internal-combustion engine, rapid combustion takes place; that is, the fuel, when mixed with the proper amount of oxygen and ignited, burns instantaneously, resulting in the production of gaseous oxides—largely carbon monoxide and dioxide—and water. These gases, being confined in a very small space, produce high pressure and consequently exert great force on the piston of the engine and thus generate power.

Oxygen, therefore, is necessary for combustion in all cases, and the chemical action taking place during the combustion of the fuel mixture in a gas engine may be represented by the following chemical equation:

$$CH_4 + 2O_2 = CO_2 + 2H_2O$$

or, for a pure liquid fuel such as octane (C_8H_{18}),

$$C_8H_{18} + 12.5O_2 = 8CO_2 + 9H_2O$$

The oxygen, in all cases, is obtained from the atmosphere; that is, a certain amount of air is mixed with the fuel before ignition takes place. Although air is only about 23 percent oxygen, the other 77 percent is largely nitrogen, which does not have any effect upon combustion.

Correct Fuel Mixtures Knowing the chemical composition of a fuel, the atomic weight of the principal elements involved, namely, carbon, hydrogen, and oxygen, and the percentage of oxygen in the atmosphere, we can readily calculate the amount of air necessary to produce perfect combustion in the gas-engine cylinder.

If pure octane is used as a fuel and the atomic weights are 12 for carbon, 1 for hydrogen, and 16 for oxygen, such a calculation would be as follows:

$$C_8H_{18} + 12.5O_2 = 8CO_2 + 9H_2O \tag{1}$$

or
$$(96 + 18) + 12.5(32) = 8(12 + 32) + 9(2 + 16) \tag{2}$$

or
$$114 \quad + \quad 400 \quad = \quad 352 \quad + \quad 162 \tag{3}$$

From Eq. (3) it is observed that 114 lb of fuel requires 400 lb of oxygen. Therefore, if air is 23 percent oxygen, a total of 1,739 lb of air is needed to supply this oxygen and the fuel-air ratio is 1,739 ÷ 114, or 15.3. In general, a fuel-air mixture of 1 to 15 by weight is considered as correct for complete and normal combustion.

A fuel mixture containing less than the required amount of air is known as a rich mixture; that is, there is not enough oxygen present to combine with all the carbon in the fuel and produce complete combustion. Hence free carbon is liberated, creating a black smoke at the exhaust. The usual indications of a too rich mixture are (1) black smoke at the exhaust; (2) lack of power; and (3) overheating of the engine.

A mixture containing more than the required amount of air is known as a lean mixture, and is best indicated by what is known as backfiring through the intake passage and carburetor. Such a mixture is very slow burning and produces (1) uneven firing; (2) lack of power; and (3) overheating. Table 10-5 shows the effect of various air-fuel mixtures on the color of the exhaust flame.

Table 10-5 Colors of Exhaust Flame

Air-fuel ratio	Color of exhaust flame	Condition of mixture
8.5 to 1	Bright yellowish-orange–black smoke	Very rich
9.5 to 1	Bright yellow	Rich
9.7 to 1	Bluish-white with faint yellow tinge	Rich
10 to 1	Light blue with trace of yellow	Rich
11.3 to 1	Light blue	Slightly rich
13.6 to 1	Intense light blue	Approaching ideal
15 to 1	Light blue of maximum intensity	Ideal
17.3 to 1	Whitish-blue of less intensity	Lean

Figure 10-9 shows a complete picture of the relationship of various air-and-fuel mixtures to the efficiency, power output and other factors in the operation of a common type of gasoline burning engine.

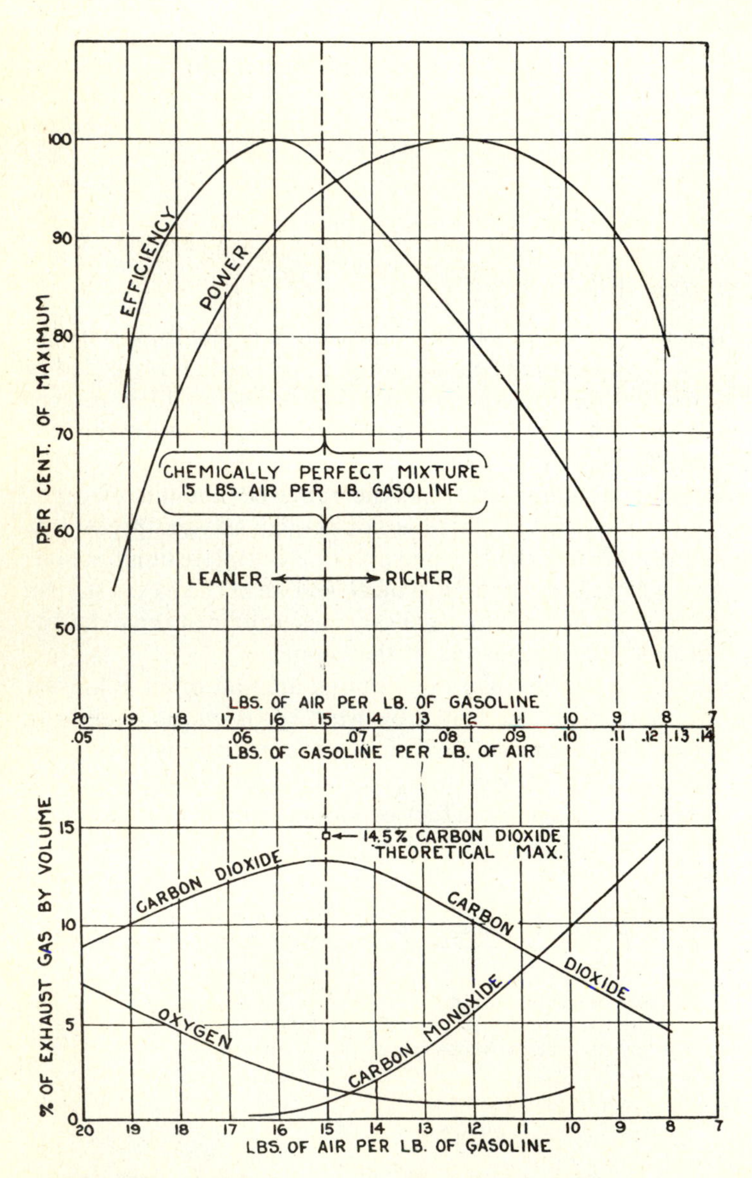

Figure 10-9 Curves showing the effect of air-fuel ratio upon power, economy and exhaust-gas analysis.

Engine Emission Control The combustion process in an engine using hydrocarbon fuels such as gasoline and diesel fuel results in the production and discharge of certain gaseous materials which have a polluting effect on the atmosphere, particularly in areas having a very high engine concentration and utilization such as, for example, in large cities with many automobiles and trucks.

As previously explained, the products of combustion of a normal engine fuel mixture include primarily carbon dioxide, water, and nitrogen in the form of nitrogen oxides. However, the conditions under which an engine operates are seldom exactly uniform. Hence, certain other harmful and undesirable materials such as carbon monoxide and unburned hydrocarbons may be present in the exhaust fumes.

Much research has been carried on concerning emissions control. The principal engine modifications recommended include the following:

1 Carburetor calibration.
2 Ignition calibration.
3 Exhaust gas recirculation system.
4 Positive crankcase ventilation.
5 Thermostatic air cleaner for controlling the temperature of air entering the carburetor.
6 Catalytic converter added to the exhaust system to reduce hydrocarbon and carbon monoxide pollutants in the exhaust gas stream. The converter contains material which is coated with a catalytic material containing platinum and palladium. Unleaded gasoline must be used for the engine as deposits from leaded gasoline reduce the effectiveness and life of the converter.

PROBLEMS AND QUESTIONS

1 Explain the refining process followed in producing gasoline.
2 Discuss the chemical composition, gravity, vaporizing characteristics, and heating value of the most common engine fuels.
3 Discuss alcohol as an engine fuel.
4 Prepare typical distillation curves for regular-grade gasoline, and diesel fuel.
5 Explain what points in a fuel distillation curve are most indicative of the fuel quality and valve. Give reasons.
6 Explain how fuels are tested and rated for antiknock quality.
7 Explain the combustion process as it takes place with a simple gaseous fuel and the relationship of this process to correct fuel-mixture control.
8 Sketch curves showing the relationship of air-fuel ratio to engine power, efficiency, and exhaust-gas composition.
9 Assuming that air contains 22.7 percent oxygen, compute the number of pounds of air required for the perfect combustion of a gallon of butane.

Chapter 11

Fuel Supply and Carburetion—Air Filters—Governing

The fuel-supply and carburetion system for an ordinary-compression carbureting-type engine consists essentially of (1) a fuel-supply container or tank; (2) a carburetor; (3) the necessary connecting lines, pump, filter, etc.; and (4) the intake manifold to conduct the mixture from the carburetor to the cylinder. The usual systems are:

1 The suction system, as used on some single-cylinder engines (Fig. 11-1).

2 Gravity feed with float-type carburetor and elevated fuel tank, as used on both single- and multiple-cylinder stationary and automotive-type engines (Fig. 11-2).

3 The force-feed system with float-type carburetor, as used on both single- and multi-cylinder stationary and automotive-type engines (Fig. 11-3).

Principles of Carburetion The functions of a carburetor are (1) to assist in properly vaporizing the fuel; (2) to mix the vaporized fuel in the correct proportions with air; and (3) to supply the engine with the proper quantity of this mixture depending upon the load, speed, temperature, and

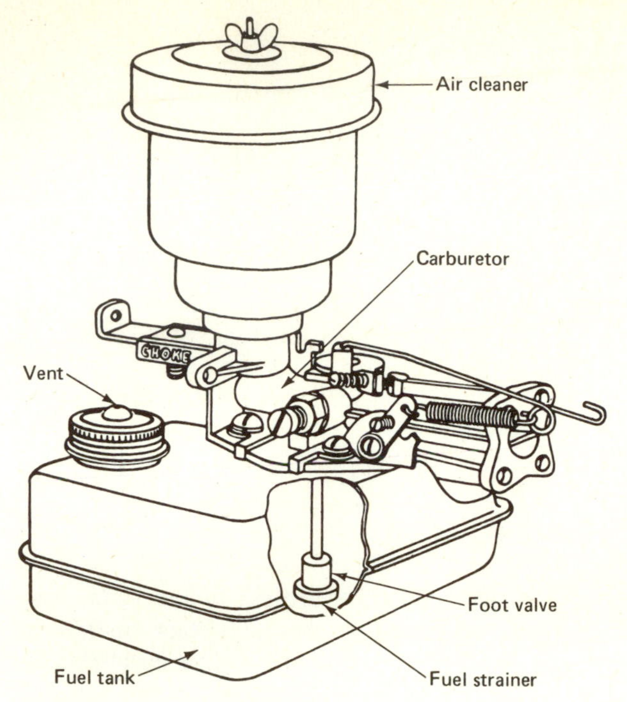

Figure 11-1 Suction system of fuel supply and carburetion. (Courtesy Briggs and Stratton Corp.)

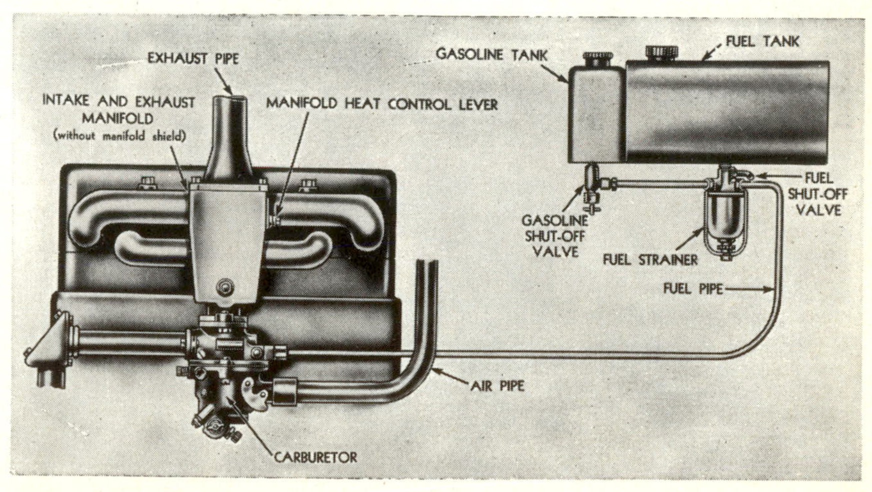

Figure 11-2 Gravity system of fuel supply for tractor.

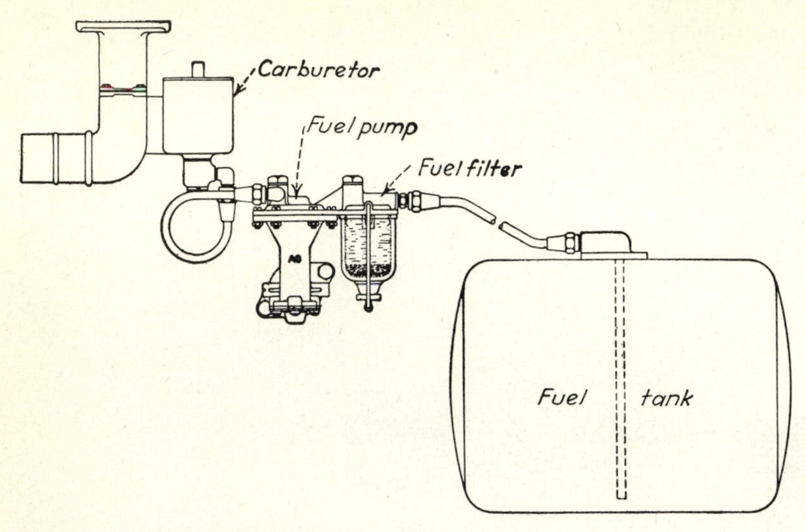

Figure 11-3 Float-type carburetor with pump feed.

other conditions present. All carburetors, regardless of the type, size, speed, number of cylinders, and other engine characteristics, operate on the same basic principles, as illustrated in Fig. 11-4. The carburetor is attached to the cylinder block by the intake manifold and, in turn, is connected to the combustion chamber by the intake valve. On the intake stroke this valve opens and the piston movement creates a low pressure in the cylinder. This results in a suction action at the air entrance to the carburetor and a high-velocity air flow past the fuel-jet nozzle in which a certain fuel level is maintained. Thus a reduced pressure is created at the nozzle and atomized fuel enters the airstream at this point and forms a combustible mixture. The air passage in the carburetor is reduced in cross section at the nozzle in order to increase the air velocity at this point and thereby create a greater reduction in the nozzle pressure than would exist

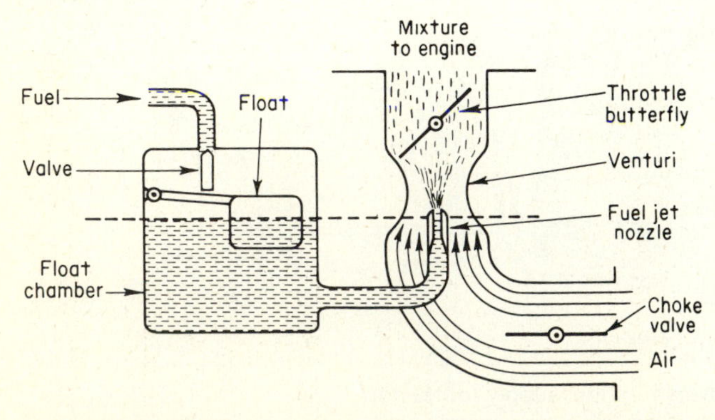

Figure 11-4 Operation of a simple jet-nozzle type of carburetor.

otherwise. This restriction in the carburetor air passage is called the venturi and is common to all types of carburetors.

The following conditions are of fundamental consideration in the proper functioning of any carburetor:

1 A constant and specific fuel level must be maintained in the nozzle.

2 There must be at least partial if not complete vaporization of the liquid fuel, regardless of surrounding temperatures.

3 The correct mixture of vaporized fuel and air must be maintained at all times, regardless of engine load, speed, temperature, atmospheric pressure, and other operating conditions.

Mixture Proportioning by a Simple Carburetor The problem of maintaining an optimum fuel mixture and supplying the correct amount of this mixture to each cylinder of an engine regardless of speed and speed range, power output, temperature, atmospheric pressure, and other operating conditions is a complex one and explains why the carburetors used on most automotive-type engines are relatively complex. For example, suppose that an engine is equipped with a simple carburetor, as shown in Fig. 11-4, and is operating at a comparatively low speed and light load, but with the proper fuel mixture. Obviously, the throttle-butterfly will be nearly closed. Now suppose that the throttle is suddenly opened with the idea of increasing the engine speed or power output. The increased throttle opening itself permits a greater suction at the nozzle, causing more fuel to be taken up by the airstream. Likewise, the velocity of the air through the carburetor increases so that it would seem likely that the correct air-fuel proportion would be maintained. However, such is not the case. Experience as well as theory has proved that the quantity of air drawn in does not increase at a rate great enough to maintain the correct air-fuel mixture, and the mixture becomes too rich. Now suppose that with the engine running at this increased speed, the fuel flow from the nozzle is readjusted and reduced so that the mixture is again correct. If the engine were again slowed down by closing the throttle, the mixture would be too lean, because the decreased suction would not permit the required amount of fuel in proportion to the air to be drawn from the nozzle. The explanation of this change and variation in the fuel mixture with changes in engine speed and load is based upon certain fundamental laws pertaining to the flow of air and liquids and the differences in their densities and other physical characteristics.

FUEL-SUPPLY AND CARBURETION SYSTEMS

The Suction System The fuel is placed in a tank that is located below the carburetor, with a fuel line connecting the two (Fig. 11-1). On the suction stroke of the piston, owing to the vacuum produced in the cylinder, fuel is drawn through the fuel line to the carburetor. At the same time, air is drawn through the air passage. The fuel mixture thus formed passes into the combustion space of the cylinder and is ignited. The correct mixture is obtained by the proper adjustment of the fuel needle valve, which controls the amount of fuel passing through the jet nozzle. With this system, as well as with all others, it is necessary to maintain a certain fuel level in the nozzle of the carburetor. This is done by placing some kind of check valve in the fuel line, which prevents the fuel from running back into the tank at the end of the suction stroke, so that, at the beginning of the next intake stroke, fuel is available at the nozzle and is immediately drawn from it. The use of the suction system is confined largely to small single-cylinder engines operating at a relatively uniform load and speed.

Gravity System The second, or gravity, system, as illustrated in Fig. 11-2, consists of (1) the tank, placed above the carburetor; (2) the fuel line; (3) a filter; and (4) the carburetor, fed by gravity. The quantity of fuel that is allowed to flow into the carburetor is controlled by a hollow metal float. This float is attached to a float valve in such a way as to allow the fuel to enter the carburetor at the same rate at which it is being consumed by the engine. This system likewise maintains an absolutely uniform level of fuel in the carburetor, regardless of the quantity in the tank, and therefore is adapted to gasoline and other liquid-fuel-burning engines.

Force-Feed System All automobile engines and some stationary power units have the fuel tank located at a level below that of the carburetor. This permits placing the tank where it is convenient for filling yet out of the way. However, a special type of pump is needed to lift the fuel from the tank to the float-type carburetor. This system is shown by Fig. 11-3.

Figure 11-5 shows the construction and operation of the usual type of pump. It is attached to the engine block in such a manner that a special cam on the camshaft operates an arm which, in turn, actuates the diaphragm and causes it to draw fuel from the tank and force it to the carburetor. The fuel is filtered by a fine screen as it flows through the pump.

Carburetor Types and Construction The suction type of carburetor, illustrated in Fig. 11-1, is the simplest possible device that can be used. It consists essentially of a metal body, the jet nozzle, a fuel needle valve, and the choke. Its use is confined almost entirely to small, single-cylinder stationary engines burning gasoline.

The narrow part of the air passage around the nozzle of a carburetor (Fig. 11-6), usually called the *venturi,* is for the purpose of creating a greater air velocity and, therefore, a stronger sucking effect at this point.

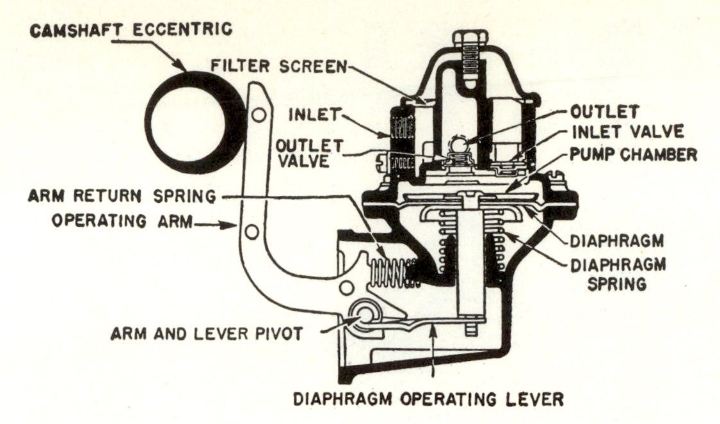

Figure 11-5 Late type of fuel pump and filter.

The purpose of the choke is to shut off the air so that a greater amount of fuel may be drawn in, when the engine is being started, because the slow piston movement provides only limited suction. Ordinarily, the choke should be used for starting only and not for adjusting the mixture during operation. The needle valve is the only means of adjusting and controlling the mixture in the suction system.

The float-type carburetor (Fig. 11-7) is a rather complex device and is usually made up of the following essential parts: (1) choke, (2) float and float valve, (3) venturi, (4) main jet, (5) main-jet nozzle, (6) main-jet needle valve, (7) idling jet, (8) idling-jet-adjustment needle valve, (9) air bleed, and (10) throttle butterfly. The float, together with the float valve, controls the fuel entering the float or fuel chamber of the carburetor from the tank and thus maintains a constant fuel level in this chamber and at the jet nozzle. The float is made of light-gauge copper or brass and is hollow.

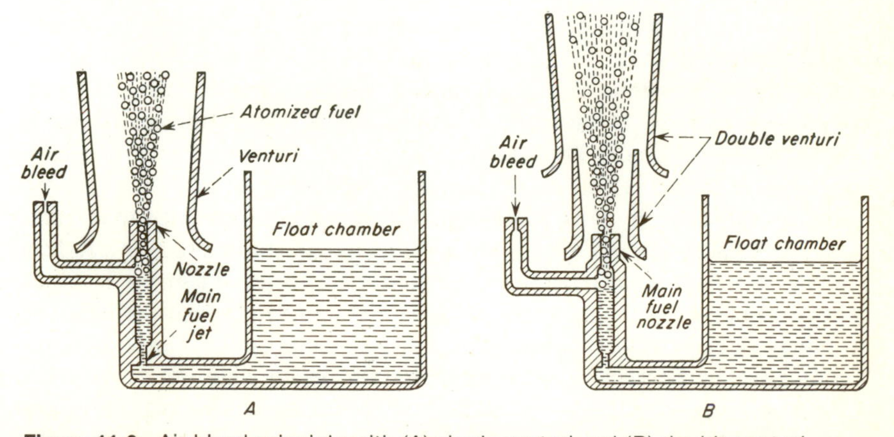

Figure 11-6 Air-bleed principle with (A) single venturi and (B) double venturi.

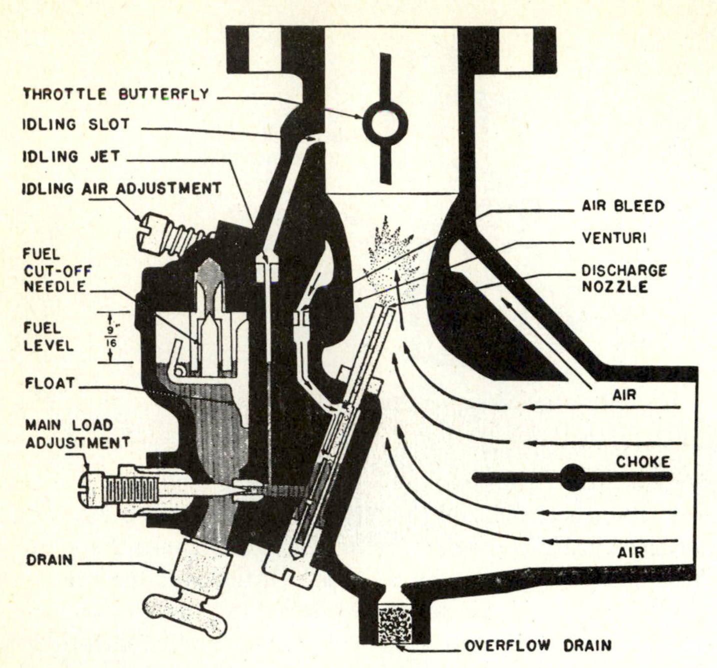

Figure 11-7 Updraft carburetor with single venturi and both load and idling mixture adjustments.

The main jet is a small opening or passage of exact size through which fuel passes from the float chamber to the main-jet nozzle. The fuel flow through the main jet may be controlled by means of a needle valve or it may be what is known as a *fixed jet.* The small, stationary, single- and multiple-cylinder engines and all automobile and truck engines are usually equipped with a fixed-jet carburetor (Fig. 11-8). Medium- and large-size tractor engines, and particularly those burning a heavy fuel, use carburetors with an adjustable main jet (Figs. 11-11 and 11-12). Some automobile-engine carburetors are equipped with a special type of main-jet control, such as a metering pin that controls and varies the fuel flow from the main jet according to the throttle position and engine speed. Such a carburetor is shown in Fig. 11-14.

The idling jet, as the name implies, supplies fuel at idling or low engine speeds only. It usually consists of a special fuel passage that leads to an opening into the airstream at the edge of the throttle-butterfly. The opening is on the manifold side of the butterfly; hence when the latter is practically closed, the manifold suction pulls the necessary idling fuel from this orifice. As the throttle opens for increased speed, the suction on the idling orifice decreases and reacts directly on the main-jet nozzle. All carburetors are equipped with a needle valve for maintaining the correct idling-mixture adjustment. In most cases this valve controls the air flow rather than the fuel, although there are a few exceptions.

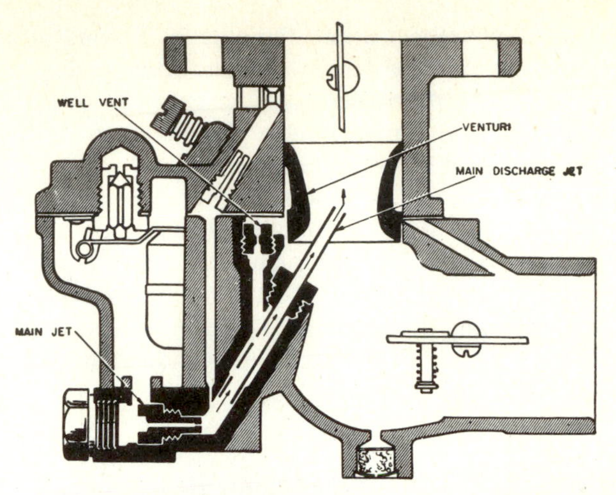

Figure 11-8 Updraft carburetor with single venturi showing fixed jet and operation under load conditions.

The term *air bleed* refers to certain types of carburetors that control the fuel flow from the main jet and the resultant fuel mixture by permitting a fine stream of air to mix with the fuel in the nozzle. This device is shown in Figs. 11-6 and 11-7. The air bubbles break up the fuel into smaller particles before it leaves the nozzle and thereby aid in producing better atomization and mixture control.

The throttle butterfly is for the purpose of controlling the quantity of mixture allowed to enter the cylinders. It therefore controls the engine speed and power. With a low speed or a light load, less fuel mixture would be required, and the throttle would be only partly open. For a high speed or a heavy load, the throttle would be wide open in order to permit the cylinders to receive a full charge. The opening and closing of this device, according to the engine speed and power output required, are effected by a foot lever, as in automobiles, or by means of a hand lever and mechanical governor, as in farm tractors.

All float types of carburetors are equipped with some sort of compensating device or arrangement or are so constructed that the mixture adjusts itself instantly to any changes in engine speed or load and thus remains properly proportioned. Several different methods are employed by carburetor manufacturers to bring about this action. The more important of these are explained in the descriptions of a number of well-known makes of carburetors.

Downdraft Versus Updraft Carburetors A downdraft carburetor is one that is mounted above the intake manifold and engine so that the air enters the upper part of the carburetor and the mixture flows downward into the manifold. An updraft carburetor is mounted below or beside the engine block and the mixture flows upward into the engine. Downdraft carburetors are used almost exclusively on automobile and truck engines and for some stationary power units, particularly V-type engines, for the reason that installation is easier and the device is more accessible and less exposed to dust, oil, and moisture. Another advantage of the downdraft carburetor is that any excess fuel caused by ''flooding'' or overchoking will run directly into the engine and thus eliminate a fire hazard. Updraft carburetors are used exclusively on tractors and many stationary power units because of more suitable installation adaptability and accessibility. There is no difference in the operating economy or efficiency of these two types of carburetors, provided they are properly designed and adapted to the engine in other respects.

Float-Type Carburetors

Zenith Model 61 The Zenith Model 61 carburetor (Fig. 11-8) is a single-venturi plain-jet updraft carburetor with a fixed main jet. It is used on some small tractors and stationary engines. Figure 11-9 shows the operation of the device under low-speed or idling conditions. The fuel for idling is supplied through the main jet to a well directly below the main discharge jet. The fuel travels from this well through the idle fuel passage to the idle jet. The air for the idle mixture originates behind the venturi. The position of the idle adjusting needle in this passage controls the suction on the idle jet and hence the idle mixture. Figure 11-8 shows the operation under high-speed and load conditions. The main jet controls the fuel delivery during the part-throttle range from about one-fourth to full-throttle opening. To maintain a proper mixture ratio, a small amount of air is admitted through the well vent into the discharge jet through the air-bleed holes at a point below the level of the fuel in the metering well.

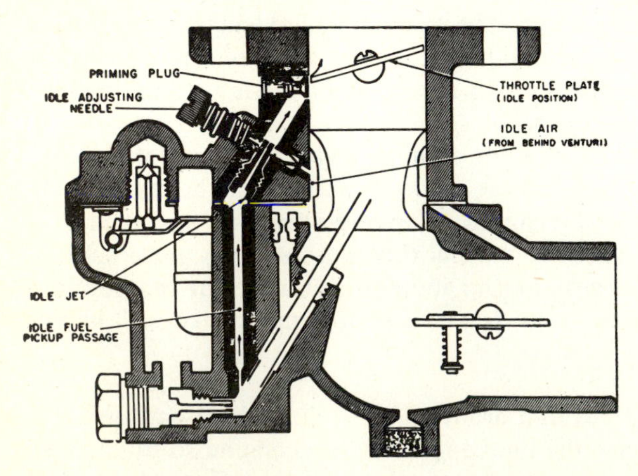

Figure 11-9 Updraft tractor carburetor showing operation under idling conditions.

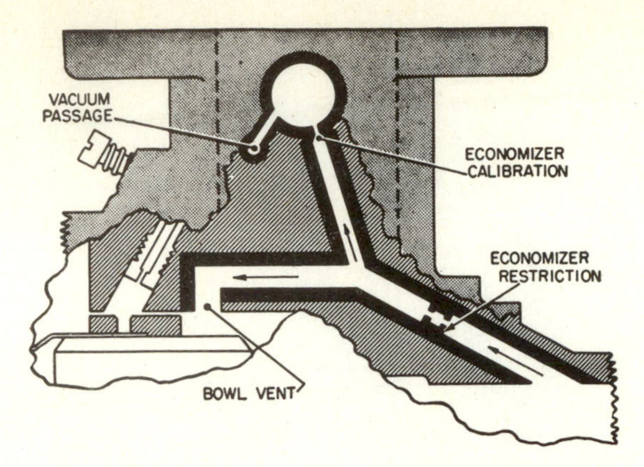

Figure 11-10 Economizer system for Zenith Model 61 carburetor.

Figure 11-10 shows the economizer system used on the Zenith Model 61 carburetor. It consists of (1) a "milled" slot in the throttle shaft, which acts as a valve to open or close the system; (2) a vacuum passage from the throttle bore to the slot in the throttle shaft; and (3) a vacuum passage from the slot in the throttle shaft to the fuel bowl. The purpose of the economizer is to provide economical fuel mixtures for part-throttle operation while still permitting the richer ratios that are needed for full-load operation. The economizer system performs its function by establishing a "back suction" on the fuel in the fuel bowl during most of the part-throttle range of operation. This "back suction" is created by manifold vacuum through the channels connecting the throttle bore with the fuel bowl. This retards the flow of fuel through the metering systems and thus permits the carburetor to operate on leaner part-throttle mixture ratios. The rotation of the throttle shaft controls the economizer system. During part-throttle operation from about one-quarter to three-quarters throttle, the passages are open and the pressure in the fuel bowl is lowered. This retards the flow through the main jet and a leaner mixture is supplied. On full-throttle opening the passages are closed and the main jet flows to full capacity to supply the richer mixture required.

Marvel-Schebler Model TSX The Marvel-Schebler Model TSX carburetor (Fig. 11-11) is used on a number of tractors and stationary engines. It is a plain-tube, single-venturi, updraft carburetor with adjustable main and idling jets.

Multiple-Barrel Carburetors

Certain multiple-cylinder, high-speed engines such as are used for automobiles, trucks, and some airplanes are frequently equipped with multiple-barrel carburetors. Such carburetors have a single-body fuel

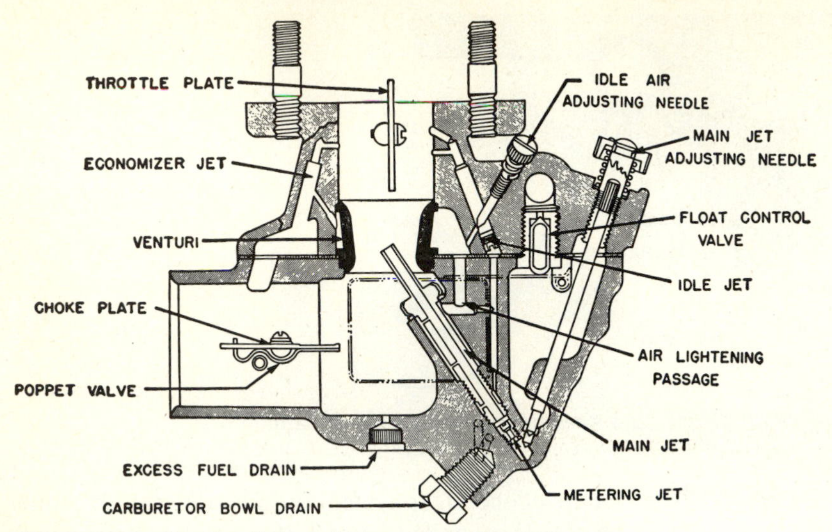

Figure 11-11 Sectional view of Marvel-Schebler carburetor.

chamber, float, air opening, and choke valve, but have multiple idling and high-speed or load circuits and throttle butterflies.

An example of a relatively simple two-barrel carburetor as used for a two-cylinder tractor engine is the Deere duplex model DLTX-75 (Figs. 11-12 and 11-13). There is a single body and fuel chamber, but there are two choke valves, two main jet nozzles, two venturis, two idling adjusting needles, and two throttle butterflies.

Four-barrel carburetors are used extensively in V-type, eight-cylinder automobile engines, largely because of the constantly changing and variable operating conditions usually existing. The four-barrel car-

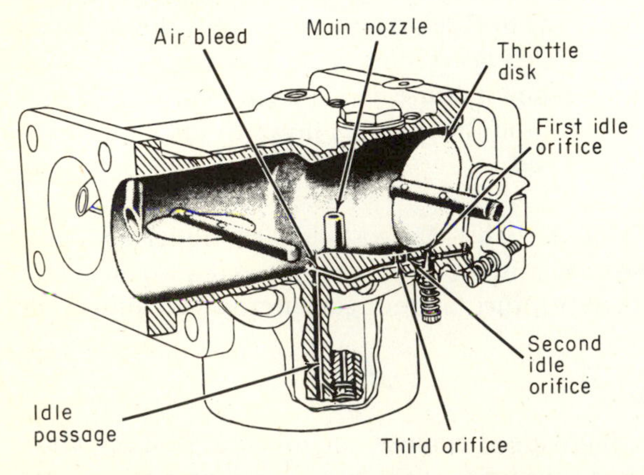

Figure 11-12 Duplex tractor carburetor showing idling-mixture circuit. (Courtesy Deere and Company.)

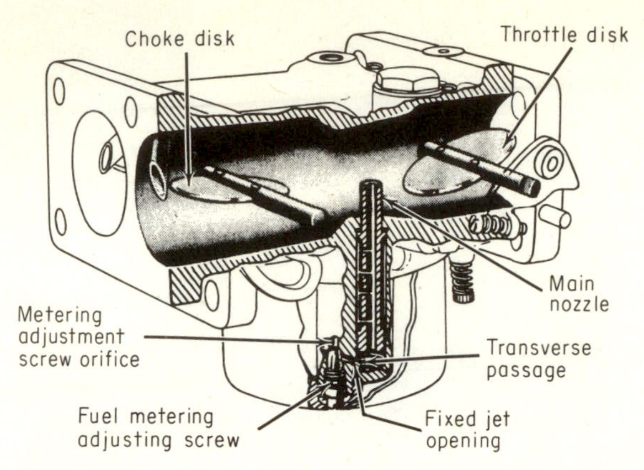

Figure 11-13 Duplex tractor carburetor showing load-mixture circuit. (Courtesy Deere and Company.)

buretor has four openings to the intake manifold, and the complete unit consists of two dual carburetors instead of four sections operating simultaneously throughout any speed or load range. The carburetor is said to have a primary side and a secondary side. In general, the primary section operates during light loads and lower speeds while the secondary section enters the picture by supplying additional fuel mixture for full-throttle operation.

Automotive-Engine Carburetors The Carter Model W-1 and Zenith Model 228 carburetors (Figs. 11-14 and 11-15) are representative of the construction and operation of carburetors used for automobile and truck engines. Both are downdraft types. The Carter W-1 has two special features for maintaining a correct fuel mixture for variable speeds and loads. The first is the triple-venturi arrangement with the main-jet nozzle feeding fuel into the airstream of the smallest venturi. The second feature is the three-step metering rod that controls the fuel flow from the chamber to the main-jet nozzle according to speed and load conditions. Figure 11-14 shows the metering rod in detail. It is connected to the throttle-butterfly control linkage in such a way that it is lifted as the butterfly is opened. As it rises in the calibrated orifice, its diameter decreases and thereby an increased fuel flow occurs. This metering rod is precision-made to an exact size, and two or more sizes are available. For example, if a rod of greater "step" diameter is used, the fuel flow through the main jet will be decreased and the mixture made leaner.

The Zenith Model 228 downdraft carburetor utilizes a vacuum-controlled jet valve to supplement the main jet in supplying the correct amount of fuel to the mixture at high speeds and heavy loads. Referring to Fig. 11-15, the main jet controls the fuel mixture at one-quarter to three-

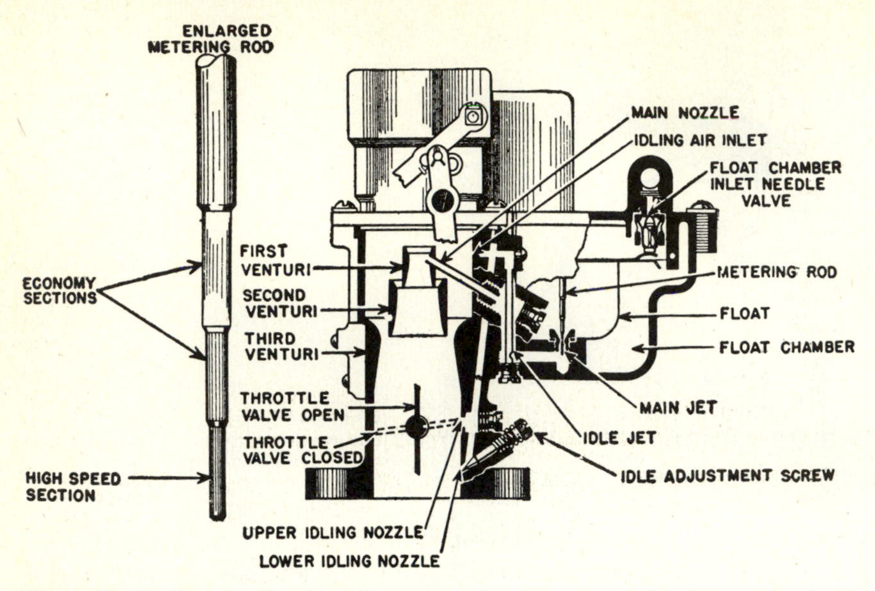

Figure 11-14 Downdraft automotive-type carburetor showing triple venturi and metering rod.

quarter throttle opening. Precise mixture control is obtained by means of the well vent or high-speed bleeder which feeds air into the main discharge jet through small holes at a point below the fuel level. As the throttle approaches a full-open, maximum-speed position, a second vacuum-controlled power jet (Fig. 11-16) comes into action and supplies some additional fuel to the main discharge jet. The vacuum piston assembly is connected to the intake manifold through a passage in the carburetor body which connects to a similar passage in the manifold. Normally, with a partly closed throttle, the vacuum effect holds the piston assembly in a raised position. However, as the throttle opening increases, the vacuum decreases and the spring pressure forces the piston assembly downward, opening the power-jet valve and feeding additional fuel to the main discharge jet. The entire system functions automatically depending upon the throttle-butterfly position as determined by the engine load and speed.

Automatic Choke All carburetors are equipped with a choke valve for the purpose of reducing the air flow and increasing the suction at the fuel jets in order to permit the cylinders to receive an effective charge of fuel mixture to ensure instant starting, particularly when the engine is cold. For stationary power units and tractors the choke is usually manually operated, but carburetors for automobile and truck engines are equipped with automatic chokes (Fig. 11-17). Their operation involves the use of manifold vacuum and a thermostatic coil spring. The spring is

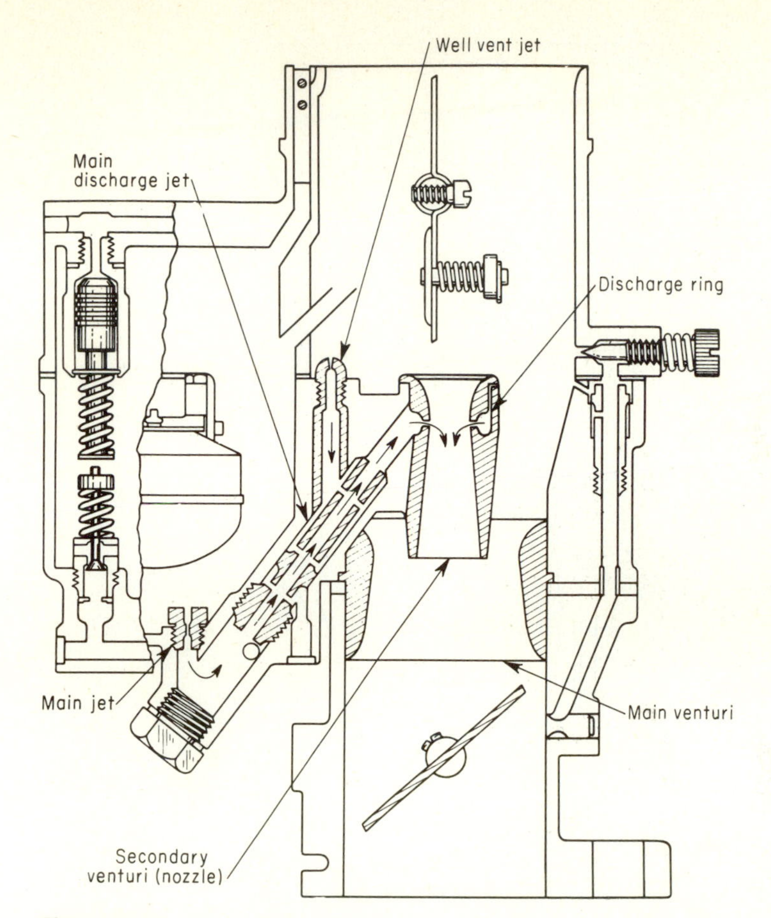

Figure 11-15 Downdraft carburetor showing medium-speed mixture control.

tightly enclosed and connected to the engine-exhaust manifold by a small tube. When the engine is cold, the spring holds the choke closed against the pull of the vacuum piston and gives the desired starting mixture. When the engine starts, heated air gradually warms the spring, causing it to uncoil and open the choke. Likewise, when the engine is running, the manifold vacuum also acts on the piston, which, in turn, assists in holding the choke valve open. A simple adjustment on the thermostatic spring housing permits adjusting the tension so as to give correct choke control under any operating conditions.

Accelerating Devices Automobile-engine carburetors are equipped with a special device to provide positive and instant fuel-mixture flow when the throttle butterfly is opened quickly to obtain quick acceleration. Without such a device there would be a lag in engine-speed pickup and a

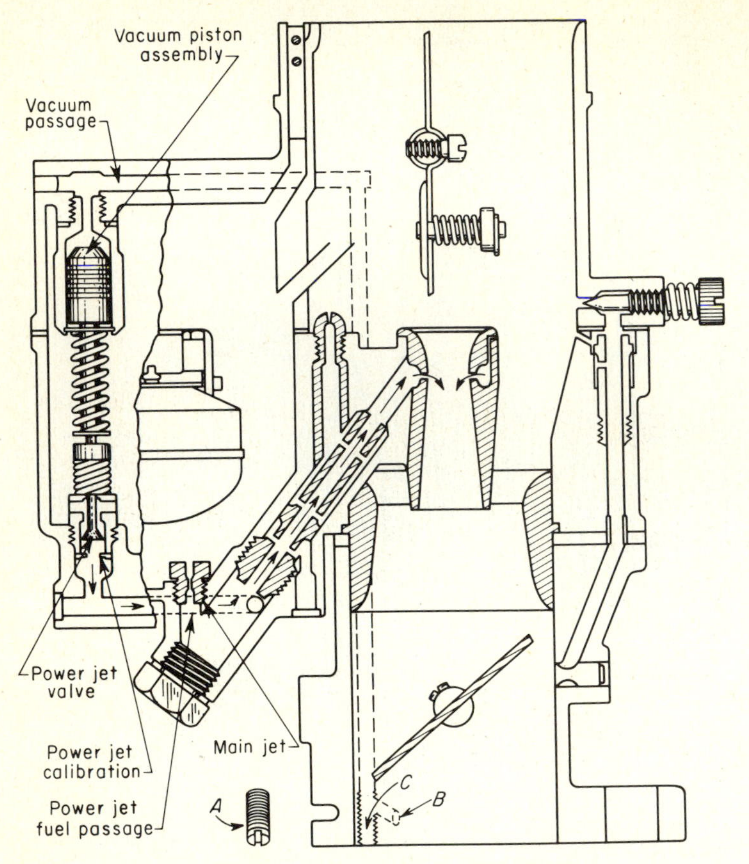

Figure 11-16 Downdraft carburetor showing high-speed mixture control.

delayed acceleration. The device usually consists of a small plunger pump that is linked to the throttle-butterfly mechanism. When the latter is opened quickly, the pump plunger forces an extra charge of fuel through a jet into the airstream. Figure 11-18 shows the operation of this device.

Superchargers A supercharger is a device that forces more fuel mixture into the cylinders than would be drawn in under normal atmospheric conditions, thus making possible a greater power output with the same piston displacement. Its use is particularly important on piston-type airplane engines operating at high altitudes in order to maintain the engine power as the atmospheric pressure decreases.

A supercharger is a gear- or exhaust-gas-driven rotary-type air pump which is inserted between the carburetor and the engine manifold. Obviously, the entire fuel supply and induction system must be specially designed so that all units work together efficiently. Figure 11-19 shows the

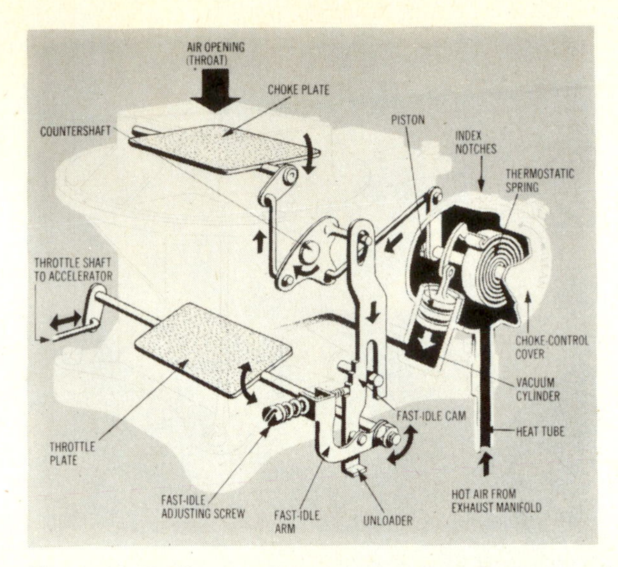

Figure 11-17 Construction and operation of an automatic choke mechanism. (Courtesy of Popular Science Monthly.)

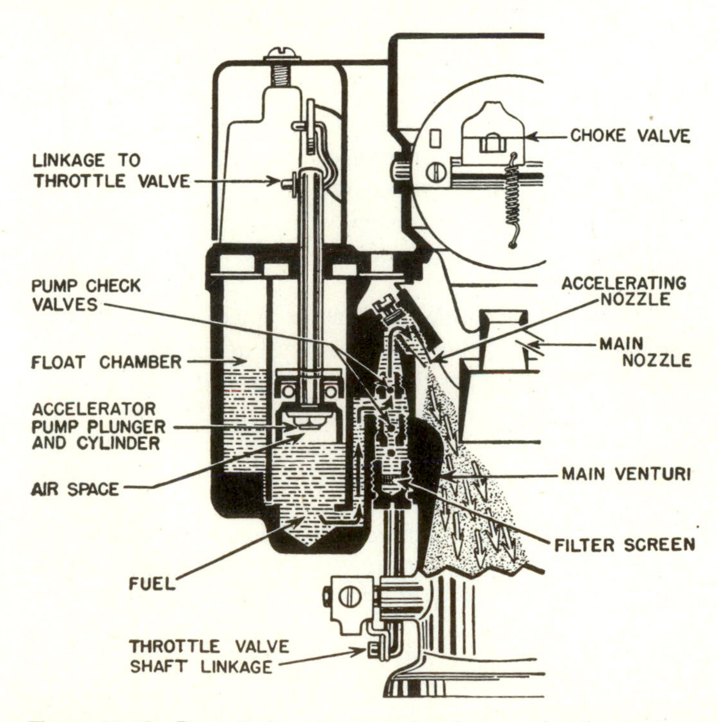

Figure 11-18 Downdraft carburetor showing accelerating pump and nozzle.

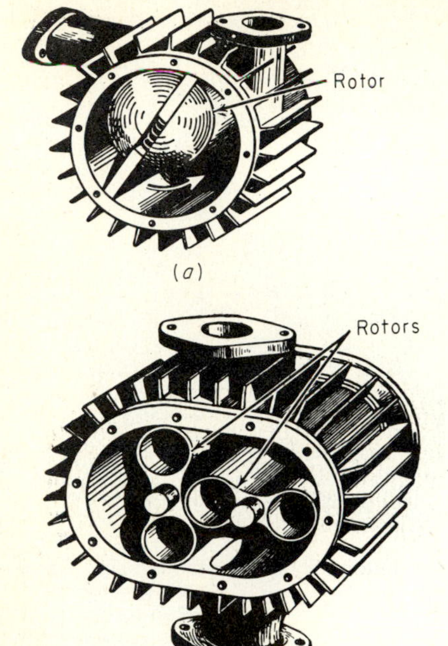

Figure 11-19 Supercharger blower types. (a) Vane, (b) roots blower.

two most common types of blowers. Figure 11-20 shows the general construction and operation of the turbocharger used on a diesel engine. The required power is derived from an exhaust-driven turbine and not directly from the engine itself. This is a definite advantage because most superchargers operate at relatively high speeds and therefore absorb an appreciable part of the generated engine power.

Carburetion of Distillate, and Special Tractor Fuels As previously discussed, such fuels are not so volatile as gasoline, and a higher engine operating temperature is essential for their successful use. The principal requirements for an engine to use these fuels satisfactorily are:

1 The engine must operate at a uniform speed and under a medium to a heavy load.

2 Provision must be made for starting on gasoline with some convenient means of changing over when the engine is warmed up. For this reason the use of such fuels is impractical if frequent starting and stopping are necessary.

3 Provision must be made for maintaining a higher intake manifold temperature than is normally required for gasoline.

4 The cooling-system temperature must be relatively high and closely controlled according to weather and load conditions by means of a thermostat valve and a radiator curtain or shutters.

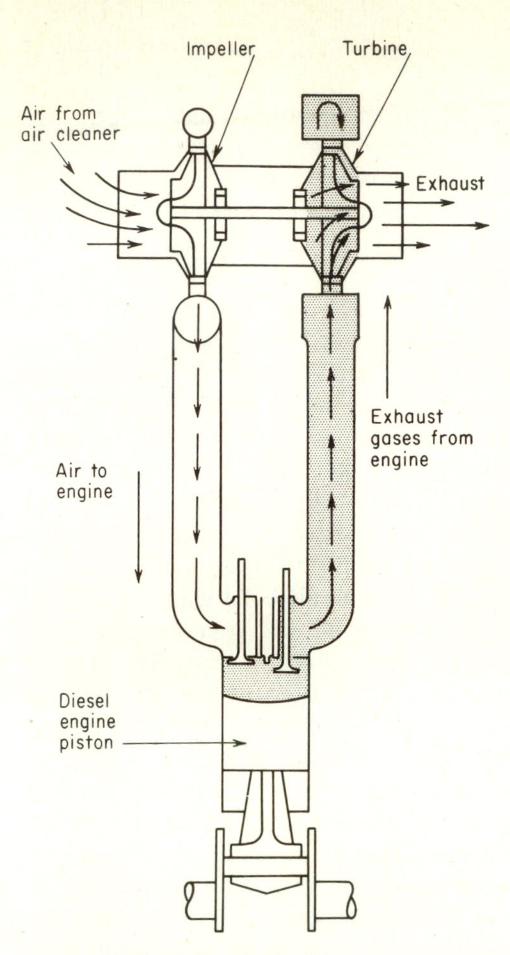

Figure 11-20 Exhaust-gas-driven supercharger for diesel engine. (Courtesy of Caterpillar Tractor Company.)

5 A relatively low compression ratio—1 : 4.5 or 1 : 5—must be used to eliminate detonation.

6 A careful and precise adjustment of the fuel mixture is essential.

Heavy-fuel carburetors are constructed and operate in practically the same way as gasoline carburetors. In fact, any engine that burns distillate successfully will also burn gasoline readily. However, owing to the lower compression ratio, maximum fuel economy will not be obtained.

Carburetion of LP Gas

Mention has previously been made of the use of commercial butane and propane as fuel for stationary engines, tractors, and trucks. These LP gases have certain advantages over gasoline and liquid fuels which make their use desirable and economical, provided there is a plentiful supply at a reasonable cost per gallon.

LP fuels for commercial use are often referred to as "butane" or "propane" and usually are a mixture of both. The mixture may vary from a nearly 100 percent propane mixture for winter use to a 70-30 percent butane-propane mixture for summer use. Some propane is desirable for the reason that it has a lower boiling temperature than butane and thereby ensures vaporization at extremely low temperatures.

A study of the physical characteristics of LP gases, Table 11-1, shows that they differ considerably from liquid fuels; hence their storage and carburetion must be handled in a different manner. First of all, the boiling temperature of LP fuels is very low and below the freezing temperature of water. This means that such fuels have a high vapor pressure at ordinary temperatures. Hence, they must be stored in a heavy, leakproof tank capable of withstanding considerable pressure. Furthermore, special equipment is needed to convert the liquid fuel to a gas, reduce the pressure, and feed it to the engine in the proper manner.

Figure 11-21 shows the make-up of a complete LP-gas fuel-supply and carburetion system, and Fig. 11-22 shows the pressure-reducing, vaporizing, and carbureting devices as they actually appear. Referring to Fig. 11-21, the liquid fuel passes through a filter to the high-pressure regulator. This device reduces the pressure to about 8 lb/in^2, and partial expansion and vaporization begin. The fuel then enters the vaporizer coils, which are surrounded by heated water from the engine cooling system, and further expansion and vaporization occur. The vapor then passes through the low-pressure regulator, which reduces its pressure slightly below atmospheric. This regulator also controls the gas supply to the carburetor and cuts it off when the engine demand has ceased. The carburetor serves as a mixer to mix the vapor and air in the correct ratio and to supply the proper quantity of this mixture to the engine at all loads and speeds.

Table 11-1 Characteristics of Commercial Butane and Propane

Characteristic	Propane	Butane
Formula	C_3H_8	C_4H_{10}
Specific gravity of liquid at 60°F	0.509	0.584
Init. boiling point at normal atmospheric pressure	−44°F	32°F
Weight per gal. of liquid at 60°F	4.24	4.86
Vapor pressure at 60°F, lb/in^2	92	12
Vapor pressure at 100°F, lb/in^2	172	37
Btu per lb	21,600	21,300
Btu per gal	91,500	103,500
Approximate octane rating	110	90

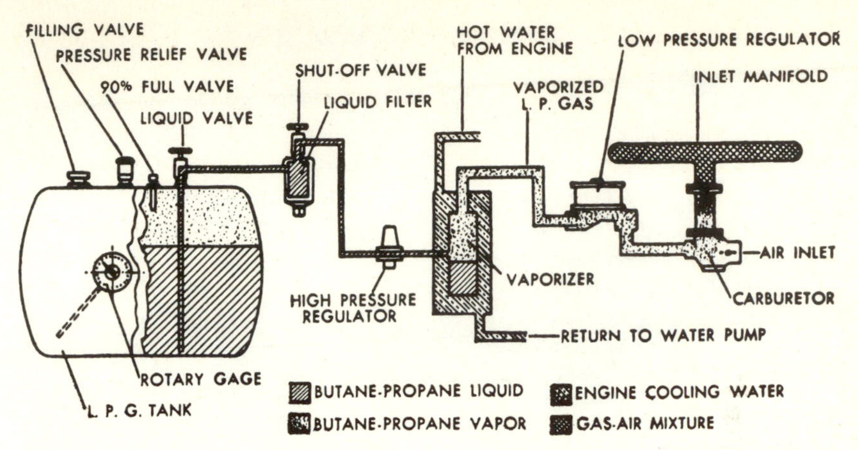

Figure 11-21 Basic parts and operation of an LP gas fuel-supply and carburetion system. (Courtesy Ensign Carburetor Company.)

A second system (Fig. 11-23), which is used to a limited extent, depends upon the gas vapor itself to flow directly from the tank through pressure regulators to the carburetor. A heat exchanger or special vaporizer is not used, and a fuel with a very low boiling point such as

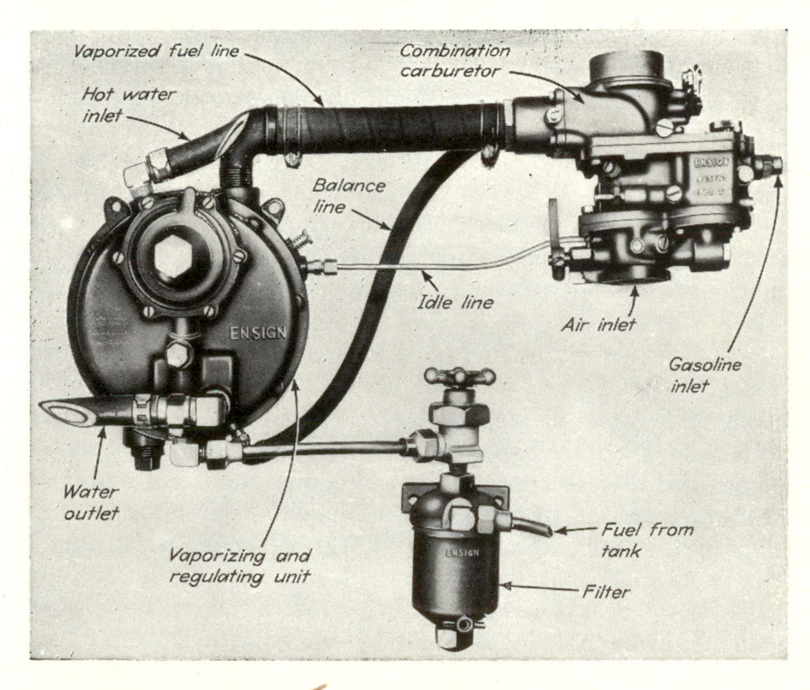

Figure 11-22 Butane-vaporizing pressure-control and carburetion equipment for tractor engine. (Courtesy Ensign Carburetor Company.)

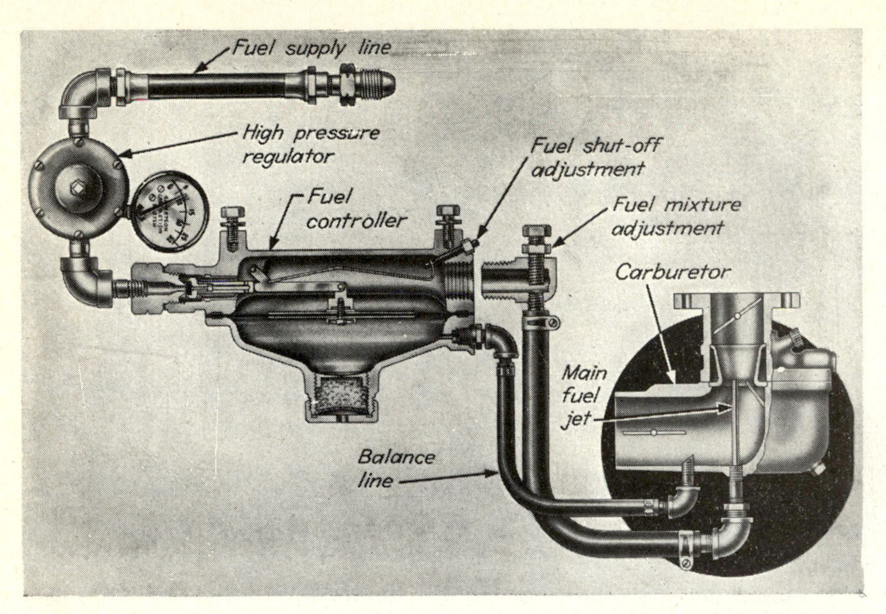

Figure 11-23 Propane pressure control and carburetion equipment for tractor engine. (Courtesy Garretson Carburetion System.)

propane gives best results, particularly in cold weather. Usually the original gasoline carburetor is used with this system, and it is possible to operate the engine on gasoline at any time if necessary.

Advantages and Disadvantages of LP Gases The principal advantage of any LP gas as an engine fuel is the reduction in engine wear and upkeep expense. Since it is a dry gas, it mixes readily with air and burns cleanly and completely without leaving residue of any kind. This means less carbon deposits, limited piston ring and cylinder wear, and no crankcase dilution.

A second advantage of this fuel is its high antiknock characteristic. LP gases have an octane rating of about 100. Hence they can be used in any high-compression, spark-ignition engine requiring a high-octane gasoline. In fact, to obtain the highest possible thermal efficiency with LP gases a compression ratio of about 8.5 : 1 is recommended. For optimum results it is also recommended that (1) a cold intake manifold be used in order to improve the volumetric efficiency; (2) the spark be carefully timed; and (3) cold spark plugs be used.

In comparing the cost of LP gases and gasoline as engine fuels, consideration must be given to their comparative weights per gallon and their heating values as well as to their actual prices per gallon. Referring to Table 11-1, propane and butane have heat energy values of 91,500 and 103,500 Btu/gal, respectively. The heat value of regular-grade gasoline is

124,000 Btu/gal. Assuming that an engine will have the same thermal efficiency when burning either LP gas or gasoline, it is obvious that gasoline will give a greater horsepower-hour output per gallon and that the cost per gallon of LP gas must be somewhat less than the cost of gasoline to offset this lower power output. On the other hand, as previously explained, if a higher compression ratio is used with the LP gas, the thermal efficiency may be increased sufficiently to partly offset the effect of this difference in heat value. The curves (Fig. 5-10) show the fuel consumption and other comparative performance characteristics when an engine is operated on LP gas, gasoline, and diesel fuel.

Handling and Storage of LP Gas Since butane and propane vaporize readily at subnormal temperatures, and since the vapor is heavier than air and settles near the floor or ground area, extreme caution must be observed in handling them. Furthermore, the high pressure existing in the storage tanks and equipment increases the possibility of leaks. Open flames, electric arcs, engine exhausts, and the like should be avoided around the storage tanks. Engines should always be stopped when butane is being transferred from one tank to another. A grounding wire or other method of carrying off static electricity should be used on tanks and hoses to prevent sparks. Hose, fittings, and safety devices should be the type recommended for butane service.

AIR CLEANERS

Mainly because early tractors were so large that their air intake was high enough to be out of the zone of dust-laden air, they were not equipped with any means or device for supplying clean air to the carburetor. Then, too, little thought was given to the harmful effects of even a very small quantity of fine dust. With the introduction of the smaller, lightweight tractors, it was soon discovered that, under most field conditions, enough dust and fine grit found its way into the cylinders through the carburetor to cause rather rapid wear. In fact, in some sections of the country, under certain conditions, enough damage could be done in even a day or two of operation to almost ruin the tractor.

The first idea of the small-tractor manufacturers to correct this was to attach a vertical tube to the carburetor air intake, extending it above the tractor far enough to obtain air that was comparatively free from the dust and grit stirred up by the machine itself. It was soon found that this arrangement, though simple, was ineffective. Some device was needed that would prevent the finest dust, as well as the more harmful coarse gritty material, from getting into the engine. Then, too, these periscopes soon worked loose and fell off or were removed by the owner in order to permit the tractor's being run under a shed.

Any type of air cleaner to give maximum satisfaction must fulfill the following requirements:

1 It must have high cleaning efficiency at all engine speeds and under all operating conditions.

2 It should offer very little restriction to the air flow, since that would reduce the power and fuel economy, nor should it interfere otherwise with the satisfactory operation of the engine.

3 It should be checked and serviced regularly in accordance with instructions.

4 It should be compact, rigid, not too heavy, and rattleproof.

Early Types of Air Cleaners The first tractor air cleaners were of the dry-filter type, consisting essentially of a sheet-metal closed container with an air opening in the side or bottom and an outlet in the top connected to the carburetor air intake. A filtering element, consisting of one or more layers of felt, eider-down cloth, wool, or similar fibrous material, was placed in the bottom of the cleaner and collected the dust as the air passed through. This type of filter soon proved unsatisfactory for several reasons. If it was dense or heavy enough to catch most of the dust, it offered too great restriction to the air flow. Also, it soon became clogged and required frequent cleaning or renewal, and it frequently got out of place as a result of backfiring.

A second simple type (Fig. 11-24) was known as an inertia cleaner. Its operation was based upon the principle that if a liquid or a solid mixture of materials of varying specific gravity is whirled violently, the heavier particles or portions will be thrown to the outside. Therefore, if the dust-laden air, on its way to the carburetor, is first made to whirl violently, the dust, grit, and other solid particles, being heavier, will be thrown to the outside

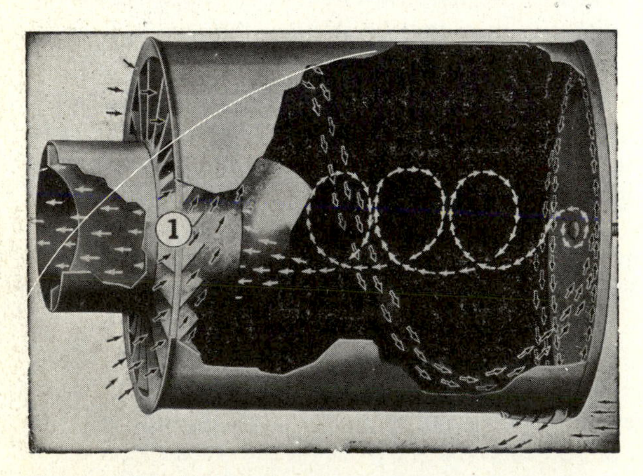

Figure 11-24 Dry inertia-type air cleaner.

and deposited or carried away, leaving the clean air to enter the carburetor at a small, central opening.

The inertia cleaner cannot be considered as efficient and effective by itself, but it is used on some tractors as a precleaner to remove the coarser material in the air before it enters the regular cleaner.

Another type of air cleaner, which was used quite extensively on tractors at one time, was the water-bath cleaner. As the name implies, the air on its way to the carburetor was drawn through water, which took up any dust and foreign material. Deflector or baffle plates in the upper part of the cleaner body prevented water being taken into the engine. Water air cleaners proved unsatisfactory for the following reasons: (1) they do not necessarily remove all the dust and dirt; (2) they require too frequent cleaning and replacement of the water, because a certain amount is taken up by the air and goes into the engine; (3) there is considerable air-flow restriction and variation with the change in water level; (4) the metal parts corrode and leak or break; and (5) there is danger of freezing in cold weather.

Figure 11-25 illustrates an early type of oil-bath air cleaner used on automobile and truck engines. It consists of a metal-gauze-filter element below which is a shallow pan containing engine oil. The entering air must pass downward and pick up a small amount of oil from the pan. This oil saturates the gauze filter and, in turn, thereby traps the dust and foreign material which gradually settles into the oil pan. The cleaner is serviced by washing the filter and pan and adding fresh oil.

Modern Types of Air Cleaners In general, present-day tractor air cleaners can be classified as either of the dry type or oil type. However, there is considerable variation in the detailed construction and operation of these two types as used on different makes, categories, and sizes of engines.

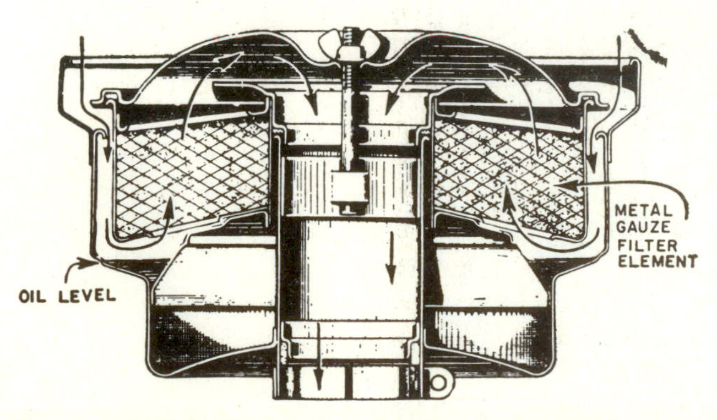

Figure 11-25 Oil-bath air cleaner for automobile and truck engines.

Oil-type Air Cleaners The oil-type cleaner was developed and introduced about 1920 and proved reasonably satisfactory and effective for many years. However, as a result of experience with air movement, dust and trash characteristics, and changes and improvements in engine design, together with their extensive utilization under many and varied operating conditions, efficient dry-type air cleaners have been developed and are being used extensively for tractor engines and similar power units.

Figure 11-26 illustrates a late model oil-bath type of cleaner. Air enters a central tube whose lower end extends into an oil cup. As the air

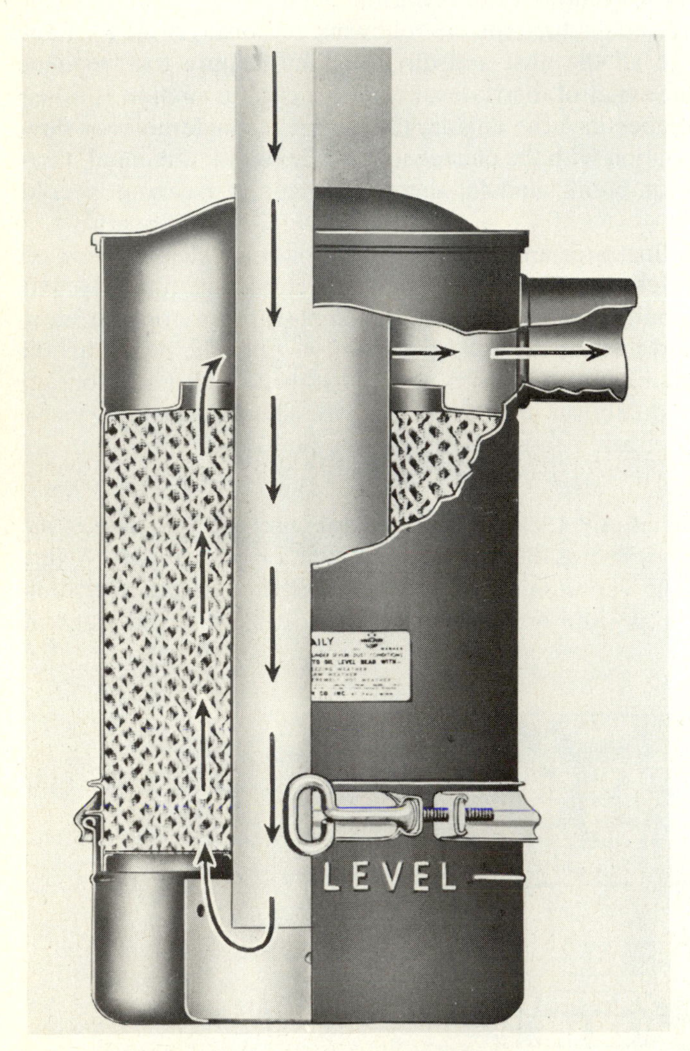

Figure 11-26 Modern type of tractor oil-bath air cleaner. (Courtesy of Deere and Company.)

passes through the oil, some dust and dirt are trapped in it. The air meantime changes direction carrying with it some of the dust-trapped oil and passes upward through a screen or metal-shavings element, which picks up the remaining foreign material. All of this material eventually settles back into the oil cup.

Servicing of the oil-bath cleaner consists primarily of inspecting the oil in the cup periodically with respect to its level, appearance, and dirt accumulation at the bottom. Eventually, depending upon the existing dust and operating conditions, the cup should be thoroughly cleaned and washed and fresh oil added to the indicated level.

Dry-type Air Cleaners Modern dry-type air cleaners are designed to filter the air through a replaceable element constructed of a specially treated filter paper. These elements are usually pleated in various forms to provide the maximum filtering area. This cleaner may be a single-stage type with paper only as a filtering medium, or combined with a precleaning section to form a two-stage cleaner.

Figure 11-27 shows a single element cleaner which utilizes centrifugal action to remove the dust and other foreign material. The entering air encounters stationary, multiple vanes, which force the air to undergo a spinning action, thereby throwing heavier dust particles outward and downward through a disk-slot into the cup. Any remaining fine particles are trapped by the paper element, while the cleaned air passes on to the

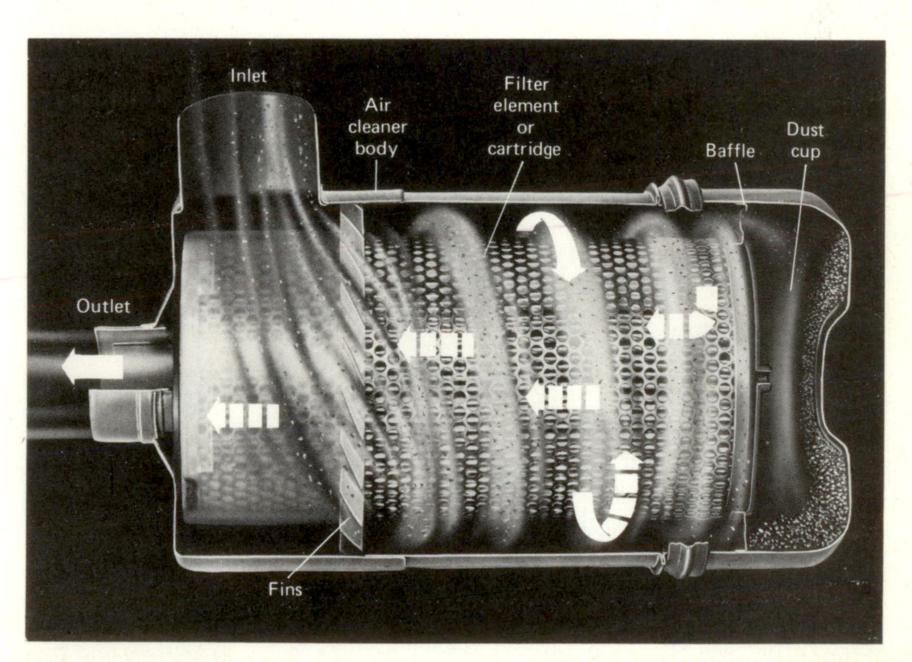

Figure 11-27 Modern dry-type tractor air cleaner. (Courtesy of Deere and Company.)

outlet and the intake manifold. An ejection valve remains closed until the engine speed or vacuum is reduced or a sufficient weight of foreign material forces the valve to open, permitting a part or all of the contents to unload.

Figure 11-28 illustrates another design of a dry-type air cleaner, which is mounted and operates in a horizontal position. It is equipped with a dust unloader that automatically opens periodically and discharges accumulated dust and foreign material. Figure 11-29 shows the construction of the dual-unit filter element. Cleaning and servicing of this unit may be done by either (1) tapping it lightly; (2) reverse-flow cleaning with compressed air; or (3) washing in water with a mild nonsudsing detergent.

Figure 11-30 illustrates a design of a dry-type air cleaner particularly adapted to heavy-duty diesel tractor engines. The air first enters the precleaner, which screens out coarse material and allows only fine dust-laden air to enter the strata tube. The dust-laden air enters the aspirated strata tube and passes through the vanes, which cause the air to swirl. The swirling action throws the heavy dust particles to the outside of the tube, where they are channeled into the scavenging system, pass through the aspirated safety valve, and are blown out of the muffler stack. The precleaned, lighter air continues on to the final safety elements and into the engine intake. The final air filter is a two-stage, dry-type with a primary and an inner safety element.

Figure 11-28 Dry-type tractor air cleaner with dual filter elements and dust unloader. (Courtesy of United Air-Cleaner Division, Halle Industries Inc.)

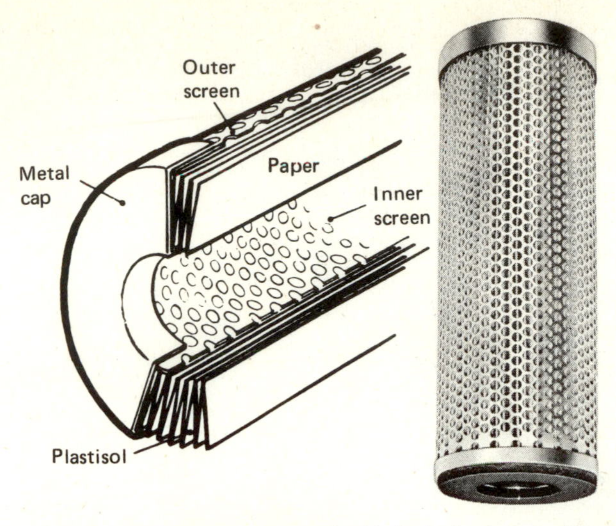

Figure 11-29 Filter elements for United air cleaner. (Courtesy of United Air-Cleaner Division, Halle Industries, Inc.)

Air-Cleaner Protective Devices The efficient and effective performance of any type of air cleaner is dependent largely upon the tractor owner or operator in religiously following the manufacturer's instructions with respect to proper servicing procedures. The mechanical condition of

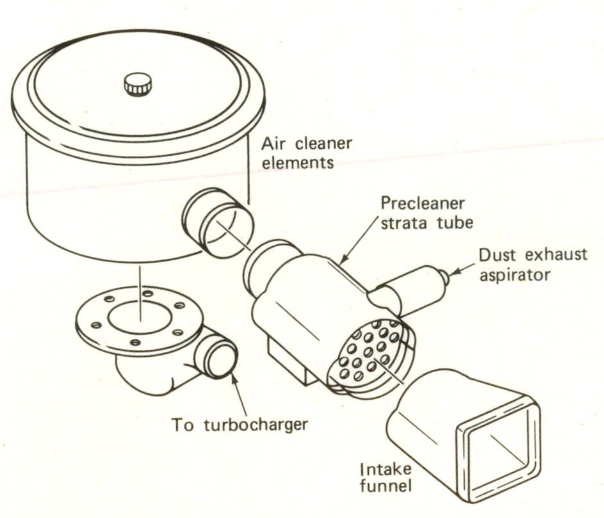

Figure 11-30 Case multistage, dry-type, heavy-duty tractor air cleaner. (Courtesy J I Case Company.)

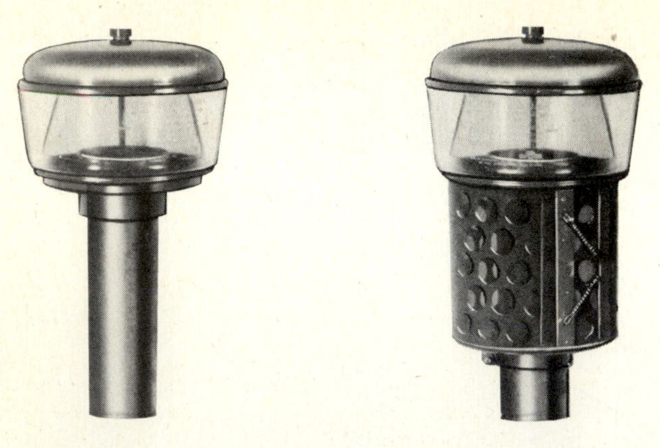

Figure 11-31 Precleaner for dry-type tractor air cleaners. (Courtesy Deere and Company.)

a tractor can be seriously damaged during a very short period of operation under most existing dust and fine-trash conditions. In order to assist the operator in avoiding such problems most tractors are equipped with one or more of four accessories: (1) a precleaner attached to the cleaner inlet, (2) a telltale light on the instrument panel or the intake manifold, (3) a dust-unloader valve attached to the primary cleaner, or (4) a restriction gauge attached to the intake manifold or mounted on the instrument panel.

A precleaner (Fig. 11-31) is attached to the air-intake pipe and extends upward above the tractor. It may be equipped with a screen that

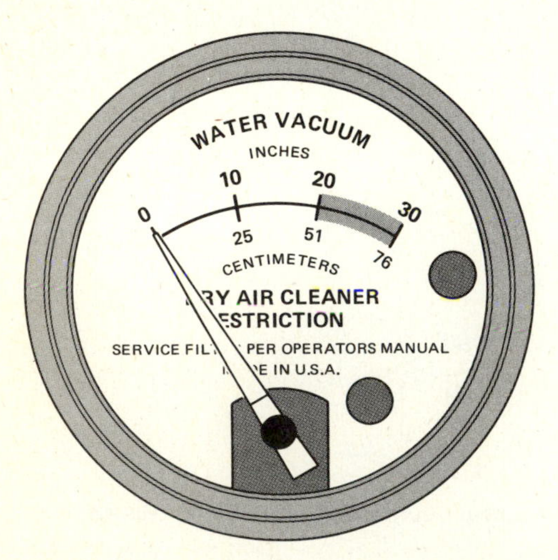

Figure 11-32 Dry-type air cleaner restriction gauge. (Courtesy of Vortox Company.)

catches coarse material before it enters the main cleaner, or it may have a collector bowl as shown.

A telltale light on the instrument panel or the intake manifold warns the operator that the air-flow is restricted, and the cleaner should be checked and cleaned.

A dust-unloader valve (Fig. 11-28), aids in collecting some dust and foreign material that is not removed otherwise, and thereby extends the operating life of the cleaner before servicing is needed. This device is equipped with a valve that opens automatically when a certain amount of dust and moisture is collected.

A restriction gauge (Fig. 11-32), measures a constant restriction condition on the air induction system and alerts the operator concerning needed service or replacement of the filter element.

ENGINE GOVERNING AND SPEED CONTROL

Most of the operations performed by a farm tractor, whether by means of a belt pulley, by power take-off gears and shaft, or by a transmission and the travel and draft action of the machine, require that the engine operate at some relatively uniform speed for any given operation. In most cases, the power requirements of the machine may fluctuate and vary considerably owing to various uncontrollable factors. Yet effective and proper performance requires that its speed remain as uniform as possible. For this reason all tractor power units are equipped with some type of mechanical speed-governing control. On the other hand, it is desirable and necessary to vary the engine speed frequently (and over a wide range) in such machines as automobiles, trucks, and airplanes in order to obtain different travel speeds. When such a machine is in use, therefore, the operator must be present to control it. The control of the speed of a tractor engine also might be held fairly constant by the manual operation of a convenient hand lever, but a mechanical governor relieves the operator of this responsibility and provides better performance.

Governor Mechanisms and Operation

The basic principle of operation of any tractor engine or similar power unit regardless of size, number of cylinders, fuel used, and other factors, is to control the flow and supply of fuel mixture to the cylinders with respect to the desired speed, power output, and existing operating conditions.

For electric ignition engines using gasoline or LP gas as fuel, the governing mechanism is connected to the throttle-butterfly of the carburetor in such a manner that it controls and varies the throttle opening according to the amount of fuel mixture needed to maintain the desired engine speed and power output. In general, the term throttle governor, is used for these. There are three types, the centrifugal governor, Fig. 11-33, the vacuum governor, and the pneumatic or air-vane governor, Fig. 11-34.

Centrifugal Governor Referring to Fig. 11-33, the mechanism is actuated and controlled by centrifugal force by means of weights rotated by a gear and shaft. Other parts include the ball-thrust collar, throttle-control arm, and throttle rod, which is connected to a hand lever. A governor-control spring is connected to the hand lever and control rod and reacts against the expansion of weights as engine speed increases.

The mechanism operates as follows: With the engine not running and the hand control in the open position, the throttle butterfly is wide open. When the engine starts and gains speed, the weights expand against the control-spring pressure, actuate the control arm and throttle rod, and gradually close the butterfly until a certain speed is established depending on the load. If the load is very small, the butterfly will be nearly closed. Now if the engine load is increased, the engine speed will tend to decrease. But this, in turn, will cause the weights to contract and thereby open the butterfly. Hence, the original speed will be maintained as long as the load does not exceed the maximum power of the engine. By placing the hand-control lever in any position between the closed or idling position and normal or rated engine speed position, any desired speed may be established and maintained by the governor. It is quite obvious that the control-spring tension must be closely correlated with the centrifugal action of the weights in order to provide sensitive speed control.

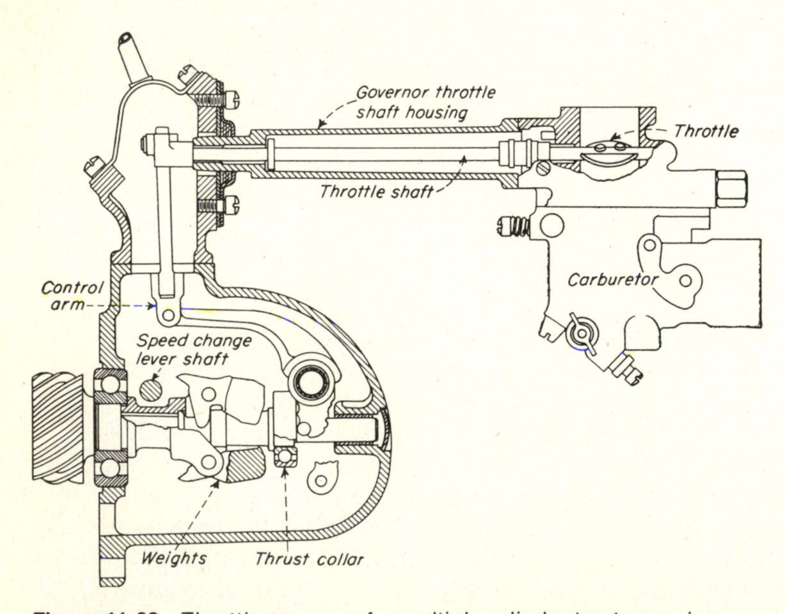

Figure 11-33 Throttle governor for multiple-cylinder tractor engine.

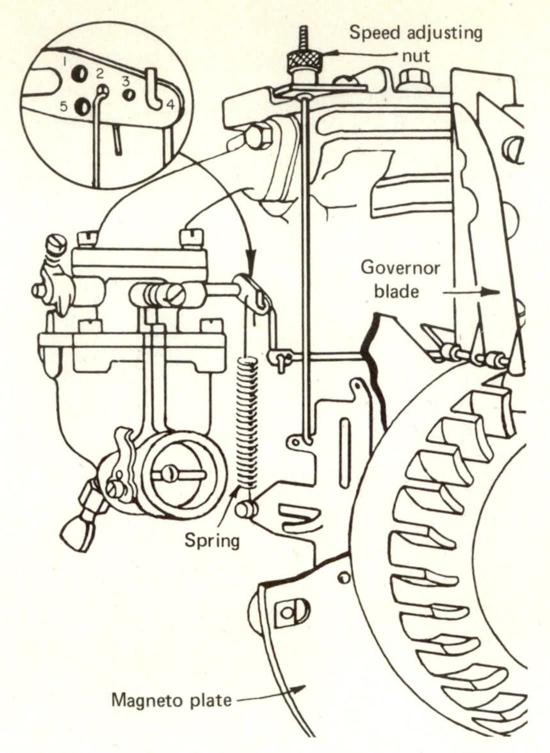

Figure 11-34 Air-vane-type governor for small single-cylinder engines. (Courtesy Briggs and Stratton Corp.)

Vacuum Governor The vacuum-type governor consists of a compact unit that includes a housing and the necessary operating mechanism as well as the throttle butterfly. The unit is mounted between the intake manifold and the carburetor and is actuated by the mixture flow in the manifold. The velocity of the mixture flow varies with the changes in engine load and power output. Hence a throttle valve located in the manifold creates a varying pressure on a spring-loaded diaphragm whose movement, in turn, controls the amount of fuel mixture needed depending upon the engine speed and power. A simple spring adjustment permits control of the desired speed range. The use of this type of governor is confined largely to stationary power units operating at a uniform speed and load.

Pneumatic or Air-Vane Governors Many small, air-cooled, single-cylinder engines used extensively on power lawn mowers and other similar types of machines are equipped with an air-vane governor as shown in Fig. 11-34. Short curved blades arranged on the side of the flywheel perimeter convert it into a blower. The air movement and pressure produced by the rotating flywheel reacts on a hinged metal blade, which, in turn, is linked to the carburetor butterfly and thereby controls the engine speed. A spring connected to the linkage reacts to control the speed and can be adjusted to provide a variable speed range.

Governor Hunting Frequently, when an engine is first started or after it is warmed up and is working under load, its speed will become uneven or irregular. It speeds up quickly, the governor suddenly responds, the speed drops quickly, the governor responds again, and the action is repeated. This is known as *hunting*. It is usually caused by an incorrect carburetor adjustment and can be corrected by making the mixture either slightly leaner or slightly richer. It is also possible for the governor itself to cause hunting by being too stiff or by striking or binding at some point so that it fails to act freely.

Governor Maintenance The dependable and successful operation of all types of governors relies largely upon precision of construction, freedom of movement of the parts, limited wear, proper installation and adjustment, and adaptability to the engine.

Diesel-Engine Governing The principles of diesel engine governing and the construction and operation of such mechanisms are fully explained in Chap. 17.

PROBLEMS AND QUESTIONS

1. Compare tractor- and automobile-engine fuel-supply systems as to make-up, fuel transfer, and carburetion.
2. Explain why most carburetors are equipped with a separate or special jet and adjustment to supply the fuel and to control the mixture for idling or low engine speeds.
3. Explain why a carburetor with a simple single jet would not give satisfactory results at all loads and speeds on any variable-speed engine.
4. Explain the meaning of mixture compensation, and list as many distinct methods of inducing compensation as shown in the carburetors described.
5. Give a suitable procedure that could be followed in adjusting a tractor carburetor having both idling and load-mixture adjustments.
6. What effect, if any, does altitude have on the fuel mixture and engine power? How can the situation be adjusted?
7. Under what conditions are multiple-barrel carburetors and superchargers desirable for internal-combustion engines?
8. If gasoline costs 60 cents, weighs 6.15 lb/gal, and has a Btu value of 20,000/lb, and if LP gas costs 50 cents, weighs 4.25 lb/gal, and has a value of 21,600/Btu

lb, which is the most economical fuel to use, assuming the same thermal efficiency in each case?

9 If, in Prob. 8, the engine developed 30 hp and used 0.60 lb of fuel/hp · h in each case, which fuel would be cheaper and what would the saving be per 10-h day?

10 What is a catalytic converter and what is its purpose? Why must unleaded gasoline be used with it?

11 Describe briefly the construction and operation of the predominant type of air cleaner used on tractors.

12 What would be your advice relative to servicing the usual type of air cleaner?

13 How would you proceed to compare the operating efficiencies of the governors on two different tractors?

14 What attention and service should be given a tractor governor to ensure proper operation?

15 Explain how the throttle hand lever of a tractor functions to control the engine speed between the idling and fully open settings.

Chapter 12

Cooling and Cooling Systems

As explained in preceding chapters, an internal-combustion engine converts only a limited portion of the total heat energy of the fuel into useful power. The unavoidable losses include (1) friction and mechanical losses, (2) cooling system losses, (3) exhaust heat losses, and (4) losses due to radiation. An analysis of the conversion and disposition of the heat energy received by an engine is known as its heat balance. This is well illustrated by Fig. 12-1, which shows the power output and heat losses for a four-stroke-cycle, spark-ignition engine under different load conditions. In general, if such an engine has a thermal efficiency of 23 percent, the losses are approximately as follows: friction 5 percent, cooling system 30 percent, exhaust gases 35 percent, and radiation and other losses 7 percent. Cooling system loss for diesel engines will be less than gasoline engines.

It is thus evident that the cylinder, cylinder head, piston, piston rings, valves and other parts must absorb and transmit a considerable quantity of heat (Fig. 12-2).

The cooling system of an automotive engine must transmit to the air enough energy to heat a six-room house on a winter day of 0°F, (−17.7°C) temperature.[1] The authors discovered that the cooling system of a 220 in^3 (3.6L) stationary gasoline power unit expelled over 100,000 Btu/h. Special provisions must be made to dissipate this heat at a reasonable rate and, at the same time, maintain an efficient engine operating temperature. Such provisions or equipment constitute the engine cooling system.

[1] *Maintenance of Automotive Cooling System*, Society of Automotive Engineers, 1947.

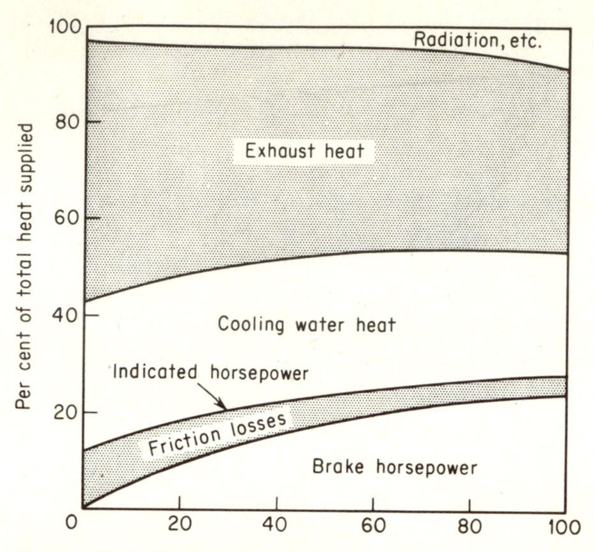

Figure 12-1 Typical heat-balance curves for a four-stroke-cycle gas engine. (Reprinted with permission from H. E. Degler, *Internal Combustion Engines,* John Wiley & Sons, Inc., New York, 1938, p. 351.)

If an engine were not equipped with some means of cooling, at least three troubles would arise as follows:

1 The piston and cylinder would expand to such an extent that the piston would seize in the cylinder, injuring the latter and stopping the engine.

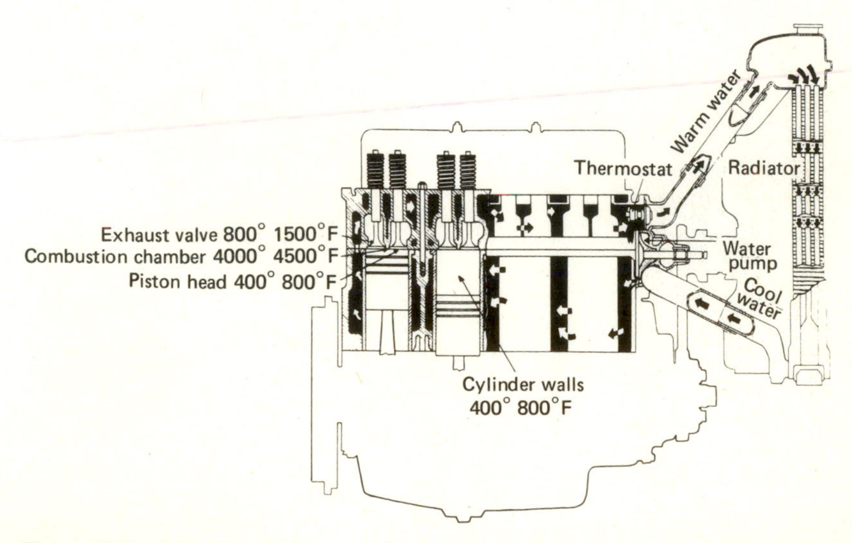

Figure 12-2 Operation temperatures of engine parts.

2 The lubricating qualities of the oil supplied to the bearings, cylinder, and piston walls would be destroyed by the high temperatures existing.

3 Preignition of the fuel mixture would take place, resulting in knocking and loss of power.

All internal-combustion engines must operate at a certain temperature to produce the best results, and they seldom give the greatest efficiency unless the temperature around the cylinder is between 170° and 200°F (76.6 and 93.3°C) (Fig. 12-3). Therefore, a cooling system that permits an excessive absorption of heat resulting in a low operating temperature is undesirable and indicates improper design.

The following is a classification of the common methods and systems of engine cooling:

1 Air
2 Liquid

Air Cooling Cooling by air alone is not used extensively but is satisfactory for certain types of engines and under certain conditions. The

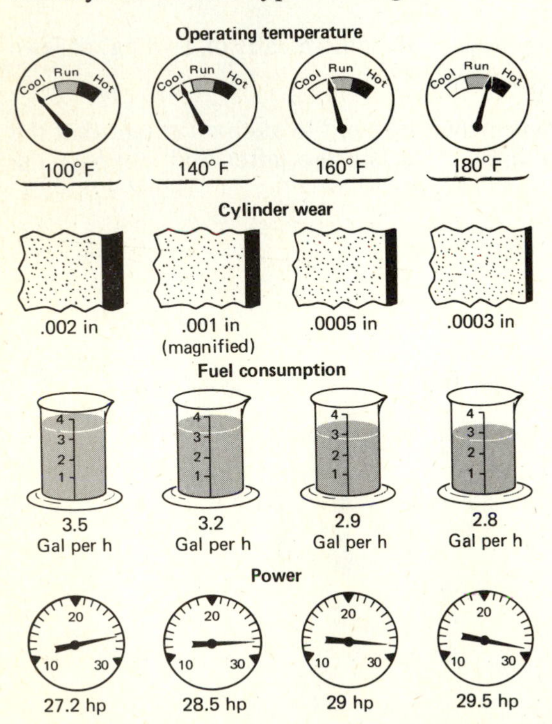

Figure 12-3 Effect of operating temperature on cylinder wear, fuel consumption, and power.

cooling effect is produced usually by means of fins or projections on the walls of the cylinder, as shown in Fig. 12-4. These fins may be placed transversely, or longitudinally with respect to the cylinder, depending upon the use of the engine and the direction of the air flow past the cylinder. Such an arrangement of fins increases the radiating surface, and therefore the heat escapes faster than it would otherwise. Figure 12-4 illustrates an engine that has the cylinder and crankcase enclosed in a sheet-metal housing and the flywheel equipped with blades so that it will create a suction of air down through the cylinder, thus producing a greater cooling effect.

Air-cooled engines are usually of small bore and stroke; that is, they have small cylinders. Multiple-cylinder air-cooled engines have the cylinders cast individually rather than in pairs or in one block, so that the maximum cooling effect will be obtained. Common examples of air-cooled engines are airplane and motorcycle engines and the power units used for lawn mowers, garden tractors, small and medium diesel tractors, and diesel power units (Fig. 12-5). Air-cooled engines have the following advantages:

1. Lighter in weight
2. Simpler in construction
3. More convenient and less troublesome
4. No danger of freezing in cold weather

The principal disadvantages of air cooling are that it is difficult to maintain proper cooling under all conditions and that it is almost impossible to fully control cylinder temperature. Air-cooled engines usually run a little hotter than water-cooled engines and require the use of heavier lubricating oil.

Liquid Cooling Cooling systems using liquids, usually water, are employed for all types of engines from the simple stationary farm engine to the most complicated multiple-cylinder high-speed types.

Water might be termed the universal cooling liquid for tractors as well as for trucks and automobiles. It has certain important advantages, among which are the following:

1. It is plentiful and readily available nearly everywhere.
2. It absorbs heat well.
3. It circulates freely at all temperatures between the freezing and boiling points.
4. It is neither dangerous, harmful, nor disagreeable to handle.

The principal disadvantages of water for cooling are:

Figure 12-4 Cutaway view of a V-type four-cylinder air-cooled engine with part of sheet metal housing removed to show fins on cylinder walls. (Courtesy of Wisconsin Motor Corporation.)

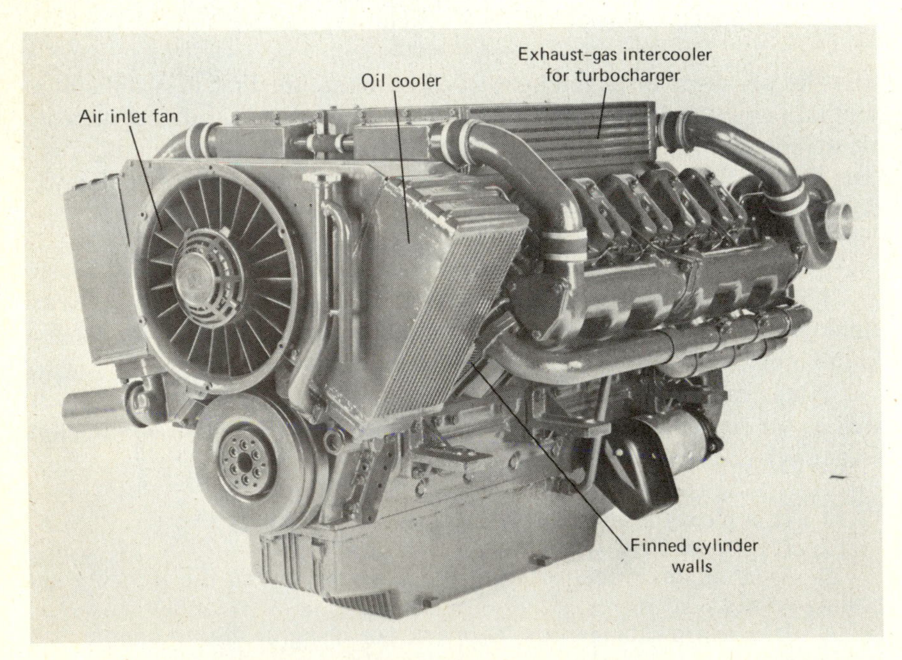

Figure 12-5 Air-cooled twelve-cylinder diesel power unit. (Courtesy of AVCO Lycoming Industrial Products Operations.)

1 It has a high freezing point.

2 It may cause excessive corrosion of the radiator and certain metal parts of the engine. Clean, pure water such as rain water gives the best results.

3 It may cause troublesome deposits in the cylinder jackets.

4 Evaporation and boiling require frequent replenishing.

The cooling system (Fig. 12-2) used on most tractors, automobiles, and trucks is a combination of liquid and air cooling. The pump and fan are mounted on the same shaft and driven by a V belt from the crankshaft. The fan draws cool air through the radiator, cooling the liquid rapidly, and also sends a blast of air past the cylinders, driving the heat away from the engine.

Radiator Construction Water-cooled tractor engines are equipped with the conventional radiator (Fig. 12-6) consisting of the core, an upper and lower reservoir, and the side members or frame pieces. Since tractors are subjected to considerable jarring and vibration and surplus weight is of little consequence, the reservoirs and frame parts may be heavier tha those found on automobile radiators.

Figure 12-6 Tractor radiator construction: (1) top tank, (2) bottom tank, (3) core, (4) fan shroud, (5) overflow tube, (6) pressure cap, (7) side members for radiator support. (Courtesy of Young Radiator Company.)

Figure 12-7 Tubular radiator construction.

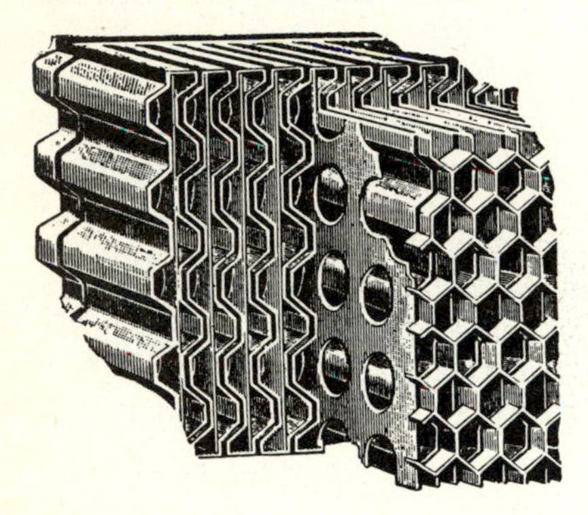

Figure 12-8 Cellular-type radiator construction.

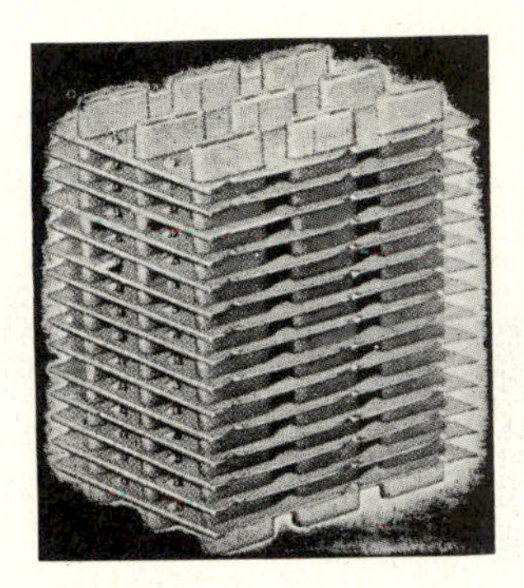

Figure 12-9 Flat-tube type of radiator construction.

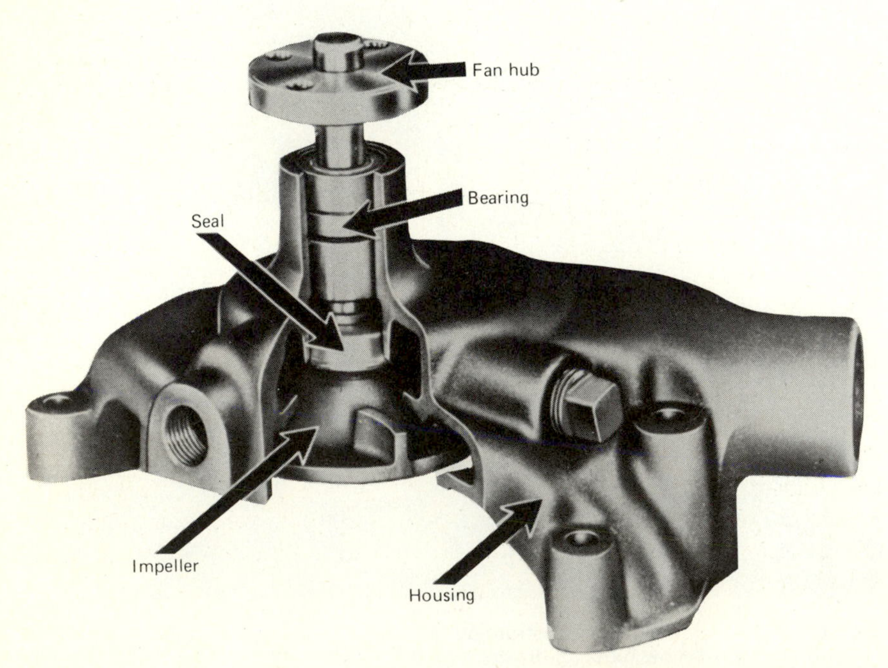

Figure 12-10 Water pump for cooling system. (Courtesy of Dana Corporation.)

There are two general types of radiator cores—the tubular type with fins (Fig. 12-7) and the cellular or honeycomb type (Fig. 12-8). The former seems to predominate, probably because of the lower manufacturing cost. The tubes are either round (Fig. 12-7) or flat (Fig. 12-9). Horizontal, thin, metal fins fastened to the tubes increase the rate of heat radiation.

Pumps, Fans, and Fan Drives Engines use a centrifugal-type pump (Fig. 12-10) to force liquid through the cooling system. Such a pump consists of a cast-iron or aluminum body, the rotating member or impeller with its curved blades, the drive shaft, and the necessary bearings, packing, and oil seals. The pump assembly is usually located on the cylinder head or block behind the radiator. Bearings are of the ball or roller type and are usually grease-packed and well sealed from water contact. The pump receives the water from the lower radiator connection and forces it through the engine block, head, and out the upper connection into the top tank of the radiator. Water-flow requirements for agricultural equipment range between ½ and 1 gallon (1.9 to 3.8 liters) per minute per horsepower developed by the engine.[1]

Fans have four to six blades and are mounted on the pump shaft. The V-belt drive has some means of tightening or adjusting the belt tension.

Pressure Cooling Most cooling systems operate under pressure by using a radiator cap (Fig. 12-11), which seals the opening and prevents leakage under low pressures. The overflow opening is also cut off by this cap. The cap is equipped with a spring-controlled valve that permits the escape of the liquid or steam if the pressure becomes too high. Another valve opens to relieve any vacuum effect caused by the cooling of the water or condensing of the vapor. The pressure system permits operating the engine at a higher temperature without boiling the water and losing it by evaporation. An increase in pressure of 1 lb/in^2 (0.07 kg/cm^2) will raise the boiling temperature of the water about 3°F (1.7°C). On the average, the boiling point will decrease approximately 1.5°F (0.8°C) for each 1,000 ft (304.8 m) above sea level.

Engine-Heat Control Since tractors are operated under a great variety of weather and load conditions, designers provide some means of control by which a uniform engine operating temperature may be maintained regardless of the varying factors mentioned. The best operating temperature is between 170° and 200°F (76.7 and 93.3°C) (Fig. 12-3). If the cooling liquid can be maintained within this temperature range at light as well as heavy load, or in cold as well as warm weather, better engine performance will be secured (Fig. 12-3).

[1] R. C. Verhaege, "Radiator Construction and Reasons," paper no. 64–635, Presented to Winter Meeting American Society of Agricultural Engineers, 1964.

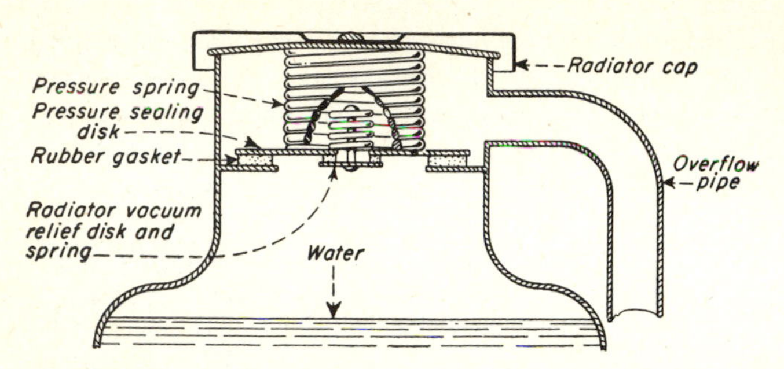

Figure 12-11 Radiator cap with valves for pressure cooling system.

The usual method of heat control is to restrict or vary the circulation of the coolant through the radiator according to operating conditions such as engine load and atmospheric temperature. Referring to Fig. 12-12, a thermostatic valve in the upper radiator connection opens and closes according to the coolant temperature. The valve is actuated by a copper bellows containing a liquid that expands when heated and contracts when cooled. The expansion of the liquid causes the bellows to expand and thereby move the valve attached to it to the open position, thus permitting circulation through the radiator. Likewise, a decrease in temperature con-

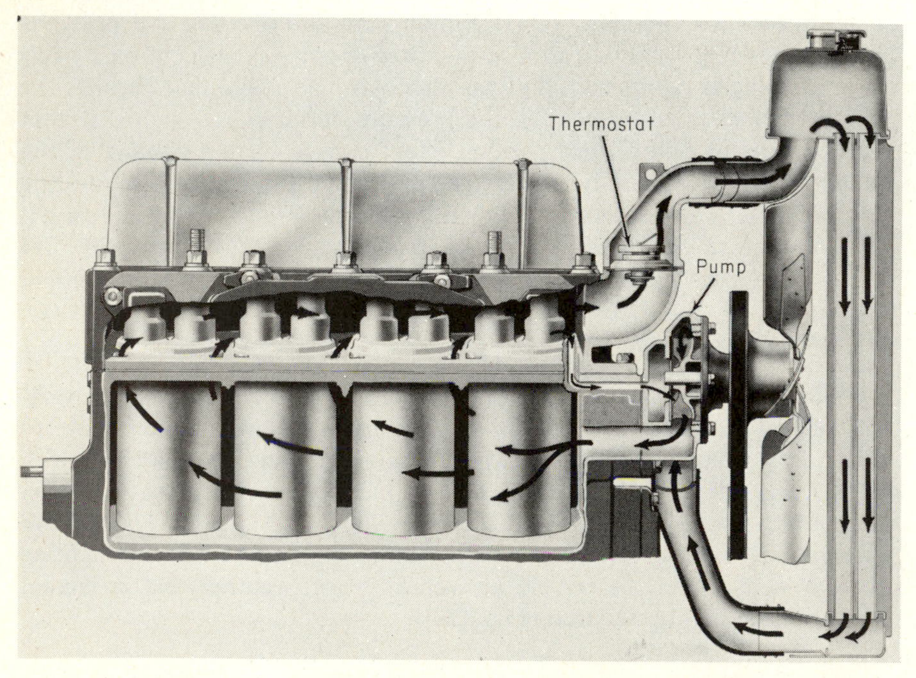

Figure 12-12 Heat-control valve in upper radiator connection. (Courtesy of Massey-Ferguson, Inc.)

tracts the liquid, and the bellows moves the valve in the opposite direction to a closed position that restricts or stops the coolant flow through the radiator. The bellows action is sufficiently sensitive to maintain the coolant temperature within a range of 10 degrees. Although some engines use a direct cutoff type of thermostat, most are equipped with the bypass type (Fig. 12-12), which prevents circulation through the radiator when the engine is cold but permits limited circulation around the cylinders. Thus the engine warms up faster and, eventually, the thermostat opens and permits circulation through the radiator.

Antifreezing Cooling Mixtures In certain sections, where freezing temperatures exist for several days or even months at a time, it is often advantageous to replace the water in a cooling system with some solution that will not freeze readily. Freezing of the cooling solution usually results in one or more troubles as follows:

1 Freezing may crack the cylinder head or block and produce either an internal or an external leak.
2 Freezing may weaken the radiator and connections or create a leak in these parts, which is often difficult to repair.
3 Freezing at a certain point in the cooling system during operation of the engine may interfere with the proper circulation of the cooling liquid and permit the engine to run too hot.

A number of liquid materials, used either alone or mixed with water, can be utilized to prevent these troubles. Only a very few, however, meet the usual requirements of a satisfactory antifreeze solution such as follows:

1 The ingredients used should be easily obtainable in operating localities.
2 The possibility of freezing should be negligible.
3 The solution should not be injurious to either the engine or the radiator through corrosion or electrolytic action, or to rubber-hose connections.
4 It should not lose its nonfreezing and noncongealing properties after continued use.
5 The possibility of fire hazard should be at a minimum.
6 The boiling point of the solution should not differ materially from that of water.
7 The viscosity should be as constant as possible through the entire temperature range involved, and the solution should remain perfectly fluid and not tend to stop up any small openings in the system.
8 The specific heat and heat conductivity of the solution should be high in order to dissipate the heat as rapidly as possible.

There are no substances satisfying all the requirements of an ideal antifreeze; however, the major requirements have been met satisfactorily by properly treated solutions of ethylene and propylene glycols. Various other substances that have been tried such as inorganic salts, sugar, alcohols, honey, glycerine, and petroleum coolants lack certain desirable qualities and are unsatisfactory.

Glycol Antifreezes Ethylene glycol has long been on the market as an automotive antifreeze compound. More recently, propylene glycol has been marketed for this purpose. The boiling points of glycol solutions are above that of water and increase with the concentration of the glycol. Protection against freezing increases with the concentration of the glycols as shown in Fig. 12-13. Because the glycols boil at higher temperatures than water, they are termed high-boiling or permanent antifreezes.

Although glycol antifreeze solutions will operate at higher engine temperatures without boiling, liquid may be lost through the overflow pipe when such solutions boil violently, in the same manner that water would be lost. Evaporation of glycol solution is practically all water. In actual service the loss of ethylene glycol by evaporation has been found to be negligible as compared to the overflow loss of liquid from violent boiling.

Antifreeze solutions composed of ethylene glycol, when properly inhibited, should, in properly maintained vehicles, protect the cooling system from corrosion for a winter's driving season. As in the case of alcohol antifreezes, the inhibitors may be weakened and depleted by hard or extended driving or through lack of proper vehicle maintenance. The

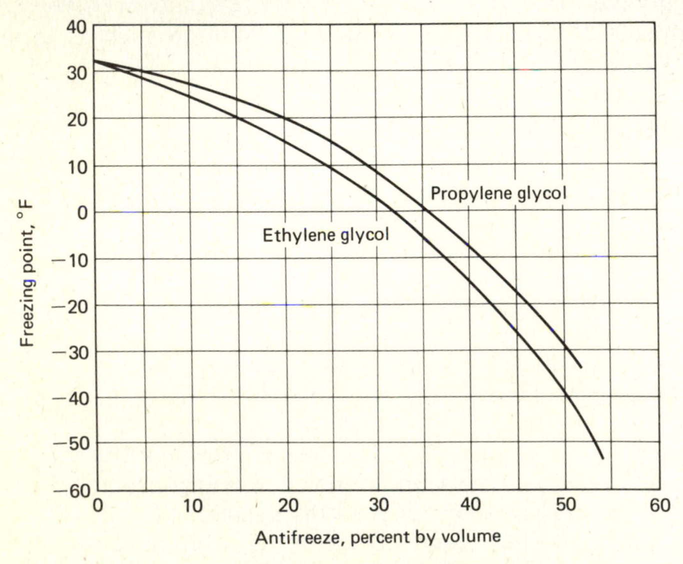

Figure 12-13 Amount of antifreeze to protect radiator.

service life of inhibitors is shortened by such conditions as high driving mileage, high engine speeds and heavy loads, air leaks into the solution (around the water-pump drive shaft, for example), combustion gas leakage into the coolant through a loose cylinder-head joint, rust deposits in the system, localized hot spots in the engine, and added contamination such as that from radiator cleaners that have not been thoroughly flushed out after use.

Corrosion The most common contaminants in water that may cause corrosion of the cooling system are dissolved minerals in the water, oxygen from the air, and acids from exhaust-gas leakage. Untreated water corrodes iron at a much faster rate than other metals. The rate of corrosion increases at higher temperature and may be even more serious in warm weather, with only water in the radiator, than in cold weather with antifreeze added to the water.

The chief constituents of scale formed on the wall of the cooling system from the use of hard water are calcium and magnesium. Rust and scale from hard water forms deposits on the cooling system wall thus reducing the transfer of heat from engine wall to the cooling system (Fig.

Figure 12-14 Cutaway showing scale formed in cooling system of an engine after 6 months of operation.

12-14). This leads to hot spots in the engine, which may cause damage without abnormally high cooling water temperature or general overheating of the engine.

Correct amount of corrosion inhibitor added to the cooling system at all times will prevent the buildup of rust and scale. Corrosion inhibitor does not remove rust and scale already formed in the system, and any necessary cleaning should be done before the inhibitor is installed. The inhibitor prevents corrosion by forming a very thin film over all the interior surfaces of the cooling system. The cooling system should be drained, cleaned, and a new inhibitor added according to the manufacturer's instructions.

PROBLEMS AND QUESTIONS

1 Name the specific heat losses encountered in an internal combustion engine, and explain how they vary with respect to the power output of the engine.
2 What are the objections to overcooling an engine?
3 What is meant by pressure cooling? What are its advantages?
4 Explain the operation of a typical thermostat as used in a tractor liquid-cooling system. Why is it important that an engine have a thermostat?
5 What are the most common contaminants in cooling water. Explain how they affect the cooling system.
6 Name the different types of antifreeze materials. Discuss their relative merits and characteristics.

Chapter 13

Fundamentals of Electric Ignition, Batteries, and Magnetism

Electricity is a broad field. In this textbook, the prime interest is in electricity as it is commonly used in mobile machines. Fundamentals of electricity and magnetism are the foundations upon which the study of electrical equipment is based. Every unit of the electrical system, whether it be an alternator, ignition coil, solenoid, or voltage regulator, operates on the same basic principles.

NATURE OF ELECTRICITY

The electron theory states that all electrical effects are caused by the movement of electrons from place to place. Electricity is the movement of electrons in a conductor. In order to understand this, let us look at the smallest unit of matter. All matter is composed of chemical building blocks called elements. There are 92 naturally occurring elements that combine into countless varied combinations to form the different kinds of matter found on earth. The smallest particle into which an element can be divided and retain its characteristic as an element is the atom.

An atom is made up of one or more negatively charged electrons rotating (orbiting) around a nucleus containing an equal number of positive charged protons and a specific number of electrical neutral particles called neutrons. Each element has its own characteristic atomic structure, but each atom of a given element has an identical number of electrons, protons, and neutrons. The electrons and protons of one element are exactly the same as the electrons and protons of any other element. The difference between elements is the number of electrons, protons, and neutrons and how they are arranged.

Hydrogen, the simplest atom, consists of one electron in orbit about its nucleus, which contains one proton (Fig. 13-1). Next in simplicity is the helium atom (Fig. 13-1), which has two electrons in orbit about its nucleus which contains two protons and two neutrons. Each electron of the helium atom follows its own individual path as it orbits about the nucleus, but the two electrons remain at the same distance from the nucleus. Each element can be listed according to its atomic number The atomic number is simply the number of protons in the nucleus of an element.

The element copper is widely used in electrical systems because it is a very good conductor of electricity. A copper atom (Fig. 13-1) contains 29 protons and 34 neutrons concentrated in the nucleus, and 29 electrons distributed in four separate shells or rings which are at different distances from the nucleus. Each electron follows its own individual path as it orbits

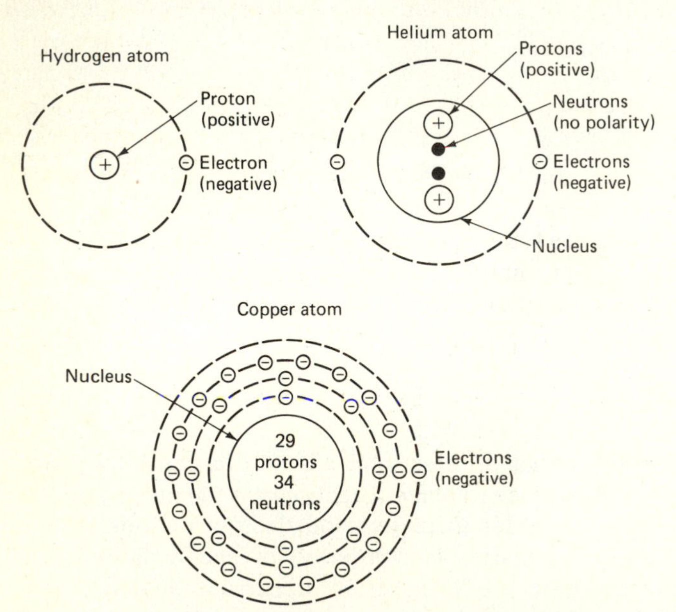

Figure 13-1 Atoms are made of negatively charged electrons orbiting around a nucleus containing positively charged protons and electrically neutral particles called neutrons.

about the nucleus, but the two electrons in the first ring remain at the same distance from the nucleus; the 8 electrons in the second but more distant ring occupy the same distance from the nucleus; and the 18 electrons in the third ring are all the same distance from the nucleus. The fourth ring is the farthest from the nucleus and contains only 1 electron. This is the secret of a good conductor of electricity. The fewer the number of electrons in the outer ring of a conductor, the more easily the electrons can be forced from their orbit to create a flow of current from atom to atom. Elements whose atoms have less than four electrons in the outer ring are generally good conductors, and elements containing more than four electrons in their outer ring are poor conductors and are called insulators. Elements containing four electrons in their outer ring are generally classified as semiconductors.

Now let us look at what happens in a copper wire when a negative charge and positive charge are located at the ends of the wire (Fig. 13-2). Copper wire contains billions of electrons and atoms but only a few atoms with only the single electron in the outer ring will be illustrated (Fig. 13-2). An electron in an atom near the positive end of the wire is attached toward the positive charge and leaves its atom. This atom, in turn, becomes positively charged because it is deficient one electron. Thus, it attracts an electron from its neighbor. The neighbor, in turn, receives an electron from the next atom, and so on. The movement of electrons is not always in one direction but is rather a haphazard drifting of electrons from atom to atom. The net effect is a drifting of electrons from the negatively charged end of the wire to the positively charged end. This flow, or current, of electrons will continue as long as the positive and negative charges are maintained at each end of the wire. This continuous flow of electrons is

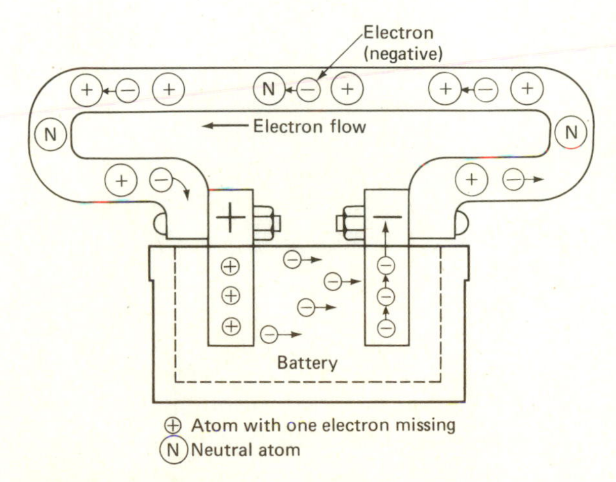

Figure 13-2 Electron movement through a circuit.

called dynamic electricity and leads us to conclude that electricity is the flow of electrons from atom to atom in a conductor.

ELECTRICAL DEFINITIONS, UNITS, AND TERMS

There are several definitions and units used in electrical ignition that are important for us to understand. The advent of transistorized ignition systems has introduced a number of electronic terms into the ignition field. Working definitions of some of the more common terms are given.

Current

The flow of electrons through a conductor is called current and is measured in amperes. One ampere (A) is an electric current of 6.28 billion billion electrons passing a certain point in the conductor in one second. Also, one ampere is that quantity of electricity that is made to flow by a pressure of one volt (V) through a circuit whose resistance is one ohm (Ω).

Voltage

Voltage is the force (pressure) that causes current (electrons) to flow in a conductor. If a conductor has an excess of electrons in one end and a deficiency at the opposite end, the result is a difference in electric pressure that causes a current to flow. Voltage is a potential force and can exist even when there is no current flow in the circuit. It is the pressure required to send a current of one ampere through a circuit whose resistance is one ohm. Electrical pressure is also called electromotive force (EMF).

Resistance

All conductors offer some measure of resistance to the flow of current. The resistance results from: (1) each atom resisting the removal of an electron due to the attraction of the protons in the nucleus and (2) collisions of countless electrons and atoms as the electrons move through the conductor. Not only do the collisions create resistance but they cause heat in the conductor. The *ohm* is the basic unit of resistance. It is the resistance that will allow one ampere to flow when the potential is one volt.

Conductors

If an element has less than four electrons in its valence (outermost) ring (Fig. 13-1), the electrons are held to the core of the atom rather loosely; they can be made to move from one atom to another easily. The movement of electrons from atom to atom constitutes the flow of electric current, and this material is a good conductor of electric energy.

Insulator

If the number of electrons in the valence ring of an atom is greater than four, the electrons are held tightly to the core and normally cannot be made to leave the atom of an element. The element is an insulator and will not conduct electric energy.

Semiconductor

A semiconductor is an electrical device that acts as a conductor under certain conditions and as a nonconductor or insulator under other conditions.

Diode

A diode is a solid-state device that allows current to flow in one direction only. It is analogous to a one-way check valve in a hydraulic system.

Transistor

A transistor is a solid-state electronic switching device in which a small electrical current can be used to switch a much larger current.

Thermistor

The thermistor is a resistor with a special characteristic. Instead of passing less current as its temperature is increased, like a normal resistor, the thermistor allows more current to flow as its temperature rises.

Zener Diode

A zener diode is one with special characteristics. In addition to passing current in a forward direction, it will, when a particular value of voltage is applied to the blocking side of the diode, also conduct current in the reverse direction. The reverse current conduction makes the zener diode adaptable for use as a voltage-sensing device.

Ohm

The unit of electric resistance is called an ohm (Ω). It is the resistance offered to the flow of one ampere under a pressure of one volt.

Ampere-Hour

An ampere-hour is the quantity of current flowing in amperes for a period of one hour. Ampere-hours = amperes $\times$ hours.

Watt

A watt is the unit of electric power or the rate at which work is performed by one ampere of current flowing under a pressure of one volt. Watts = volts $\times$ amperes.

Ohm's Law

As previously stated, the current flowing in an electric circuit is dependent upon the resistance of the circuit and the pressure or voltage. Ohm's law is expressed as follows:

$$\text{Current in amperes } (I) = \frac{\text{pressure in volts } (E)}{\text{resistance in ohms } (R)}$$

or, using the common symbols,

$$I = \frac{E}{R} \quad \text{or} \quad E = IR \text{ and } R = E/I$$

Circuits

The electrical system of a tractor is a combination of interrelated circuits. One requirement for any circuit is to have a voltage supply and a complete path through which current can flow. There are three types of circuits: (1) series; (2) parallel; and (3) parallel-series.

Series Circuit A series circuit has only one path for the current to flow. Batteries (cells) arranged in series are connected with the negative of the first to the positive of the second, negative of the second to the positive of the third, etc. The series connection produces a total voltage equal to the sum of the voltages of the batteries and a total amperage equal, approximately, to the average amperages of individual batteries. Cells in lead acid batteries are connected in series.

Parallel Circuit A parallel circuit provides more than one path for current to flow. A parallel arrangement of batteries would have all positive terminals connected by one wire and all negatives by another. The voltage of the set would be about the average of the voltage of individual batteries. Used for lighting and internal circuits of many components.

Parallel-Series Circuit The parallel-series circuit is what its name implies. Part of the circuit is connected in series and part is connected in parallel. The internal circuits of many components are connected in parallel-series.

There are two ways to describe current flow in a circuit. The *conventional theory* arbitrarily chose the direction of current flow to be from positive terminal of source, through external circuit, to negative terminal of the source. The discovery of the electron led to the electron theory of current flow, which is from negative terminal of source, through external circuit, to positive terminal of the source. Since either theory can be used, we will use the conventional theory in the remainder of this book because of its wide use in industry.

SOURCES OF ELECTRICITY

There are two basic sources of electricity used by modern tractors. They are: chemical and mechanical. The battery is an electrochemical device for converting chemical energy into electrical energy. It is not a storage tank for electricity as often thought but, instead, stores electrical energy in chemical form.

Chemical Generation of Electricity

Chemical devices for generating an electric current are known as cells or batteries. Correctly speaking, a cell is a single unit, and a battery consists of two or more cells connected together. Such cells for generating electricity are made up of four fundamental parts: (1) positive material, (2) negative material, (3) electrolyte, and (4) container. A simple cell (Fig. 13-3) consists essentially of two dissimilar materials immersed in a solution called the electrolyte. If these materials are connected externally by a good conductor of electricity, such as copper wire, chemical action takes place between the solution and these materials, and an electric current flows through the wire; that is, chemical energy is converted into electrical energy.

Only certain combinations of materials will generate electricity in this manner. Some of the more common ones and the voltage produced are

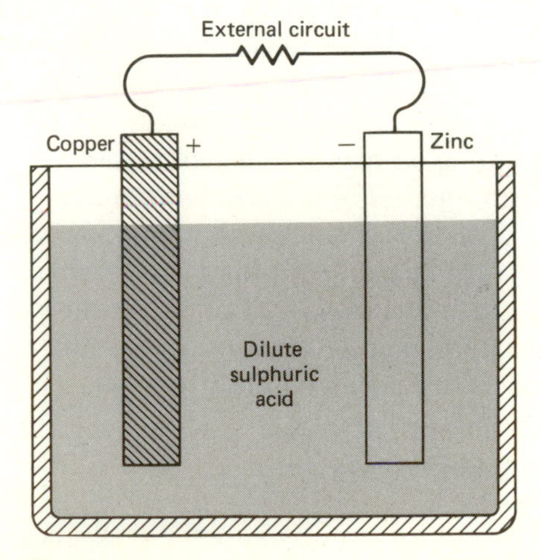

Figure 13-3 A simple cell for generating electricity.

Table 13-1 Chemical Combinations Used in Different Types of Cells

	A Daniell cell	*B* Leclanché cell or dry cell	*C* Lead-acid storage cell
Positive material	Copper (Cu)	Carbon (C)	Lead dioxide (PBO_2)
Negative material	Zinc (Zn)	Zinc (Zn)	Lead (Pb)
Electrolyte	Sulfuric acid (H_2SO_4)	Ammonium chloride (NH_4Cl)	Sulfuric acid (H_2SO_4)
Voltage	1.0	1.5	2.0

given in Table 13-1. Combinations *A* and *B* form what are known as primary cells. Combination *C* is the one used in the ordinary lead-acid automobile and tractor batteries and forms what is known as a secondary cell or battery.

Primary and Secondary Cells A primary cell is one in which the chemical action, taking place as the cell discharges, changes one or more of the active materials—particularly the negative element and the electrolyte—in such a way that when the cell is completely discharged or "dead," it can be restored to its original condition only by renewing the materials that have been so changed. For example, in Fig. 13-3 the zinc plate is gradually consumed, the zinc replacing the hydrogen in the sulfuric acid. This hydrogen collects on the positive copper plate, which remains unchanged. Consequently, the negative zinc and the electrolyte are gradually broken down and eventually must be replaced.

In a secondary cell the active materials also undergo a chemical change during the discharging process. When the cell becomes completely discharged, a replacement of these materials is unnecessary; it can be restored to its original condition by sending an electric current through it in a direction opposite to that of discharge. This process is commonly known as recharging. Of course, this charging and discharging action cannot be carried on indefinitely because these secondary or so-called storage cells gradually lose their strength and efficiency for other reasons. Their lifetimes vary from 2 to 10 years or more, depending upon their type, construction, quality of materials, use, care and other factors.

LEAD-ACID BATTERY CONSTRUCTION AND OPERATION

Plates The active positive and negative materials in the lead-acid cell are in the form of rectangular plates. A number of positive plates are connected together to form what is known as a positive group or element (Fig. 13-4). In a similar manner, a negative group is made up of several negative plates (Fig. 13-4). The two groups are then placed together (Fig. 13-4) so that the positive and negative plates alternate and form an element. Since it is necessary to have a negative on each side of a positive, there must be one more of the former, and the total number of plates per cell will be an odd number. Separators or insulators, as described later, are inserted between all plates.

Plate Construction and Manufacture Since the active materials in both the positive and negative plates are of a brittle nature and have little mechanical strength, a framework made of some neutral metal is necessary to hold them in place. The cellular frame (Fig. 13-4) made of a

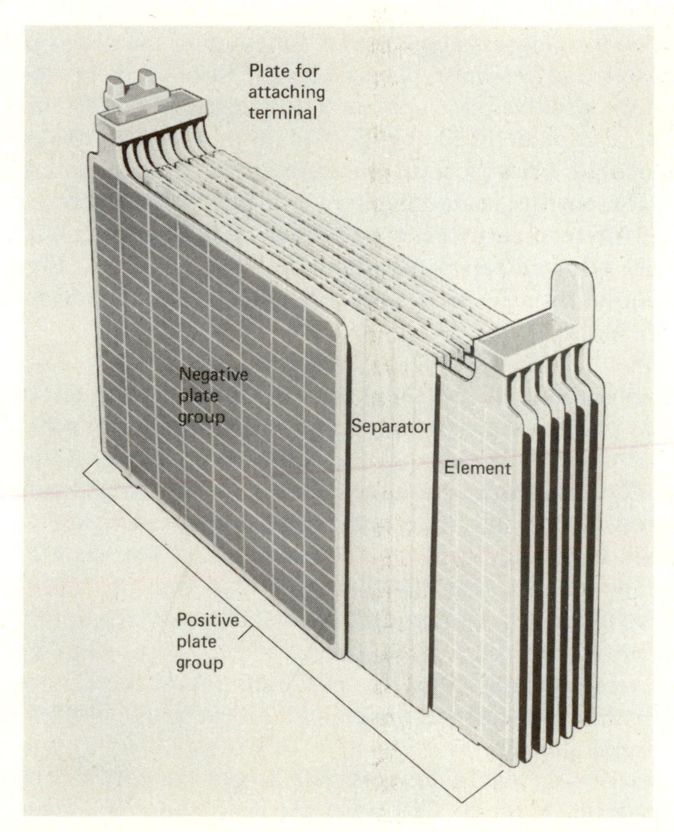

Figure 13-4 Positive and negative groups of plates partly assembled with separator to form a cell for lead acid battery. (Courtesy of Delco-Remy Division, General Motors Corporation.)

lead-antimony alloy, is known as a grid. There is considerable variation in the grid construction as used by different manufacturers. In all cases, the purpose of the grid is the same—namely, to prevent the active materials from cracking and falling out of the plate (thereby reducing the cell output) and to conduct the electric current to and from the active materials of the positive and negative plates. Maintenance-free batteries use grids containing other metals such as calcium or strontium.

The first step in manufacturing positive and negative plates is to paste the active material onto grids. The paste is a mixture of lead oxide, sulfuric acid, and water. Paste for the negative plates has expanders added to prevent the negative material from contracting in service and reverting to a dense, inactive state that would greatly reduce the performance of the battery. When pressed into place, the paste dries and hardens like cement and adheres to the ribbed structure. At this stage, the positive plate has a light brown color; the negative plate has a slight gray color due to the expander in the material. The plates are then subjected to an electrochemical process known as forming; this consists of submerging the plates in sulfuric acid and water and sending a direct current through them (the current flowing in at the positive plate and out at the negative plate). This process converts the lead oxide of the positive plate into lead peroxide, which is a dark chocolate brown color, and converts the negative into gray, spongy lead. The positive material is now a highly porous material that enables the electrolyte to easily penetrate the plate. The spongy lead of the negative plate allows deep penetration by the electrolyte, thus permitting more material than the plate surface to take part in the chemical reaction.

Separators The purpose of the separators is to act as insulators between the plates to prevent them from making contact with each other and creating an internal short circuit. At the same time, the separator must be a porous material that will permit the free passage of charged ions of the electrolyte between the negative and positive plates. Originally, separators were made from chemically treated wood (Fig. 13-5), such as cypress, redwood, or fir. Present-day batteries use porous, resin-impregnated cellulose fibers or acid-resistant synthetic rubber separators alone or in combination with perforated rubber sheets (Fig. 13-6) or glass-fiber mats, as retainer walls. The retainer walls retard the loss of active material from the plate surface under high vibration application, thus improving the separator life.

Separators have ribs on one side; this side is placed next to the positive plate with the ribs vertical. This type of construction minimizes the area of contact between plate and separator, provides greater acid volume next to the positive plate, aids acid circulation, and permits gas to rise to the top of the cell. Some batteries use a porous envelope separator

Figure 13-5 A wood separator for a lead-acid cell.

Figure 13-6 A thread-rubber separator for a lead-acid cell.

around each individual plate, which is sealed on two sides and bottom but open at the top to allow gases to escape. This type of separator allows elimination of the sediment space in a battery and prevents shorts due to misalignment of separators and plates.

Electrolyte The electrolyte used in the lead-acid cell consists of a mixture of about 2 parts chemically pure, concentrated sulfuric acid (H_2SO_4) to 5 parts distilled water, by volume. The concentrated acid will have a specific gravity of 1.835, but the diluted mixture of the above proportions will drop to about 1.300 at 70°F (21°C).

The usual method of determining the strength of the solution is to test it by means of a hydrometer syringe, as shown in Fig. 13-7. The hydrometer is placed inside the syringe so that it is only necessary to insert the latter in the cell, draw out some of the liquid, make the reading, and squirt the liquid back into the cell without removing the syringe nozzle. The ordinary battery hydrometer is graduated from 1.100 to 1.300, which is the maximum range of variation of the electrolyte between a completely discharged and a fully charged condition. Table 13-2 gives the conditions present in a battery at different stages of charge, assuming the surrounding temperature to be 70°F (21.1°C).

Container Containers for the lead-acid batteries are made either of polypropylene (Fig. 13-8), hard rubber, or other plasticlike materials. The tops of the plastic containers are usually one piece, while tops for hard-rubber containers may be one piece or consist of individual cell covers sealed with a tarry, acid, and waterproof compound. The one-piece battery cover is sealed to the container by a high temperature and pressure process or by epoxy resin. The battery top usually has a vent-well for each

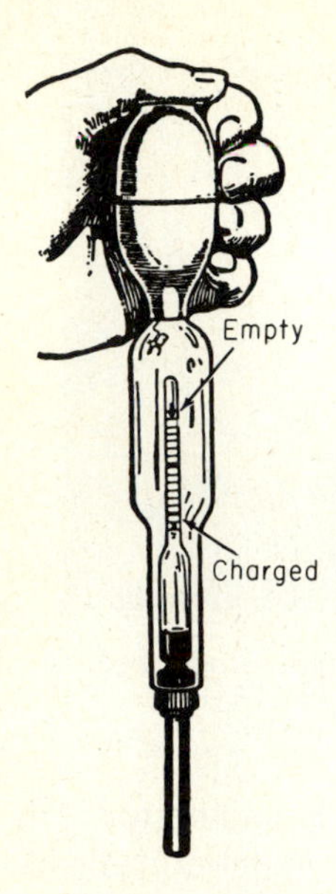

Figure 13-7 Hydrometer syringe for determining the specific gravity of the electrolyte.

cell that permits testing of the electrolyte, refilling the cell, and providing the proper airspace above the electrolyte. The vent plug is provided with one or more small holes that allow the escape of gas formed during the charging and discharging action, without forcing the electrolyte from the battery.

Chemical Action in the Lead-Acid Battery As already stated, the active materials in a lead-acid cell of a battery, when fully charged, consist of: (1) sponge lead (Pb), or negative material; (2) lead peroxide (PbO_2), or positive material; and (3) dilute sulfuric acid solution, or electrolyte (H_2SO_4). If the terminals of such a battery are connected by a conductor so that an electric current will flow, the battery is said to be discharging and the chemical action takes place as in Figure 13-9.

Table 13-2 Characteristics of Lead-Acid Cell at Different Stages of Charge

Condition of battery	Specific gravity	Cell voltage	Freezing point, °F
100% charged	1.230–1.300	2.2	−90
75% charged	1.200–1.250	2.1	−60
50% charged	1.190–1.220	2.0	−20
25% charged	1.175–1.190	1.9	0
Completely discharged	1.150 or less	1.8 or less	20

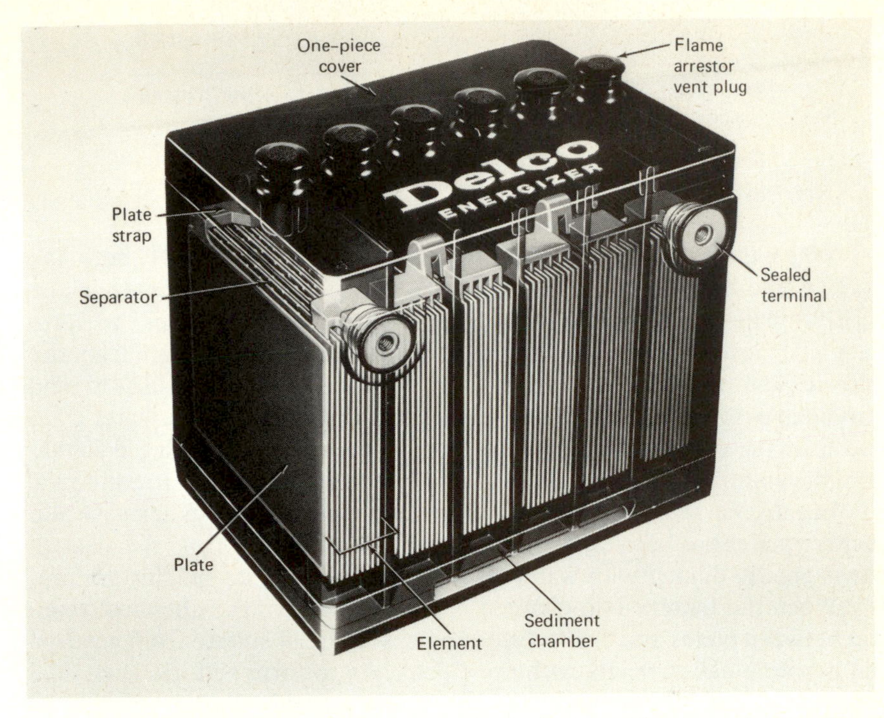

Figure 13-8 Construction of a 12-volt lead-acid battery. (Courtesy of Delco-Remy Division, General Motors Corporation.)

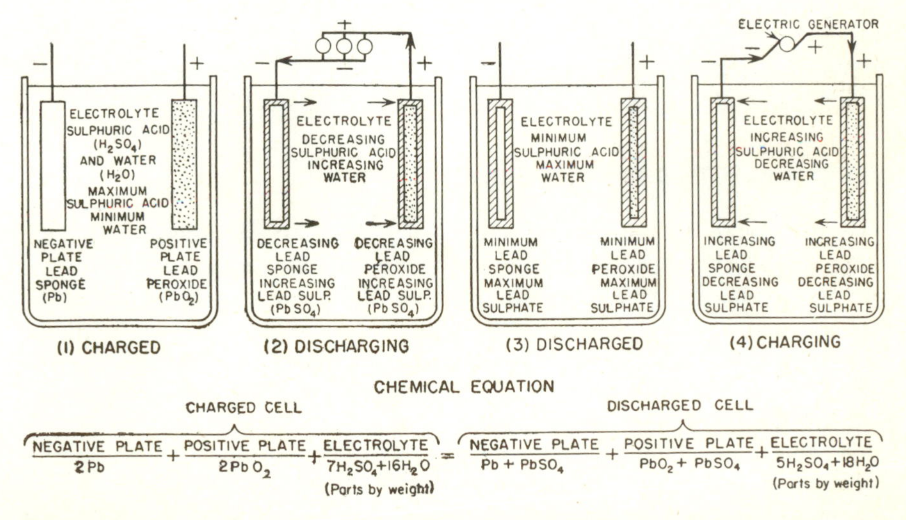

Figure 13-9 Action taking place in a lead-acid cell during a cycle of discharge and charge.

$$\underset{\text{Negative plate}}{2Pb} + \underset{\text{Positive plate}}{2PbO_2} + \underset{\text{Electrolyte}}{7H_2SO_4 + 16H_2O} \underset{\text{Recharging}}{\overset{\text{Discharging}}{\rightleftarrows}}$$

$$\underset{\text{Negative plate}}{Pb + PbSO_4} + \underset{\text{Positive plate}}{PbO_2 + PbSO_4} + \underset{\text{Electrolyte}}{5H_2SO_4 + 18H_2O}$$

When the battery discharges, the electrolyte reacts with both the positive and negative plates . . . oxygen from the lead peroxide in the positive plates combines with hydrogen from the sulfuric acid to form water . . . lead from the lead peroxide combines with the sulfate from the sulfuric acid to form lead sulfate . . . hydrogen from the sulfuric acid combines with oxygen from the lead peroxide to form more water . . . lead from the sponge lead in the negative plates combines with the sulfate from the sulfuric acid to form lead sulfate . . . and electric current flows.

In a discharged battery, most of the active material from negative and positive plates has been converted to lead sulfate ($PbSO_4$), and the electrolyte is greatly diluted with water (H_2O), thus lowering its specific gravity.

When the battery is recharged by the alternator, the chemical reaction between plates and electrolyte is reversed. Lead sulfate from positive and negative plates reacts with the electrolyte to form sulfuric acid. The removal of sulfate from the negative plates restores sponge lead as its active material. The oxygen from the water recombines with the lead in the positive plates to form peroxide. Thus, the chemical action is reversed, the strength of the battery is restored and the specific gravity of the electrolyte increases to about 1.300.

Since there is a marked and uniform drop in the specific gravity of the electrolyte, owing to the formation of water and the absorption of the acid during the discharge process, the measurement of the specific gravity serves as the most convenient, accurate, and satisfactory means of determining the state of charge of a cell or battery (Table 13-2).

VEHICLE ELECTRIC REQUIREMENT

There are three main functions of the tractor battery. Its major responsibility is to supply current to start the engine. Current required to crank an engine varies widely from engine to engine. It depends on the engine bore, stroke, compression ratio, number of cylinders, engine/starter cranking ratio, temperature, oil viscosity, resistance in electric circuit, and type of transmission. All these factors are considered when the design engineers select the original equipment battery for a specific engine.

The second function of a battery is to supplement the vehicle electrical load requirement whenever it exceeds the maximum output of the charging system. This occurs when the vehicle engine operates at "idle"

Table 13-3 Power Requirement for Vehicle with Average List of Accessories and 12-V System

Electrical system	Amperes required
Ignition	3–4.2
Radio	.4–2
Headlights	
Low beam	8–14
High beam	10–18
Blower	
Heater (no air conditioner)	6–9.3
Heater (with air conditioner)	16–19
Air conditioner (summer)	17–23
Summer starting	150–250
Winter starting	225–400

speed. Table 13-3 lists typical current requirements for cars or tractors. A load of 40 A is not unusual in summer, with air conditioning and no lights, higher with lights.

A third function of the battery is to act as a voltage stabilizer in the charging system. High transient voltage may be generated occasionally when a circuit is broken. The battery helps to absorb this peak voltage and protect components such as diodes, from damage.

CAPACITY RATING OF BATTERIES

The Society of Automotive Engineers (SAE) and the Battery Council International (BCI) adopted new capacity ratings for batteries in 1971. The two standards are known as the Cold Cranking Rating and Reserve Capacity Performance.

Cold Cranking Rating The primary function of the battery is to provide electrical power to crank the engine. Cranking an engine involves a large discharge of amperes over a short span of time. The Cold Cranking Rating is defined as the discharge load in amperes that a battery at 0°F (−17.8°C) can deliver for 30 s and maintain a voltage of 1.2 V per cell or higher (7.2 V for a 12-V battery; 3.6 V for a 6-V battery). Batteries for diesel must discharge rated amperes for 90 s and maintain voltage of 1.0 V per cell.

Reserve Capacity Performance This rating gives the number of minutes a new, full-charge battery at 80°F (26.6°C) will deliver 25 A continuously while maintaining a voltage of 1.75 V per cell. The 25 A are considered the power required for ignition, lights and normal accessories to keep the machine or vehicle operating if the charging system should fail sud-

Table 13-4 Effect of Number of Plates on Cold Cranking and Reserve Capacity of Lead-Acid Batteries

Battery size, volts	Plates per battery	Cold cranking at 0°F (−17.8°C), amps	Reserve capacity at 80°F (26.6°C), minutes	Cold cranking power, watts	Battery dry weight, lb
6	33	325	85	1,950	17.4
6	45	440	145	2,640	21.5
6	63	660	250	3,960	35.1
6	99	975	420	5,850	47.5
12	42	235	45	2,820	22.6
12	54	310	83	3,720	30.7
12	66	460	150	5,520	50.8
12	90	640	230	7,680	71.0
12	150	750	290	9,000	70.0

denly. Reserve capacity is always expressed in minutes, the time available to seek help. Thus, the larger the number of minutes, the greater the margin of safety.

Battery Voltage and Amperes Each cell in a lead-acid battery has a potential of approximately two V. Cells in lead-acid batteries are connected in series; thus, voltage equals number of cells times two. A 6-V battery has three cells; a 12-V battery will have six cells.

The principal factors affecting the voltage of a lead-acid battery are: (1) strength of the electrolyte, (2) charging or discharging rate, and (3) existing temperature.

Nearly all lead-acid batteries consist of a number of cells connected in series; the ampere capacity of the battery is the same as the capacity of the individual cells. Amperes produced depend primarily upon the size and number of plates per cell (Table 13-4), because the greater the size or number of plates, or both, the greater the total plate surface area and the quantity of active material available for generating electricity. Other factors affecting the ampere capacity are thickness of plates, porosity of plates and separators, quantity and density of electrolyte, and temperature.

SELECTION OF REPLACEMENT BATTERIES

If comparable performance is to be obtained, it is necessary to select a replacement battery at least equal in size to the original equipment battery. The design engineers selected with great care the original battery for

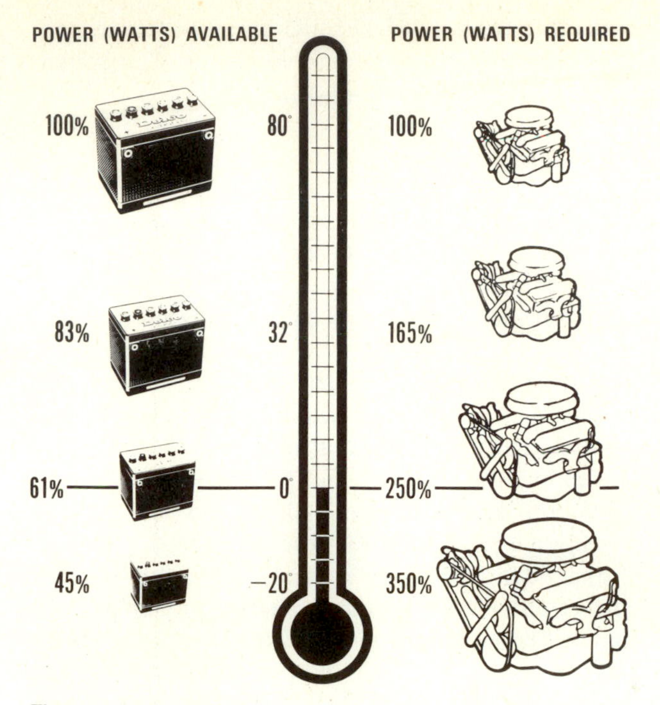

Figure 13-10 Effect of temperature on available power from a battery and power required to start an engine. (Courtesy of Delco-Remy Division, General Motors Corporation.)

the vehicle. Thus, replacing the original battery with one smaller in capacity may result in poor performance and short life.

Figure 13-10 shows why a battery of sufficient electrical size is essential if satisfactory cranking of an engine is to be achieved at low temperatures. At sub-zero temperatures, the capacity of a battery at full charge is approximately 45 percent of its capacity at 80°F (26.6°C). At the same time, the load imposed on the battery by the engine is approximately 3½ times the normal cranking load at 80°F (26.6°C).

DRY-CHARGED BATTERIES

A dry-charge battery is manufactured with charged plates and then each cell element is thoroughly washed, completely dried, and assembled into the battery case. The electrolyte is not added, the advantage being that the battery does not deteriorate with nonuse but remains in a fully charged condition as long as moisture is prevented from entering the cells. Activation consists of adding electrolyte of 1.265 specific gravity to each cell. Dry-charge batteries should be given at least a limited charge after activation before placing them in a vehicle.

"MAINTENANCE-FREE BATTERIES"

The terms "maintenance-free" and "no-maintenance" batteries, used so widely within the past 10 years, have lost some of their meaning. However, a new generation of batteries has been developed, incorporating changes in chemistry, manufacturing processes, and materials of construction (Fig. 13-11).

Antimony has been used for years in the grid of batteries to strengthen them. When even a small amount of antimony is present, the battery gases lose electrolyte and use water that must be replaced. The antimony has been replaced by other metals, such as calcium, cadmium, and strontium. Batteries with grids made from lead-calcium used 82 percent less water than conventional batteries. The new grids, produced from lead-calcium strips, are expanded into shape at room temperature rather than casted. A self-discharge characteristic during long periods of inactivity is another significant difference between lead-antimony and lead-calcium batteries. In lab tests, lead-calcium batteries retain sufficient cranking ability to start an automobile engine after 12 months of storage at 80°F (26.6°C), while conventional batteries dropped below recommended

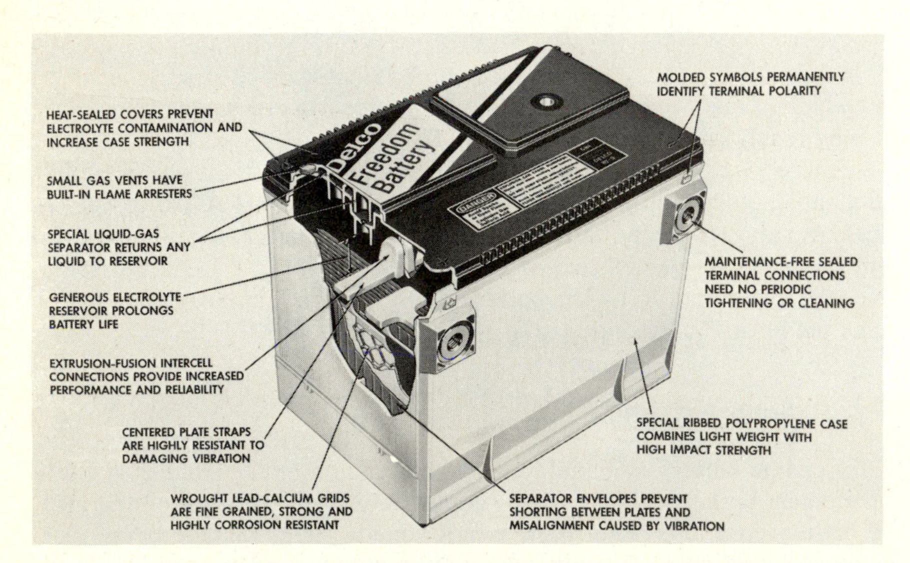

Figure 13-11 Construction of a "maintenance-free" battery. Note large reservoir of electrolyte above plates. (Courtesy of Delco-Remy Division, General Motors Corporation.)

cranking capacity in 3 to 4 months after manufacture. Separators made like an envelope, with the plates inside, prevent lead from falling out of the grid and also prevent shorting between plates and misalignment caused by vibration.

BATTERY TROUBLES

Sulfation The life of any storage battery is dependent upon the care and attention it receives. Maximum service should not be expected from a storage battery of any kind if it is not inspected and checked periodically. Perhaps the most common cause of short life is sulfation, which results from a protracted discharged condition. As already mentioned, during the normal discharging process, the active material in the plates is changed to lead sulfate ($PbSO_4$) which, in turn, is converted back into lead peroxide and lead when the battery is recharged. However, if the cell or battery is permitted to remain in a discharged condition for any length of time, this sulfate gradually hardens and is not readily converted to its former state. The longer the cell remains uncharged, the more difficult it is to recharge. A badly sulfated cell or battery can be recharged only with difficulty, if at all, and then seldom regains its original effectiveness.

High Charging and Discharging Rates Excessive charging or discharging rates for periods exceeding a few seconds or minutes are injurious to a storage battery in that heat generated causes warping or buckling of the plates, thereby loosening the active material and reducing its life and efficiency. As previously stated, it is important that a battery be charged or discharged only at a certain rate based upon its capacity in Ah.

Other Causes of Battery Troubles Other common causes of battery troubles include: (1) failure to keep electrolyte at proper level and over tops of plates; (2) use of undistilled or impure water for replenishing electrolyte; (3) dirty, badly corroded, or loose connections and terminals; and (4) not keeping battery top clean.

MAGNETS AND MAGNETISM—INDUCTION

Nature of Magnetism It has been found that under certain conditions a piece of iron or steel or some iron alloy possesses the property of attracting or repelling other pieces of iron or materials containing this metal. This peculiar action or form of energy is known as *magnetism,* and the metal possessing it is called a *magnet*. Other metals and nonferrous alloys, such as aluminum, copper, brass, bronze, tin, lead, zinc, and so on, do not have this property and are therefore termed *nonmagnetic*.

Soft iron of a more or less pure form absorbs or takes up magnetism

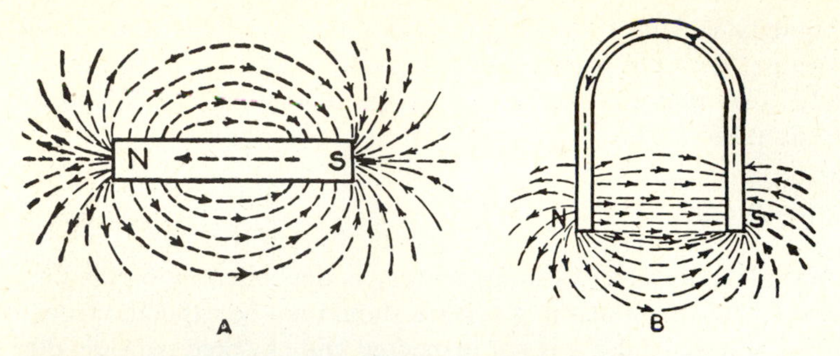

Figure 13-12 Steel bar and horseshoe magnets showing magnetic field.

readily and in greater quantity, so to speak, but retains this property only while the magnetizing force or medium is very near or acting upon it. In other words, soft or pure iron forms what is known as a *temporary magnet*. Steel and alloys containing iron are not magnetized so readily but, when once so treated, retain this property to a certain degree even after the magnetizing medium has been removed. The common bar and horseshoe magnets (Fig. 13-12) are, therefore, made of hard steel and are known as *permanent magnets*. The well-known magneto magnets are likewise permanent magnets and are made of a very high-grade steel alloy.

Characteristics of Magnetism If a steel-bar magnet is held under a piece of cardboard (Fig. 13-13), on which has been placed a thin layer of iron filings, and the cardboard is tapped gently, the filings will assume a rather definite arrangement and form more or less distinct lines, which radiate from one end of the magnet to the other. The attraction appears to be greater at the ends or poles of the magnet but also exists in the space around it between the poles. This space about a magnet in which there exists more or less attraction for other pieces of iron is known as the *magnetic field*. When iron filings are held over a U-shaped or horseshoe magnet (Fig. 13-14), a similar effect is produced and a magnetic field likewise exists.

If a pocket compass is brought near an ordinary bar magnet, the end of the needle which ordinarily points north will be attracted by one pole of this magnet but will be repelled by the other pole, as shown in Fig. 13-15. The end of the magnet attracting the north end of the needle is called the *south pole* and the end repelling the north end of the needle is known as the *north pole* of the magnet. It is likewise observed that the lines of force about the magnet flow from north to south.

If the north poles of two steel magnets are brought close to each other (Fig. 13-16) and iron filings are sprinkled on a cardboard held over them, the filings will arrange themselves as shown, indicating that there is a repulsion or repelling effect. On the other hand, if a north pole of one

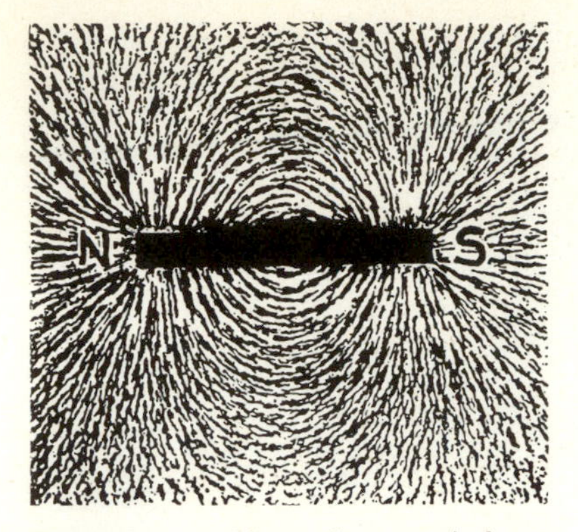

Figure 13-13 Lines of magnetic force around a bar magnet as shown by iron filings.

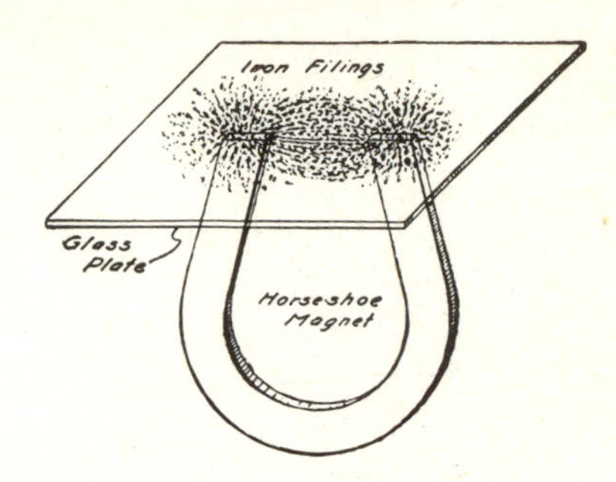

Figure 13-14 Effect produced by a magnet held under a glass plate which is covered with iron filings.

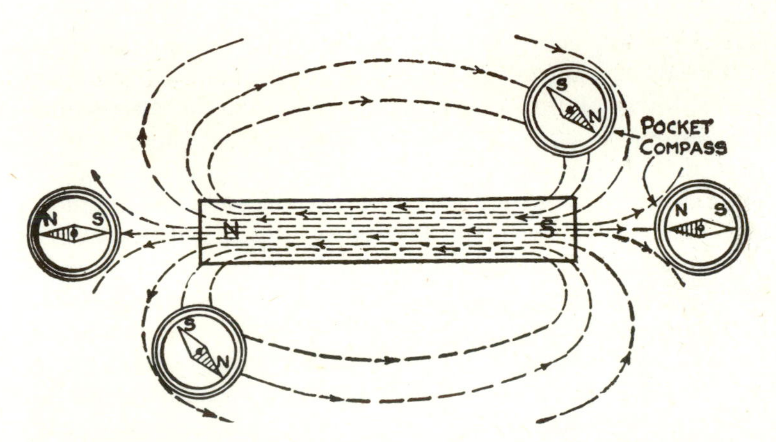

Figure 13-15 Polarity of a bar magnet as indicated by a compass.

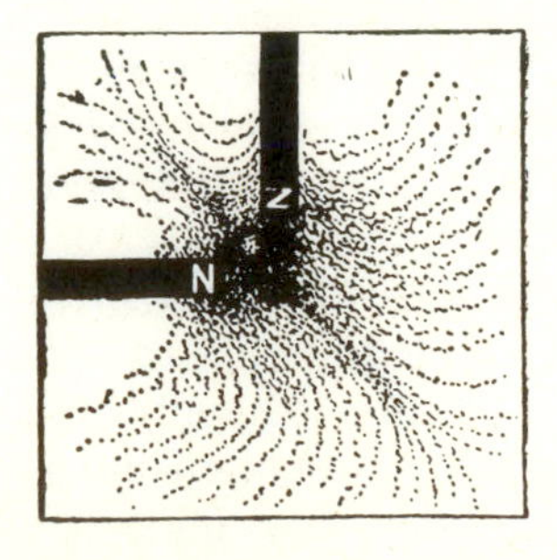

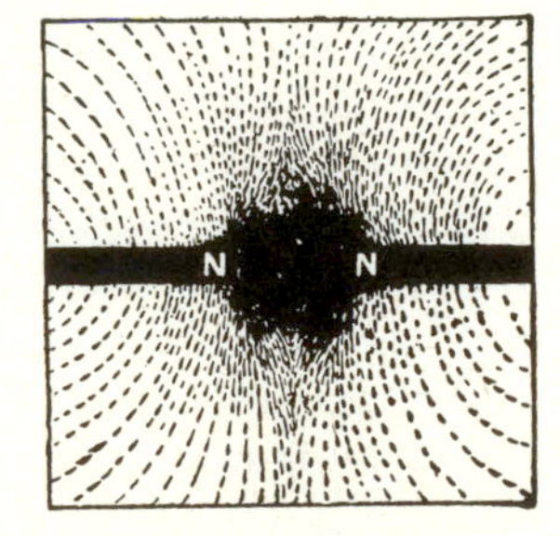

Figure 13-16 The repelling effect of like poles.

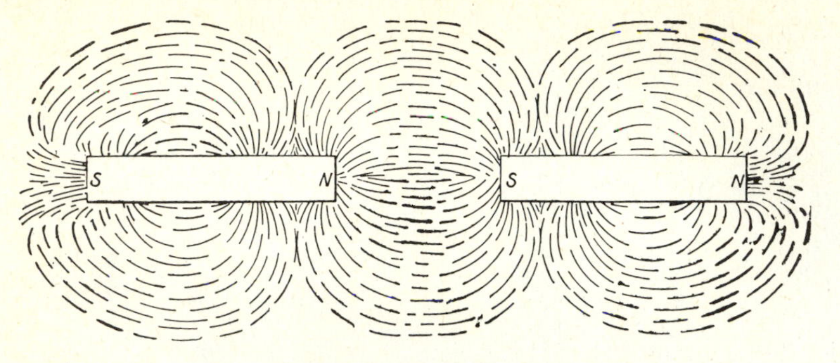

Figure 13-17 The magnetic attraction produced by unlike poles.

magnet is brought close to the south pole of the other (Fig. 13-17), there is a very strong attraction and the filings arrange themselves differently. In other words, it can be said that like poles repel, and unlike poles attract each other.

If a piece of soft iron is brought near to a permanent steel magnet, the soft iron itself becomes a magnet and the lines of force tend to concentrate themselves in it. If the soft iron is removed, however, it does not retain its magnetism. If, instead of a piece of soft iron, a piece of unmagnetized soft steel is placed near a permanent steel magnet, as previously described, the piece of soft steel, upon removal from the magnet, will, perhaps, retain a small amount of magnetism, known as *residual magnetism*. The ability to retain this residual magnetism is known as *retentivity*.

A piece of very hard steel, unmagnetized, when brought close to a magnetized piece of steel or a strong magnet, will not take the magnetism so readily as soft iron, but once it becomes magnetized, it retains a considerable amount and is said to have better retentivity than soft iron or soft steel. These characteristics of soft iron, soft steel, and hard steel are very important and useful in connection with the successful operation of numerous electrical devices such as coils, magnetos, and electric generators.

Electromagnetism Although magnetism and electricity are two entirely different forms of energy, they are more or less connected or related. For example, if an electric current is passed through a piece of copper or iron wire, and a portion of the latter dipped in iron filings (Fig. 13-18), they will be attracted to it and the wire will have the properties of a magnet. When the circuit is broken, the filings drop off. This shows that the conductor of electricity has a magnetic field about it. If an ordinary compass is placed in different positions about a wire carrying an electric current, and the action of the needle observed in these different positions, it will be found that the lines of force form concentric circles about the wire, as shown in Fig. 13-19. Likewise, it has been found that the direction

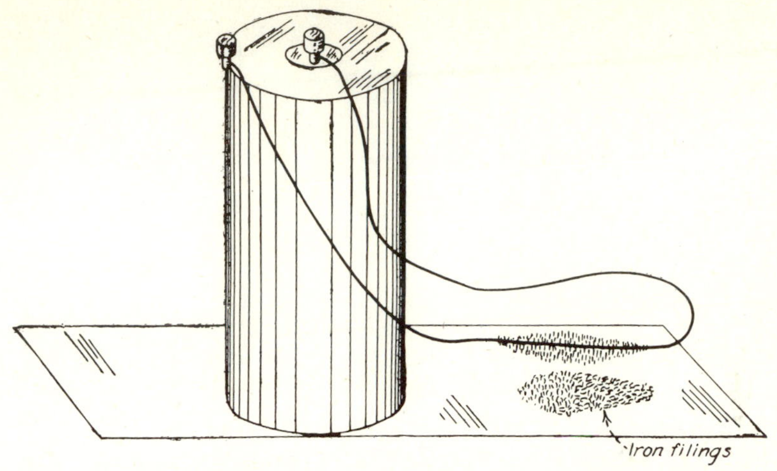

Figure 13-18 The magnetic field about a conductor of electricity as indicated by iron filings.

taken by the lines of force is determined by the direction of the current through the wire, as shown in Fig. 13-20.

If a certain length of wire is made into a coil, as in Fig. 13-21, and the current passed through it, the lines of force about each turn of wire in the coil will combine to form a concentrated or strong magnetic field about the entire coil. Such a coil of wire through which an electric current is flowing is known as a *solenoid* and possesses the same properties as a piece of magnetized steel; that is, it has a north and a south pole and a magnetic field around it. If the current is suddenly broken or ceases to flow, the magnetism likewise disappears.

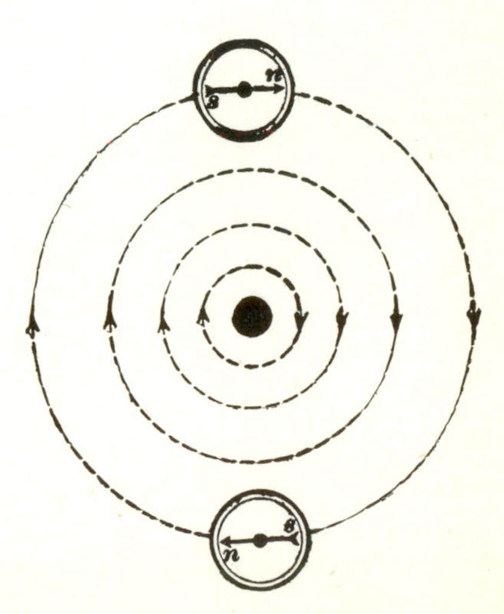

Figure 13-19 Direction of lines of force about a conductor of electricity as indicated by a compass.

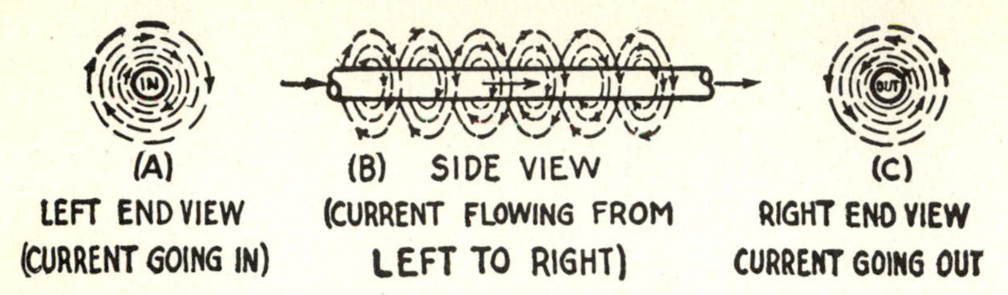

Figure 13-20 Effect of direction of flow of current on lines of force.

The polarity of a solenoid depends upon the direction of flow of the current through the wire, and, if the direction of flow is known, it may be determined as follows: Grasp the coil with the right hand, with the fingers pointing around the coil in the direction of the current flow. The thumb will then point toward the north pole. The polarity can likewise be determined by means of an ordinary compass in the same manner as that in which the polarity of the steel magnet is determined (Fig. 13-15).

The Electromagnet If the end of a piece of soft iron is brought near a coil or solenoid, consisting of many turns of insulated wire, and a current is passed through the coil, the iron will be attracted and drawn or sucked into the center of the solenoid (Fig. 13-22). The soft iron thus becomes magnetized and is known as an electromagnet. It possesses the same characteristics as an ordinary steel-bar magnet. Since the soft iron is extremely magnetic and readily magnetized, the lines of force tend to concentrate in this iron core, and the coil becomes much stronger, magnetically, than without the iron core. If the core of soft iron is withdrawn, this iron will lose its magnetic properties. A steel core would not produce so strong an electromagnet because steel does not take up magnetism as readily as soft iron.

The strength of such an electromagnet depends upon two factors, namely, the number of turns of wire around the core making up the coil, and the quantity of current flowing in the wire in amperes; that is, the greater the number of turns of wire and the greater the current flow in

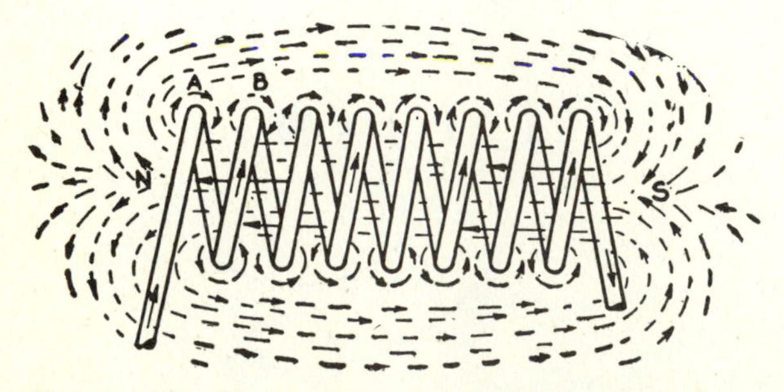

Figure 13-21 The magnetic field about a coil of wire carrying an electric current.

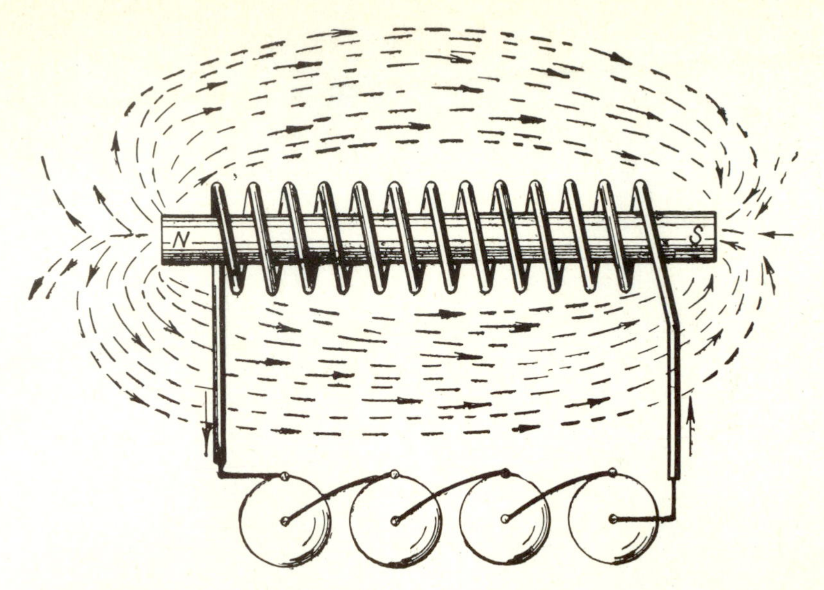

Figure 13-22 A simple electromagnet.

amperes, the stronger the electromagnetic field and the electromagnet produced.

Induction In the preceding discussion it has been shown that a conductor of an electric current has a magnetic field existing about it as long as the current flows. Now, if the action is reversed and, in some manner, a magnetic field is suddenly set up about a conductor of electricity, an electric current will flow in the conductor just at the instant the field builds up. Again, if a magnetic field existing about a conductor is suddenly reduced in strength or completely broken down, a current will flow in the conductor at the instant the change occurs. This peculiar relationship or interaction of electricity and magnetism is known as *induction*.

It should not be assumed that if a conductor is merely held in a magnetic field of uniform strength a current will flow. Such a current is generated only by a change in the number of lines of force about the conductor. This change may be from a low to a high field strength or from a high to a low field strength.

Electromagnetic Induction The phenomenon of induction is well demonstrated and illustrated by means of an ordinary single-winding coil (Fig. 13-23). If an ordinary 6-V dry cell battery is connected to such a coil as shown and the circuit is closed, a strong magnetic field is built up and exists as long as the current flows. Some of the electrical energy of the battery becomes magnetic energy. Now, if the circuit is quickly broken at some point, an electric spark of considerable size is observed where the

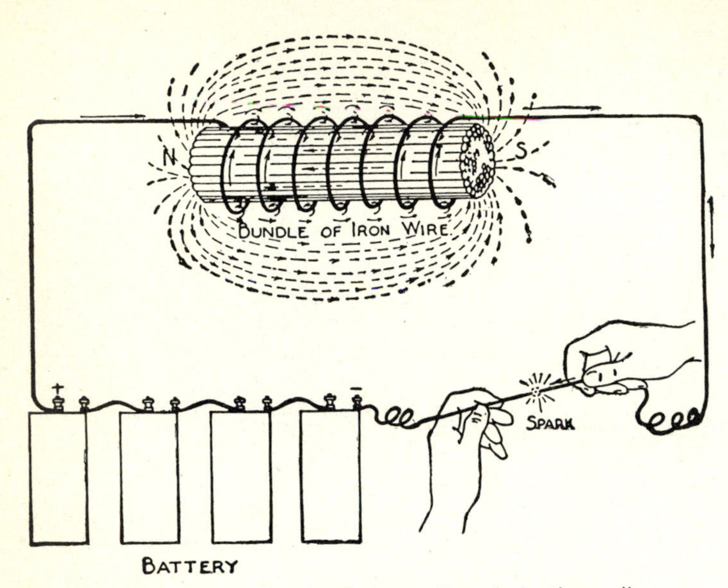

Figure 13-23 Action of a simple low-voltage induction coil.

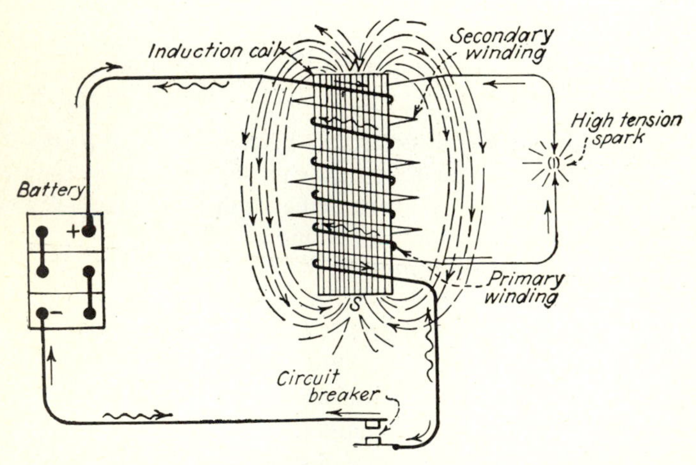

Figure 13-24 Construction and action of a two-winding high-voltage induction coil.

Figure 13-25 Principle involved in generating an electric current by induction.

break occurs. Bearing in mind that an electric spark is electricity flowing through air and that the latter is a very poor conductor, it is evident that this sudden breakdown of the magnetic field about the coil sets up a high voltage in the circuit, which produces an instantaneous flow of current. The magnetic energy is thus dissipated and converted into electrical energy again, and this action is known as *electromagnetic induction.*

The double-winding coil (Fig. 13-24) demonstrates this action in a somewhat more conclusive manner, perhaps, than the single-winding coil. Suppose that the winding attached to the battery, ordinarily known as the *primary,* is made of reasonably coarse wire and has fewer turns than the other, *secondary* winding, which is finer and has many turns. If the primary circuit is closed, current flows through it and sets up a magnetic field, which also surrounds the secondary winding, since it is on the same iron core. Now, if the primary circuit is suddenly broken, a spark will jump the small gap left between the ends of the secondary. Thus the magnetic field which existed about this winding, although created by the primary winding, generates for an instant a current flow in the secondary winding, due to its sudden breakdown or dissipation. This again is called an *induced current.*

The phenomenon of induction, produced in a somewhat different manner than previously described but fundamentally the same in principle, is shown in Fig. 13-25. If the permanent steel horseshoe magnet is first held so that the coil of insulated copper wire is between its poles, a magnetic field exists about the coil. Now, if the coil is quickly moved in the direction indicated, a voltage and resultant current flow will be generated for an instant, as indicated by a sensitive galvanometer. Likewise, if the coil is quickly returned to the first position, a current will flow at the instant the coil enters the field. In this device it will be observed that the movement of the magnet and its field away from the coil, or the movement of the coil away from the magnet, results apparently in the lines of force being cut or broken. In fact, the action is often so expressed, and the maximum voltage and current flow occur when the greatest number of lines of force are cut.

The voltage and amount of current induced in an electric circuit depend upon three factors, namely, (1) the strength of the magnetic field about the conductor, (2) the number of turns making up the conductor involved, and (3) the speed or rate at which the magnetic field changes or breaks down. In other words, the stronger the magnetic field, the greater the number of turns of wire in the field, and the quicker the field strength changes with respect to the conductor, the stronger the voltage and current induced. The principles of electromagnetic induction, as just described, are important and fundamental to a clear understanding of the action and operation of all ignition coils, magnetos, electric generators of all kinds, transformers, doorbells, alarms, and many similar devices.

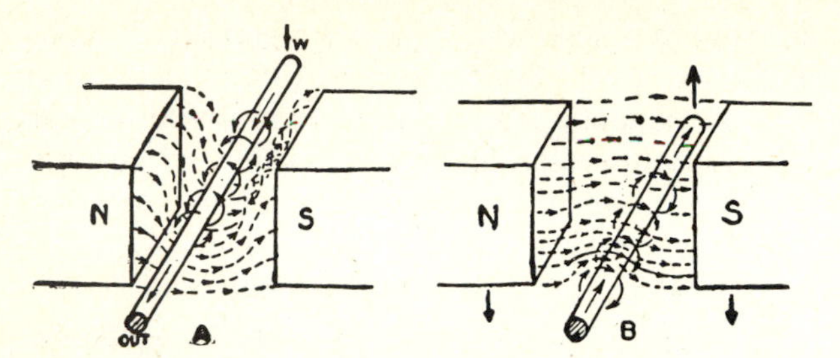

Figure 13-26 Effect of polarity and direction of motion of conductor on direction of current flow.

Direction of Flow of Induced Current Just as there is a definite relation between the direction of the current in a solenoid or coil and the polarity of the electromagnet so formed, so there is also a definite relation between the arrangement of the poles, the direction of movement of the conductor through the magnetic field, and the direction of flow of the induced current. For example, referring to Fig. 13-26A, it will be observed that if the conductor is moved downward through the field between the poles of the magnet, the lines of force are bent or distorted and assume a more or less circular form about the conductor as the latter passes through the field. The direction of the induced current is thus determined by the direction of the circular lines of force about the conductor. In Fig. 13-26B the conductor is shown moving upward through the field, so that the lines of force pass around it in the opposite direction; therefore the induced current flows the other way.

A rule for determining any one of the three conditions involved—namely, the direction of flow of the magnetic lines of force, the direction of motion of the conductor, and the direction of flow of the induced current—when two of them are known is called the *right-hand three-finger rule* and is illustrated by Fig. 16-27.

PROBLEMS AND QUESTIONS

1 Explain the electron theory of current flow as related to electricity and state how electricity flows in a copper conductor.

2 Explain the following electronic terms that have been introduced into the

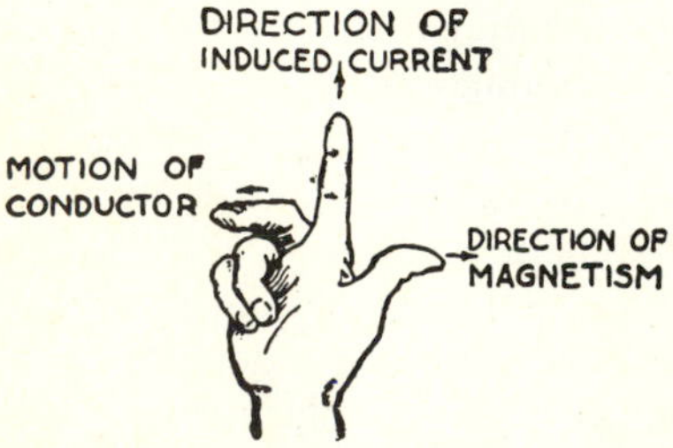

DIRECTION OF INDUCED CURRENT

MOTION OF CONDUCTOR

DIRECTION OF MAGNETISM

Figure 13-27 Right-hand three-finger method for determining direction of induced current.

ignition field: a) semiconductor, b) diode, c) transistor, d) thermistor, and e) zener diode.

3 Explain the difference between primary and secondary cells. Explain the chemical action that takes place when a secondary cell is discharged; when it is recharged.
4 Explain the three main functions of a tractor battery.
5 Explain the Society of Automotive Engineers and the Battery Council International capacity ratings for batteries.
6 Explain the meaning of the term electromagnetic induction in the secondary winding of a two-winding coil.
7 Explain the relationship of polarity, current flow, and direction of motion of the conductor for a simple electric generator.

Chapter 14

Battery and Magneto Ignition Systems

All electric ignition systems for internal-combustion engines are either one of two types; namely, (1) the conventional breaker-point system and (2) the electronic systems. The current for either system may be generated chemically, that is, by means of a battery, or mechanically, by means of a magneto or similar electric generator.

The essential functions of any electric ignition system are (1) the generation of a large, hot spark in the cylinder and (2) the production of this spark at the right instant in the travel of the piston. In other words, if a good spark is produced in the cylinder at the right time, the combustible mixture of fuel and air should be properly ignited.

Voltage Required for Ignition An electric spark is nothing more than an electric current flowing through air. In any electric ignition system such a spark is produced by making the current flow between two points separated by a small fraction of an inch and forming a part of an otherwise complete electric circuit. Since air and other gases are very poor conductors of electricity, an extremely high voltage is necessary to produce the desired spark in the cylinder of any engine. Furthermore, at the time that the spark is required in the cycle, the gaseous mixture surrounding the sparking points is under compression. Consequently, this mixture is considerably more resistant to the passage of an electric current than if it were at a lower or ordinary atmospheric pressure. The curves (Fig. 14-1) show the relationship of gap voltage to compression pressure under ideal conditions.

Again, the greater the width of the spark gap, the greater the voltage necessary to make the current flow across the gap and create a desirable spark. This is clearly shown by reference to the curve in Fig. 14-1. Therefore, to obtain the best results, it is important to use as narrow a gap as possible in the spark plug and to maintain the highest possible voltage in the circuit.

It has already been explained that the voltage of most lead-acid batteries used in mobile equipment is 12 V. In view of this, and since voltages ranging from a few thousand to several thousand volts are required if any electrical ignition system is to function properly, it would seem impractical to attempt to utilize this source of electric energy for ignition purposes. That is, to do so would apparently require a large number of batteries connected in series, which would mean excessive cost, weight, and perhaps, considerable trouble. However, since the spark is needed in the cylinder at a certain instant only, and need not continue flowing for any length of time, even for a second or less, the source of the electrical energy is not required to supply a steady flow of current. Therefore it is possible to utilize the principle of electromagnetic induction, as previously ex-

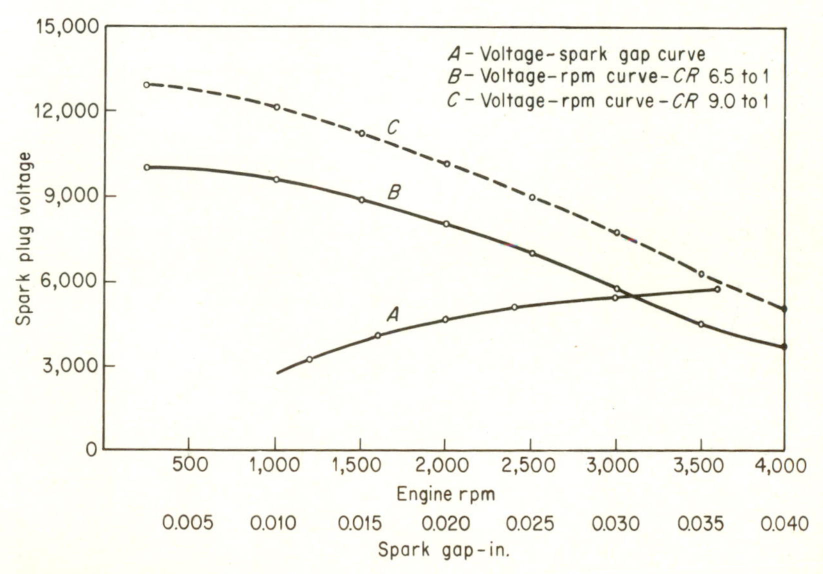

Figure 14-1 Curves showing relationship of spark-plug voltage to gap size and compression ratio under ideal conditions.

plained, to increase or "step up" the low voltage of an ordinary lead-acid battery to the desired degree.

BATTERY IGNITION

There are several types of battery ignition systems used on tractors, multiple-cylinder power units, small stationary farm engines, and light-weight high-speed engines. These include:

1 Conventional breaker-point system
2 Transistorized-assist breaker-point system
3 Magnetic-pulse ignition system
4 Capacitor discharge system

All these systems utilize the principles of electromagnetic induction to increase the battery voltage to the 25,000 to 40,000 V required to jump the spark-plug gap in the cylinder. Two electric circuits, the primary and secondary, make up all ignition systems.

Conventional Breaker-Point Ignition System

A schematic diagram of a conventional breaker-point ignition system, as shown in Fig. 14-2, has a primary and secondary circuit. The primary circuits contain the battery, switch, primary coil winding, breaker mechanism, and condenser. The secondary circuit includes the secondary coil winding, distributor, and spark plugs. The system operates as follows: When the cam rotation allows the breaker points to close, current from the battery flows in the primary winding of the coil, establishing an elec-

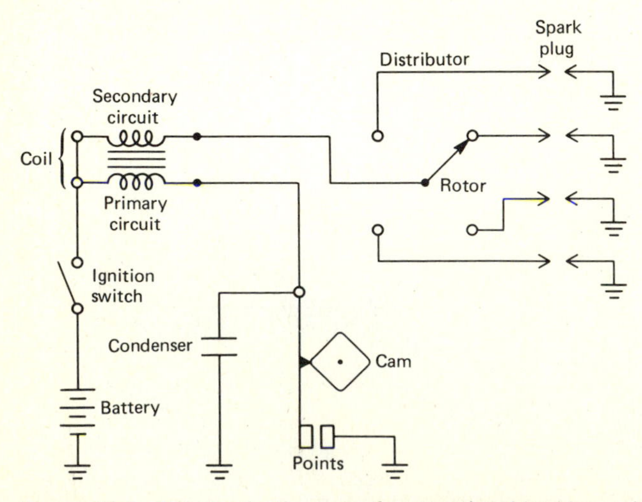

Figure 14-2 Schematic diagram of conventional breaker-point ignition system.

tromagnetic field. When the points open, the electromagnetic field collapses, inducing a high voltage in the secondary winding of the coil. This voltage is conducted to the appropriate spark plug, where it ignites the fuel-air mixture in the cylinder. Voltage produced is a function of both the strength of the magnetic field, the number of turns of wire in the coil, and the rate at which it collapses.

Coil The coil is made up of a soft-iron core consisting of a bundle of soft-iron metal strips, a primary winding of about No. 18 insulated copper wire, and a secondary winding consisting of many thousands of turns of very fine insulated copper wire (Fig. 14-3). In the manufacture of such a coil, the wire itself that makes up the winding is well insulated. The different turns of wire are insulated from each other, and the layers forming each winding are well insulated one from the other. This thorough insulation is necessary in order to prevent the high voltage generated in the coil winding from breaking through and creating an internal short circuit. In addition to thorough insulation, the coil is surrounded by a moistureproof and waterproof compound and encased in a metal casing. The ends of the windings are brought through the case and fastened to the terminal pieces or posts.

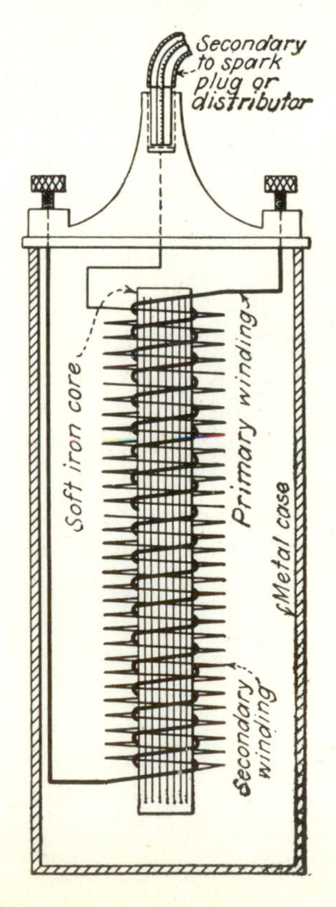

Figure 14-3 Construction of a typical coil for an ignition system.

The operation of an ignition coil is as follows: When the battery is connected to the primary circuit by the ignition switch (Fig. 14-2), the electric energy flows through the primary coil and forms an electromagnetic field, as previously described. If the flow of electric energy from the battery is suddenly stopped, the electromagnetic field collapses very quickly, thus inducing a high voltage electric charge in the secondary winding, and also in the primary winding, since they are both in the same magnetic field. It is not necessary that the secondary winding have an actual electrical connection with the primary winding in order to have electric energy induced in it by the breakdown of the magnetic field. Since the secondary winding consists of many more turns than the primary winding, the voltage induced is very high, and the electric energy flows across the spark-plug gap and produces a spark. Therefore, if a coil of this kind can be connected to a battery and to the spark plug of the engine, and if some means can be devised by which the primary circuit can be broken just at the instant the spark is needed in the cylinder, the fuel charge will be ignited.

Breaker Mechanism The device that is used to interrupt or break the primary circuit is known as a breaker or interrupter mechanism. It consists essentially of three parts: (1) a movable insulated point, (2) a stationary grounded point, and (3) a rotating cam. A lobe on the cam strikes the rub bar on the movable point and separates it from the stationary point at the instant the spark is desired. The breaker mechanism is usually a rather small device with a spring that holds the movable point against the stationary one when contact is desired. The contact points themselves are usually made of some hard and highly heat-resistant material such as a tungsten alloy. The breaker cam for a four-stroke-cycle engine, with any number of cylinders, will rotate at camshaft speed or one-half crankshaft speed. The points must open once in two revolutions of the crankshaft to produce the correct number of sparks for a single-cylinder, four-stroke-cycle engine. Therefore, if the cam is properly timed, and if it opens the points at the right time in the compression stroke, the ignition system will function correctly.

For multiple-cylinder engines, the breaker cam as shown in Fig. 14-4, has as many lobes or projections as there are cylinders, and opens the points that number of times per revolution. The advancing or retarding of the spark is usually produced by shifting the cam with respect to the breaker points. The points are located on the breaker plate inside the distributor housing as shown in Figs. 14-4 and 14-5.

Condenser When the breaker points open, a current is induced in the primary as well as in the secondary winding, and a spark will occur at the points. Since this sparking will soon pit or roughen the points and

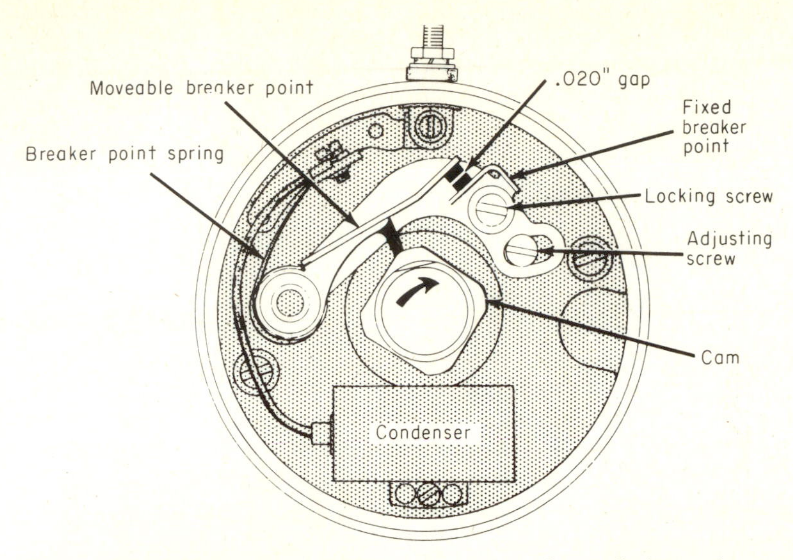

Figure 14-4 Breaker-point ignition assembly for a four-cylinder engine.

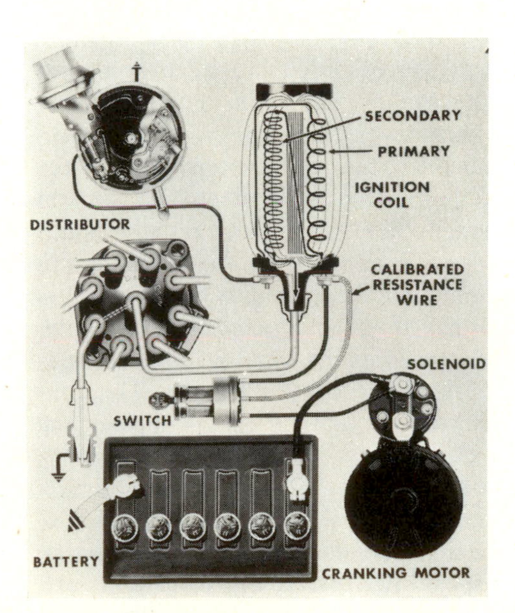

Figure 14-5 Typical ignition system for an eight-cylinder engine. The coil is shown schematically with magnetic lines of force indicated, distributor in top view with its cap removed and placed below it. (Courtesy of Delco-Remy Division, General Motors Corporation.)

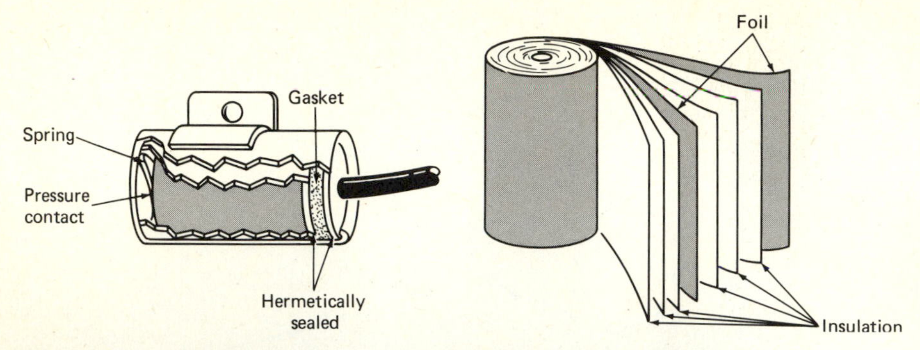

Figure 14-6 Condenser assembled in case and with insulation and aluminum foil unrolled. (Courtesy of Delco-Remy Division, General Motors Corporation.)

eventually render them inefficient or inoperative, it is desirable that it be eliminated. This is done by means of a condenser (also called a capacitor), which consists of two layers of tinfoil separated and insulated from each other by waxed paper. As shown in Fig. 14-6, it is not possible for an electric current to flow through a condenser as it would through a wire or a coil. The condenser is connected in the circuit as shown in Figs. 14-2 and 14-4. That is, it might be said that it is connected across the breaker points or bridges them.

The condenser functions as follows: The current induced in the primary circuit at the time of separation of the points, instead of dissipating itself in the form of a spark at the points, surges into the condenser but not through it, then surges outward into the circuit again in the opposite direction through the primary coil, then reverses again and surges backward and continues until it finally dies out. This action not only eliminates the arcing almost entirely, but in addition, helps to demagnetize the iron core and break down the electromagnetic field much more rapidly than it would be, without the condenser. Consequently, the current induced in the secondary circuit is of a much higher voltage, and a better spark is produced. In other words, we might say that the condenser protects the points by eliminating the arcing and intensifies the voltage in the secondary winding, producing a better spark at the spark-plug gap. The condenser is normally located on the breaker plate (Fig. 14-4) near the breaker-point mechanism.

Distributor A distributor (Fig. 14-7) is necessary in any multiple-cylinder ignition system to distribute the secondary voltage to each spark plug. The entire mechanism consists of two parts, the rotor and (distributor) cap, with its contact and terminals for the wires to the spark plugs. The rotor is always located just above the breaker cam and is turned by the same gear and shaft. As the rotor rotates a metal spring and strip (on the rotor) connect the center terminal of the cap with each

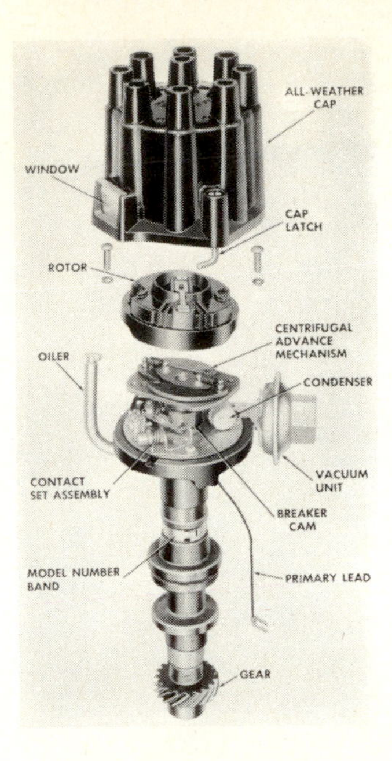

Figure 14-7 Complete breaker-point and distributor assembly. (Courtesy of Delco-Remy Division, General Motors Corporation.)

outside terminal in turn so that the high voltage surges from the coil are directed to each spark plug according to the firing order. The cap also serves as a cover to protect the rotor, points, and condenser from dirt, dust, moisture, and oil accumulation.

Spark Plugs The spark plug is the device used in spark-ignition systems that provides the gap for the high voltage electric surge to jump and ignite the charge in the cylinders. Numerous types and styles of spark plugs are manufactured for many different types of engines that use an electric ignition system. The spark plug (Fig. 14-8) is made up of the following parts: the steel outer shield, which is threaded and screws into a tapped hole in the cylinder head; the insulator, usually made of porcelain; the copper-glass gaskets that provide the seal between the insulator and the shell to prevent loss of compression and leakage; the electrodes or points; and the terminal.

The insulator is the most important part of the spark plug and must be made of high-grade material to withstand the high voltage and prevent leakage of the electric surge from the coil by a short circuit. It must also withstand the high temperatures and pressures of the combustion chamber. Porcelain is used almost exclusively as the insulating material.

The spark-plug electrodes must be made of material that will withstand high temperatures and will not be burned away rapidly by the arcing and heat to which they are subjected. They are usually made of

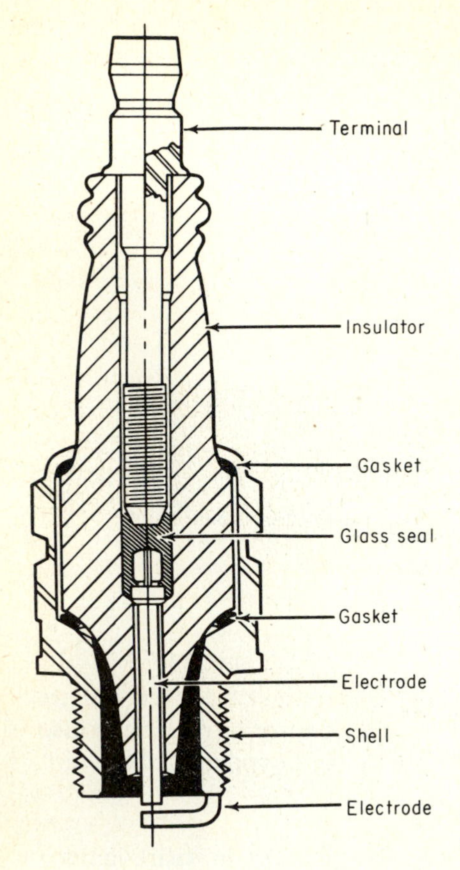

Figure 14-8 Typical spark-plug construction.

some special nickel or tungsten or chromium alloy, with a small percentage of iron. The gap at the electrode is very important in the effective and efficient operation of the spark plug in an ignition system. In general the width of the gap depends upon the engine design and conditions under which the engine operates, but will vary from 0.020 to 0.080 in (0.508 to 2.03 mm).

Size and Reach of Spark Plugs The size of spark plugs is determined by the diameter of the thread end of the shell. There are five common sizes: the ⅞ in (2.22 cm) SAE plug, and four sizes of metric plugs. The ⅞ in (2.22 cm) SAE plug has a thread similar to that used on a ⅞ in (2.22 cm) bolt, with 18 thread/in (2.54 cm). The metric size of plugs are 10, 12, 14, and 18 mm.

Operating conditions often determine the size plugs used by engine-design engineers. Plugs installed in high-abuse applications require more breathing area. For this reason, the 18 mm plug is used in tractors, snowmobiles, and industrial power units. The 14 mm is frequently used in most four-stroke-cycle automobiles, lawn mowers, snowblowers, and power saws. The 10 and 12 mm plugs are used where space is a critical factor as in motorcycles and racing engines.

The "reach" of a spark plug is the distance from the gasket seat to the end of the threads. Different reaches are necessary because of the variation in cylinder head design and thickness. The most common reaches are $^3/_8$ (9.5), $^7/_{16}$ (11.1), $^1/_2$ (12.7), and $^3/_4$ in (19.1 mm). Many aluminum engines use the longer reach plugs, $^1/_2$ in (12.7 mm), and $^3/_4$ in (19.1 mm), to assure a better and stronger head fit. It is extremely important to install plugs with the reach specified by the engine manufacturer in order to prevent severe engine damage.

Spark Plug Heat Range Improvements in design and the use of higher compression ratios and faster burning fuels have resulted in higher combustion-chamber temperatures in automotive and tractor engines. Efficient operation of these engines depends greatly upon the temperature of the spark plugs. If the plug dissipates heat rapidly and remains too cool, combustion is likely to be sluggish and the plug might become fouled. If the plug dissipates heat slowly and remains too hot, preignition and knocking might occur and the points can be burned away rapidly. For this reason, not only must any engine use a certain size and reach of plug, but it must be of the correct type as related to heat control. The operating temperature of a spark plug is determined by the distance the heat must travel, as shown in Fig. 14-9. The greater this distance, the higher the plug temperature, the shorter this distance, the cooler the plug operates.

In general, high-compression engines operating at high speeds and heavy loads , require "cold" plugs, and engines operating at low speeds and light loads for short periods of time use "hot" plugs. The selection of the correct type of spark plug from the standpoint of heat range for any engine is very important.

As observed, the heat travels from the combustion chamber through the spark plug and is transmitted to the head at the plug shoulder, where it seats (Fig. 14-9). Most plugs use a copper gasket as a seal at this point to

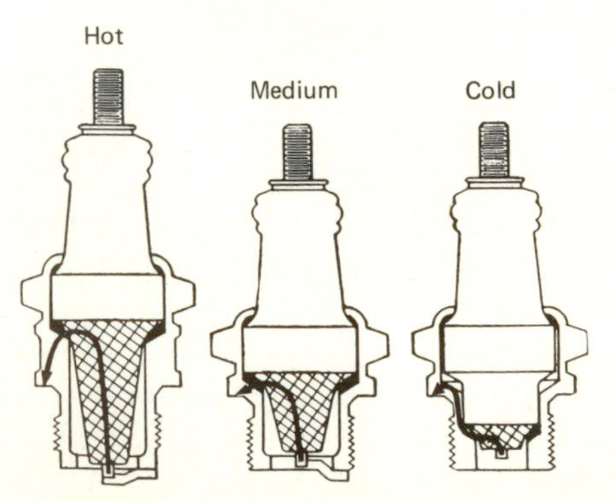

Figure 14-9 Spark-plug construction control heat flow.

improve heat dissipation. However, some engines are equipped with spark plugs having a beveled shoulder, which bears directly on the beveled ground seat in the cylinder head. Thus, the gasket is eliminated without any reduction in heat dissipation.

Transistorized-Assist Breaker-Point System

Arcing, which causes the points to burn, and wearing of the rub block on the movable point that changes ignition timing, are the two most common maintenance problems of the conventional ignition system. An early attempt to overcome arcing of the point was the addition of a power transistor in the primary circuit to minimize the current carried by the breaker points. With this system, the points turn the transistor on and off and carry only about 300 mA. The majority of the coil primary current is switched by the power transistor. Larger primary currents can be used without excessive point burning, thus producing larger secondary voltages than these in a conventional system. Obviously, coils for use with this system must be designed to handle higher primary currents. No condenser is required in this system.

Magnetic-Pulse Ignition System

This system utilizes a magnetic-pulse distributor, which outwardly resembles a standard unit, and is mounted and driven in the same way as a conventional distributor. The cam, breaker points, and condenser are eliminated and replaced by a magnetic-impulse generator, which serves the same function. The magnetic pickup coil (Fig. 14-10A), which provides the signals to the amplifier (Fig. 14-11), contains a permanent magnet with a pole-piece on top of it. The pole-piece has the same number of teeth pointing inward as there are cylinders in the engine. There is a pickup coil containing many turns of wire inside the permanent magnet under the teeth on the pole-piece.

An iron timer-core, which has the same number of teeth as cylinders in the engine, is mounted on the distributor shaft. When the shaft rotates one revolution, the teeth on the timer-core and the pole-piece align eight times (for an eight-cylinder engine). Each time the teeth align, a magnetic field is established through the pickup coil (Fig. 14-10). Further shaft rotation moves the teeth apart, thus breaking the magnetic path that causes the magnetic field to quickly collapse through the coil. The moving magnetic field induces a voltage in the pickup coil that sends a signal to the amplifier (Figs. 14-11 and 14-12A). This pulse triggers the transistorized amplifier, which interrupts the flow of electric current in the primary circuit, thus producing the same effect as the opening of the breaker points.

The operation of the pulse amplifier is shown in Figs. 14-12A and 14-12B. The amplifier consists of two circuits separated by capacitor C-1. When the ignition switch is closed, a voltage bias is put on transistor TR-2

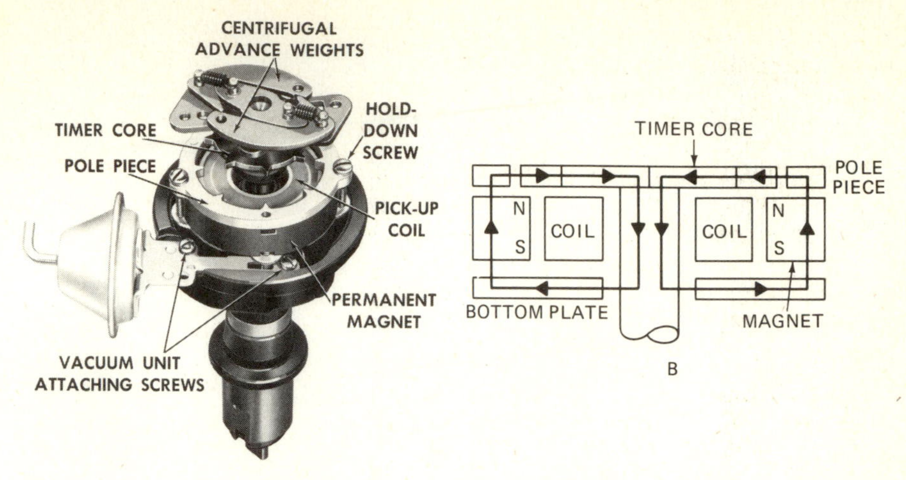

Figure 14-10 Magnetic-impulse generator shown complete (A) and in cross section (B). (Courtesy of Delco-Remy Division, General Motors Corporation.)

and causes current to flow from the battery through resistors R-7 and R-1, through emitter base circuit of TR-2, and through R-2 to the ground. The current flow also triggers TR-2 and flows through the emitter collector circuit of TR-2 and on to the ground through R-5. The electrical potential between R-7 and R-1, for a 12-V system, is about 8 V and slightly lower after R-1 and even lower after the emitter-base circuit of TR-2. The left side of capacitor C-1 also has a charged voltage of about 8 V.

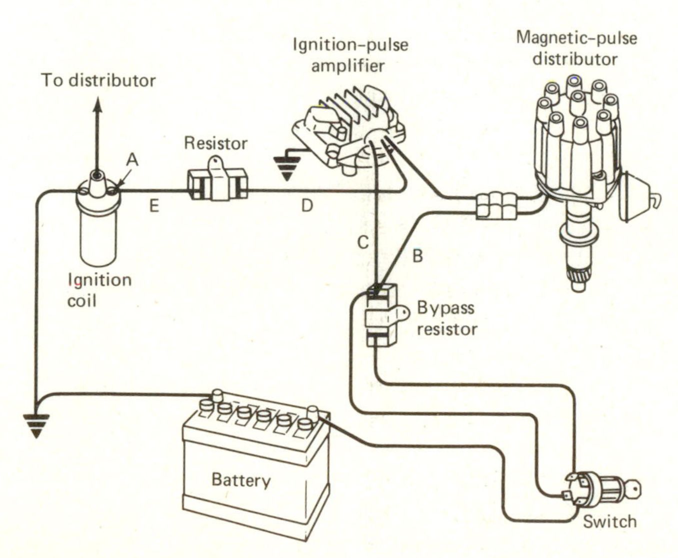

Figure 14-11 Wiring diagram of magnetic-pulse ignition system. (Courtesy of Delco-Remy Division, General Motors Corporation.)

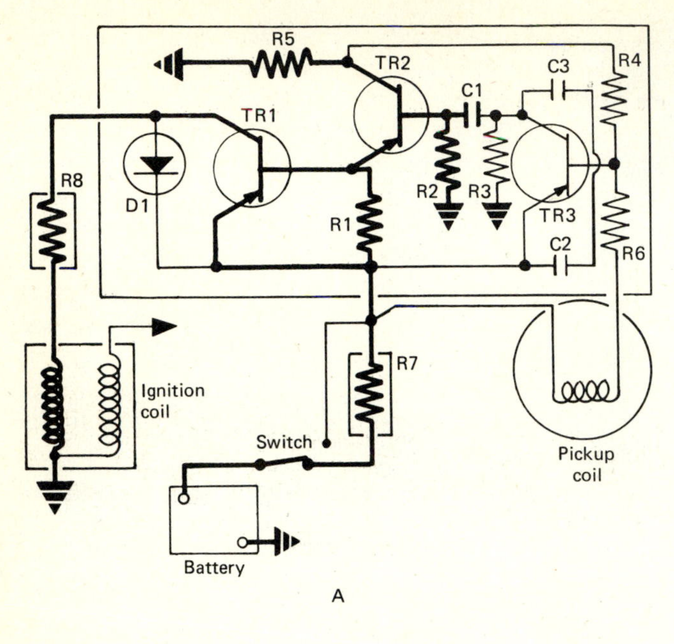

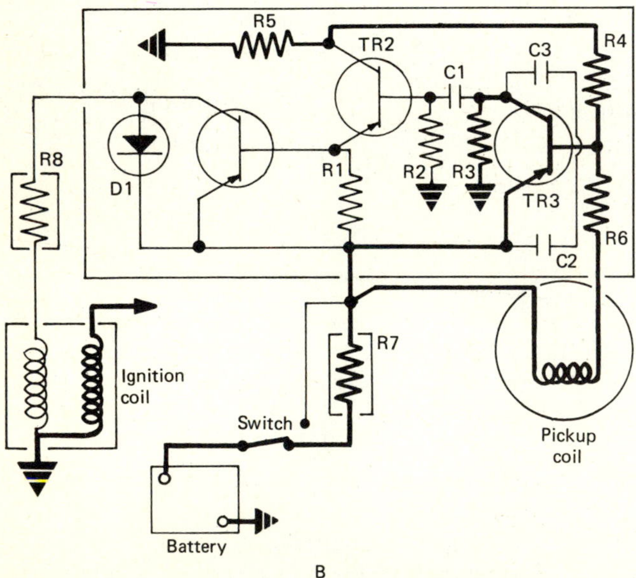

Figure 14-12 (A) Electrical flow shown in heavy lines from battery through ignition-pulse amplifier and coil during build up of electromagnetic field in primary circuit. (B) Electrical flow shown in heavy lines when pickup coil signals ignition-pulse amplifier to turn off the flow of electricity in the primary circuit thus producing a high voltage surge in the secondary circuit by electromagnetic induction. (Courtesy of Delco-Remy Division, General Motors Corporation.)

The voltage bias across the emitter-base circuit of TR-1, because of the voltage drop across R-1, triggers TR-1, thereby causing current to flow from the battery through R-7, through emitter-collector of TR-1, through R-8 and the primary winding of the ignition coil and on to the ground. This builds up an electromagnetic field in the ignition field.

When it is time for a spark plug to fire, a signal from the pickup coil (Fig. 14-12), causes transistor TR-3 to become conductive by supplying the transistor base with current carriers. With TR-3 conductive, current carriers are drained away from the base of TR-2 causing it to become a nonconductor. With no current flowing through TR-2, there is a reduced voltage drop across resistor R-1, and the base of TR-1 becomes approximately the same voltage as the lower TR-1 connection. Thus transistor TR-1 becomes, in effect, a diode that stops the current flowing from the battery through TR-1 to the coil primary winding. With the stoppage of current, the electromagnetic field collapses, inducing a high voltage in the ignition coil secondary winding.

In this type of system there are no breaker points to adjust or wear out. Everything is done electronically. The only adjustment required is timing the ignition system to the engine.

Within the last few years, General Motors has introduced the High-Energy Ignition System (H.E.I.). The ignition coil for this unit is built into the distributor cap (Fig. 14-13) and it produces voltage of up to 35,000 V. The wiring is simplified with only one lead from the battery. The only other leads are the high-tension cables that go to the individual spark plugs from the distributor. This ignition system cannot use standard spark plugs since it requires a wider gap—as much as 0.080 in (0.20 cm). Special plugs are made for this system.

Capacitor Discharge System

The capacitor-discharge ignition system is another variation of the conventional breaker-point system in which additional steps are taken to increase the voltage across the spark plugs. A transistorized oscillator power circuit increases battery voltage to 250–350 V of direct circuit, which is stored on a capacitor (condenser) (Fig. 14-14). A timing device, cam-actuated breaker points, discharges the capacitor into the primary winding of the coil, which is a pulse transformer. The higher voltage from the capacitor produces a strong electromagnetic flux field, which rises sharply for a longer period of time than that in a conventional system, thus inducing higher current in the secondary. The extremely fast rate at which the voltage rises at the spark plug (about 100 times faster than other systems) and the longer duration of the arc, offers the best ability to fire partially fouled spark plugs. The spark plug fires as the magnetic field is rising rather than collapsing as in other systems.

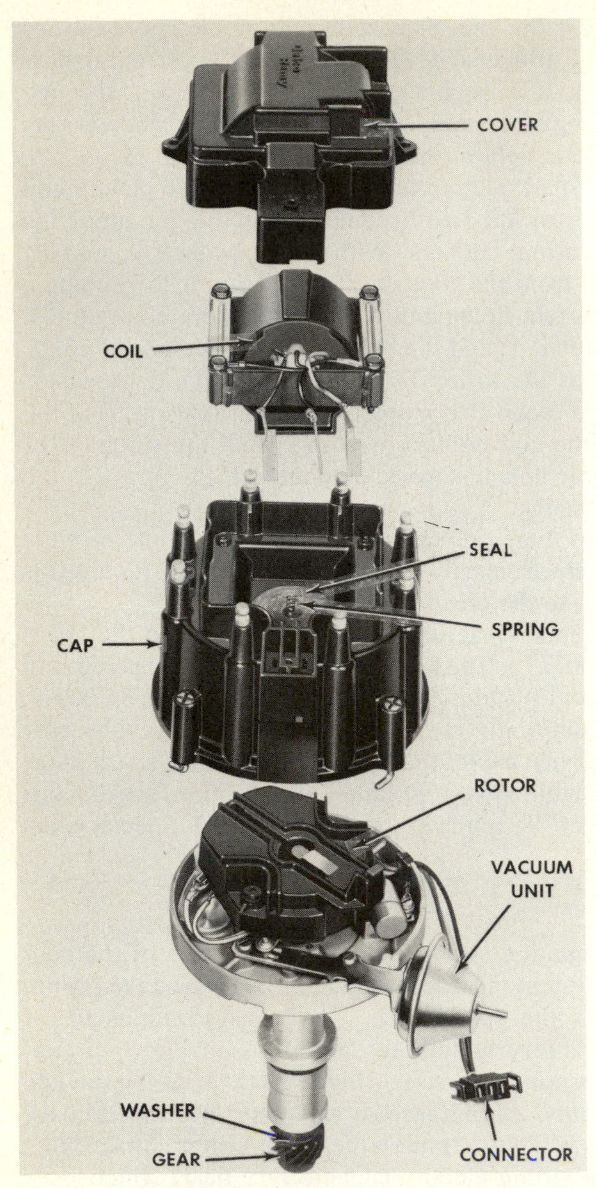

Figure 14-13 High-energy distribution. (Courtesy of Delco-Remy Division, General Motors Corporation.)

A magnetic-impulse generator could replace the cam and points as a timing device to discharge the capacitor. Another variation of this system uses a light pulse received by a photocell to trigger the discharge of the capacitor into the primary coil. The light-pulse system could operate down to zero speed and has no theoretical upper speed limit.

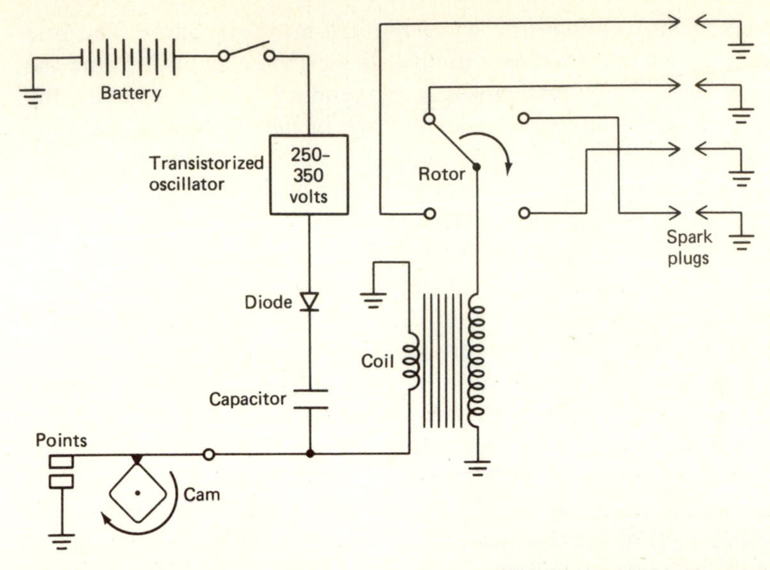

Figure 14-14 Schematic diagram of capacitor-discharge battery-ignition system.

Complete Voltage Picture

We have described what happens in the ignition system to produce the high-voltage surge required to fire one spark plug. A light beam picture or pattern of the electrical impulses that occur in an ignition system can be produced by an oscilloscope. Let us now look at a typical oscilloscope pattern showing the complete voltages in the secondary circuit during the production of one spark (Fig. 14-15) in a conventional breaker-point ignition system.

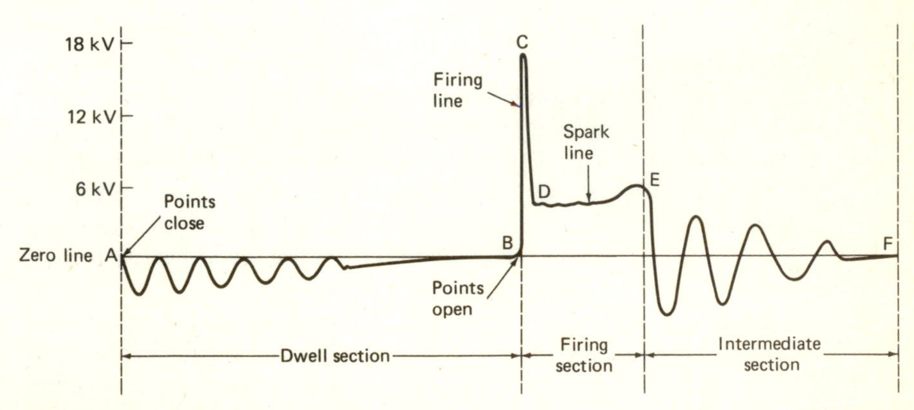

Figure 14-15 Pattern or trace of the voltage in a secondary circuit for one complete spark-plug firing cycle.

It is convenient to divide the wave pattern into four parts. The first section is *A* to *B*, which represents the dwell section or length of time the points are closed. The points close at *A*, sending current through the primary winding and inducing a small voltage in the secondary winding. This voltage oscillates and decreases in value until, at point *B*, the secondary voltage is zero. A magnetic field builds up in the coil's primary circuit as the points remain closed from *A* to *B*.

The points open at *B*, and the secondary voltage immediately increases to *C*. This is the voltage required to start the spark at the spark plug. *B* to *C* is called the firing line. After the arc starts between the electrodes in the spark plug, less voltage is required to maintain current flow than was required to start it initially, thereby the voltage drops down to *D*. The spark continues from *D* to *E* and is called the spark line. The duration of line *D* to *E* may be .002 s for a typical eight-cylinder engine operating at 3,000 rpm or about 20° rotation of the crankshaft. At E the spark terminates because of insufficient energy in the coil, but there is still some energy left in the coil, which is dissipated throughout the circuit in the form of a damped oscillation between *EF*. The alternating voltage between points *E* and *F* is due to the inductive-condenser effect of the coil and distributor condenser. The points again close at *F* and the cycle repeats to fire the next spark plug. A complete cycle would require about .005 s for an eight-cylinder engine at 3,000 rpm.

Variations from the normal pattern are usually an indication of trouble in the ignition system. Most oscilloscopes can be adjusted to show the wave forms of all cylinders side by side, thus providing a comparison between cylinders, or show wave form for individual cylinders.

Ignition Timing One of the more important factors in the correct operation of any internal-combustion spark ignition engine is the time ignition occurs. This is particularly true in multiple-cylinder and variable-speed engines. That is, the combustible mixture of fuel and air must be ignited so that the piston will receive the pressure of the burning fuel-air mixture as it completes the compression stroke and is ready to start on the power or working stroke. If the spark comes too early, ignition will take place before the piston reaches the end of the compression stroke, and the result will be a knocking effect and appreciable loss of power. If the spark occurs too late, ignition will be delayed and the maximum effect from the burning mixture will not be received by the piston, resulting in considerable power loss and overheating.

Obviously, the time for the development of maximum pressure in the combustion chamber is determined by the time of occurrence of the spark. Furthermore, it should be kept in mind that the occurrence of the spark and the development of maximum pressure from the burning mixture are two distinct and separate actions and are not simultaneous as it might appear. This, perhaps, can be best understood by bearing in mind that the fuel mixture when ignited by the spark is "set on fire," so to speak, at one point only. Before maximum pressure can be developed, the flame front must spread throughout the entire mixture. A very short, but nevertheless definite, period of time is required for the propagation of the flame front through the mixture. The rate of travel is dependent on factors such as the type of fuel used, the compression pressure in pounds per square inch, the load on the engine, the speed, the cylinder temperature, and fuel-air ratio.

In order to provide for this lapse of time between the ignition of the mixture and its complete combustion by the time the piston has reached the end of the compression stroke, the spark is advanced as illustrated by Fig. 14-16; that is, it is made to occur before the piston reaches the end of the stroke by an amount necessary to give the best results. Spark advance is usually designated in degrees of crankshaft travel; that is, an advance of 30° means that the spark occurs when the crankshaft still lacks 30° of being on compression dead center. A late spark is known as a *retarded spark* and usually occurs on, or very near, compression dead center (Fig.

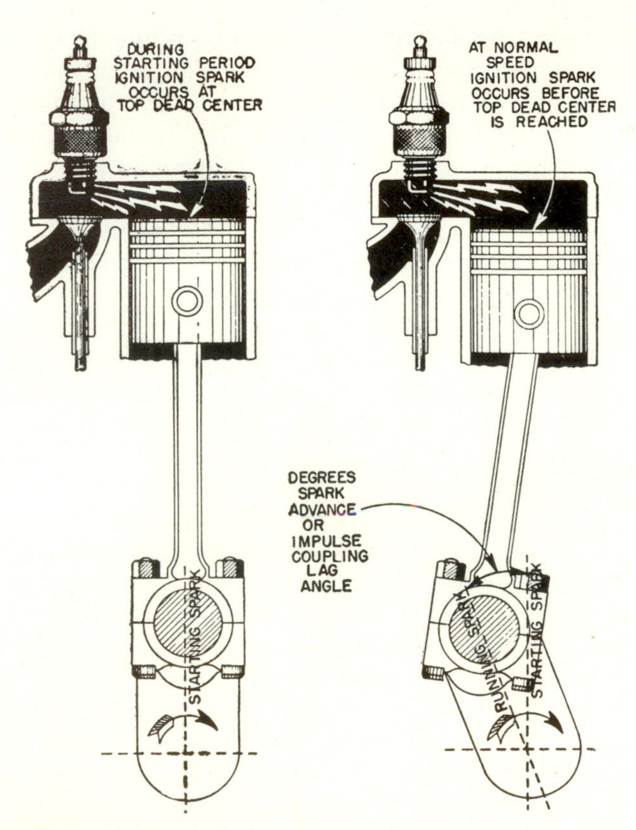

Figure 14-16 Spark timing for starting and running.

14-16). A retarded spark is for cranking the engine only, to prevent its kicking back. As soon as the engine is started and reaches its normal speed, the spark should be advanced the necessary amount to allow for smooth operation.

As previously stated, the amount of spark advance varies with a number of conditions, the most important of which are engine crankshaft speed, kind of fuel, quality of fuel mixture, compression pressure, fuel-air ratio, and cylinder temperature. It is quite evident that with a given fuel and compression pressure, the greater the speed of the engine, the greater the spark advance must be up to a certain point. That is, if the fuel mixture requires a certain fraction of time to be ignited and burned, the piston will travel farther in a high-speed engine during this brief interval than it will in a low-speed engine. Again a lean fuel mixture, or a fuel mixture at a low or medium pressure, will not burn as rapidly as a dense mixture or one under higher compression. Consequently, low-compression engines running under light loads require more spark advance than those operating at higher compression pressures and heavy loads.

Spark-Timing Mechanisms In the spark ignition system, the time at which the spark occurs is determined by the time at which the breaker points separate. As seen in Fig. 14-7, these points and the distributor rotor are located on the same shaft, which is gear-driven by the camshaft. Usually some provision is made for adjusting the spark timing by slightly rotating the distributor head and breaker points about the breaker cam and shaft. Also, the gap setting of the points can affect the timing of the spark. Therefore, they should always be adjusted for the recommended gap setting before the spark timing is checked or changed.

Ignition-Timing Procedure Most engine manufacturers recommend the use of a timing light for checking and adjusting the spark setting for high-speed engines. A mark on either the flywheel or the crankshaft fan-belt pulley of the engine indicates the idle-speed spark setting. When the engine is running at recommended low idle rpm, the timing light should flash just as the mark passes a stationary pointer or indicator. If this idle-speed setting is correct, the operating-speed spark setting will be taken care of by the automatic spark-advance mechanism.

Automatic Spark Control Multiple-cylinder engines using battery ignition are usually equipped so that the spark timing is controlled automatically according to the engine speed and other operating conditions. The most common arrangement consists of a pair of weights held together by two springs and mounted on the distributor shaft just under the breaker points. The weights react against the advance cam, which is a part of the breaker cam, and cause the latter to rotate in the direction it normally turns. At low speeds, the springs hold the weights together as shown in

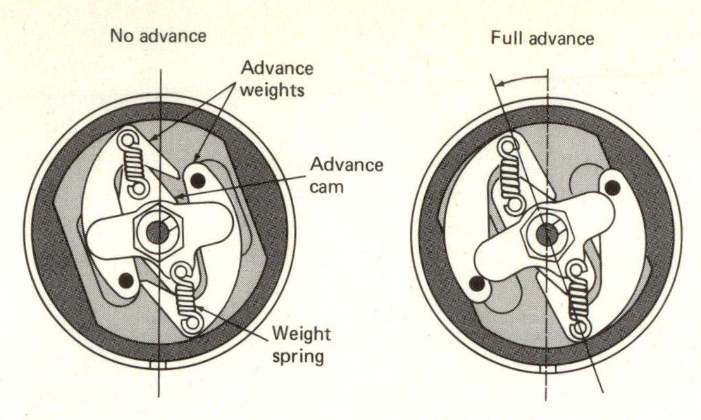

Figure 14-17 Centrifugal weight method of spark advance.

Fig. 14-17, so that there is no spark advance. As the speed increases, centrifugal force causes the weights to expand and act on the advance cam, which in turn rotates the breaker cam and thereby opens the breaker points earlier, thus creating an advanced spark.

Tractor engines using battery ignition are usually equipped only with the centrifugal automatic spark-control device. However, automobile and truck engines use a combination of centrifugal and vacuum control. The vacuum mechanism is actuated by manifold suction and advances or retards the spark according to the engine load as determined by the throttle-butterfly position. A spring-loaded diaphragm is linked to the breaker-point plate and rotates the points with respect to the cam. Figure 14-18 shows how these two types of spark-control mechanisms function with respect to each other and according to engine speed.

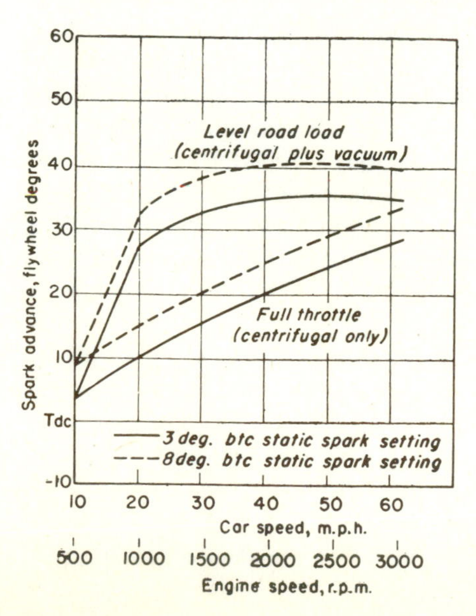

Figure 14-18 Spark timing for automobile engine equipped with centrifugal and vacuum control.

MAGNETO IGNITION

Mechanical devices for generating electric energy are termed either electric generator or magnetos. Such devices require a magnetic field which, as we have seen, can be produced either by a permanent magnet or by means of an electromagnet. Magnetos are distinguished from other electric generators in that they utilize permanent steel magnets for providing the magnetic field.

Use of Magnetos for Ignition The principal use of magnetos is for the mechanical generation of electricity for spark ignition in internal-combustion engines. For many years magnetos served as the means of ignition for tractor engines. Magnetos are still used on some stationary farm engines, larger multiple-cylinder power units; and many small, lightweight, high-speed engines.

Magneto Types and Construction Magnetos for ignition purposes may be classified as high-tension conventional or solid-state capacitor discharged, and of either unit- or flywheel-type construction. A high-tension magneto generates a very high-voltage current such as is needed for a spark-plug ignition system. Most unit-type magnetos are gear-driven, and the armature or rotor rotates at some specific speed, usually crankshaft speed. The conventional magneto is a complete, self-contained ignition device, with the exception of the spark plugs, and may be readily attached or detached from the engine; however, many small, single-cylinder, high-speed engines have the magneto built into the flywheel in such a manner that certain essential parts are attached to it and rotate with it. This arrangement provides simplicity and compactness.

Although there is considerable variation in the detailed construction of the different makes and types of magnetos, the basic construction and design of all magnetos are essentially the same. The fundamental parts are (1) a base and frame structure of some nonmagnetic material, usually an aluminum alloy; (2) a permanent alloy steel magnet or magnets; (3) soft-iron field or pole-pieces; (4) an armature or rotating member; (5) a winding or windings; (6) breaker points and condenser; (7) a distributor with gears and rotor (if it is a multiple-cylinder engine).

Permanent Steel Magnets Magneto magnets must be made of a steel alloy because pure soft iron will not retain magnetism. An aluminum-nickel-cobalt alloy known as Alnico has been developed and is being used extensively. Its advantages are greater magnetic energy per volume of material and greater retentivity or stability. It is claimed that Alnico is highly resistant to such demagnetizing forces as stray fields of opposite polarity, high temperatures, and excessive mechanical vibration.

Inductor-Type Magnetos Most present-day magnetos are so constructed that the windings are fixed or stationary, and the magnets are usually a part of the armature and rotate. Such magnetos are termed inductor types. Figure 14-19 shows the general construction and operation of such a magneto. The winding and its soft-iron core (Fig. 14-20) are mounted on the upper ends of the pole-pieces. Since the permanent magnet is a part of the rotor, the magnetic flux or field is forced to change its path of flow through the pole-pieces and coil-core during rotation as indicated. A change occurs every half revolution, and the maximum induced current impulse comes just at the instant the flux lines through the core are at their minimum value, as in Fig. 14-19. A complete high-tension magneto of the inductor type is shown by Fig. 14-21. All current models of magnetos use this type of construction.

Conventional Unit-Type Magneto

The important parts of a magneto are as follows: (1) nonmagnetic base and frame; (2) permanent steel magnets; (3) armature; (4) soft-iron pole-pieces; (5) primary winding; (6) secondary winding; (7) breaker points; (8) condenser; and (9) safety gap. For convenience and clarity, so that the construction and operation of the high-tension magneto will be readily understood, the electrical circuits and parts involved are shown by Fig. 14-21.

The operation of the magneto is as follows: As the armature rotates and passes through the position at which the lines of force change their path through it, a current is induced in the primary winding of coarse insulated wire. Just as this current is induced in the primary winding, the breaker points, which are in the primary circuit, are separated by a cam, interrupting the current and breaking down the strong electromagnetic field set up by it. This results in the instantaneous induction of a very high voltage in the secondary circuit, which consists of many thousands of turns of very fine wire. Therefore, if the breaker points separate at the correct time in the travel of the engine piston, and the secondary winding can be connected to the spark plug, a spark will be produced and will ignite the mixture.

Breaker Points The breaker points (Fig. 14-21) are similar to those used in battery-ignition systems. They consist of a stationary noninsulated point and a movable insulated point and arm that comes in contact with the cam. These points are located near one end of the armature, and the cam that separates them will rotate with the armature.

The contact points themselves are made of a high-grade alloy steel so that they will not pit or become rough too quickly. One point is usually adjustable so that the space between them, when they are separated, can be adjusted. This space should be 0.015 to 0.020 in (0.038 to 0.051 cm).

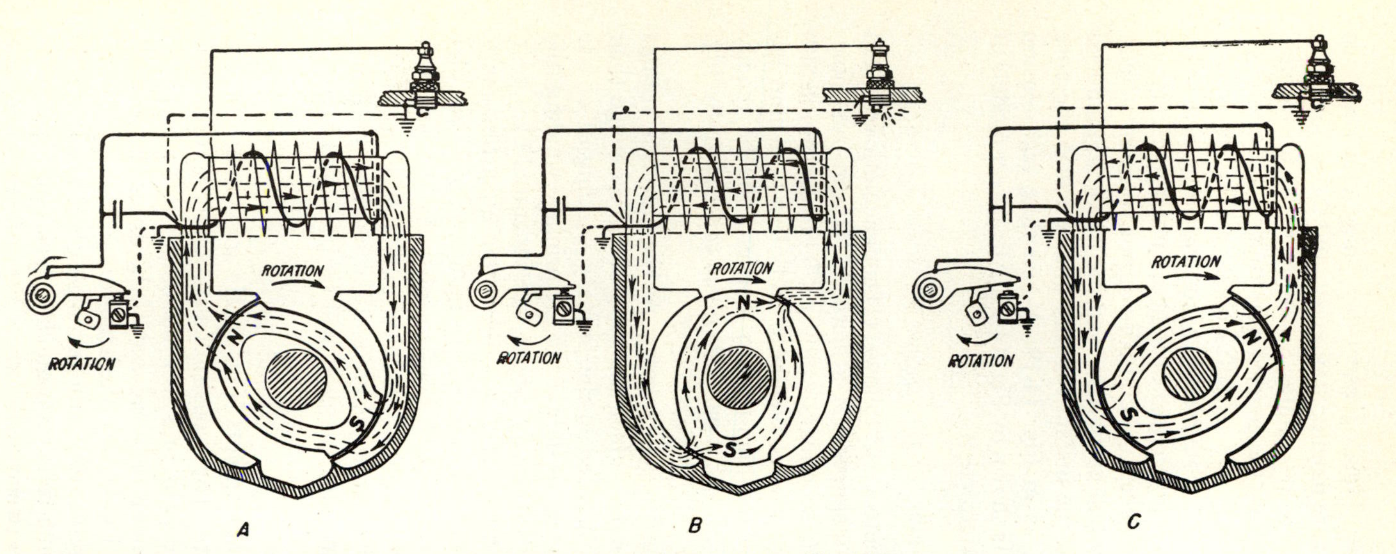

Figure 14-19 Generation of electrical impulse by an inductor-type magneto. (Courtesy of International Harvester Company.)

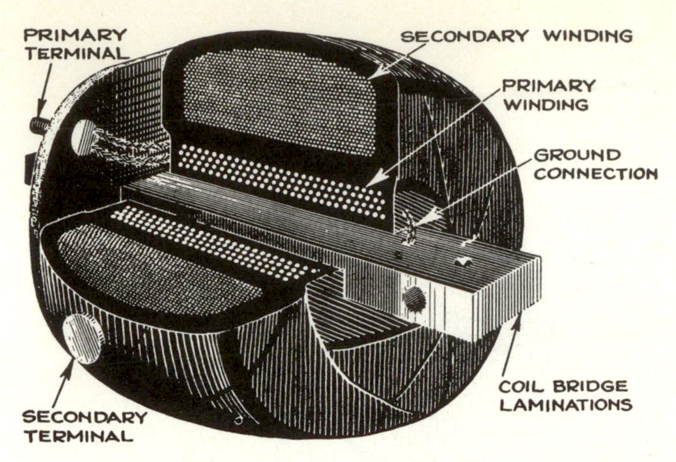

Figure 14-20 Core and windings for a magneto. (Courtesy of Fairbanks Morse and Company.)

Condenser The magneto must be equipped with a condenser (similar to that used in a battery-ignition system), for the purpose of eliminating the arcing at the breaker points and intensifying the voltage in the secondary circuit. The condenser is connected in the primary circuit as shown in Fig. 14-21.

Safety-Spark Gap If the wire that carries the current from the magneto to the spark plug should become broken or disconnected at any point, it is obvious that the high-voltage current induced in the secondary winding could not take its regular path and, therefore, might dissipate itself by breaking through the insulation in the winding on the armature. This would create a permanent short circuit and, therefore, make the magneto useless until it could be rewound. To eliminate the possibility of such a breakdown in case of a broken secondary circuit, some unit-type magnetos are equipped with what is known as a safety-spark gap. This gap consists of a point fastened to that side of the secondary winding that leads to the spark plug or the collector ring, and another point exactly opposite the first and grounded to some metal part of the magneto. The gap must be about ¼ to ⅜ in (.67 to .95 cm) wide, which is somewhat greater than the spark-plug gap, otherwise the current would have a tendency to jump at this point at all times, rather than at the spark plug.

Distributor A magneto for a multiple-cylinder engine, in addition to the parts named, requires a distributor cap and rotor to distribute the secondary current to the different spark plugs in the correct order. This is akin to the battery-ignition system, which requires a distributor to distrib-

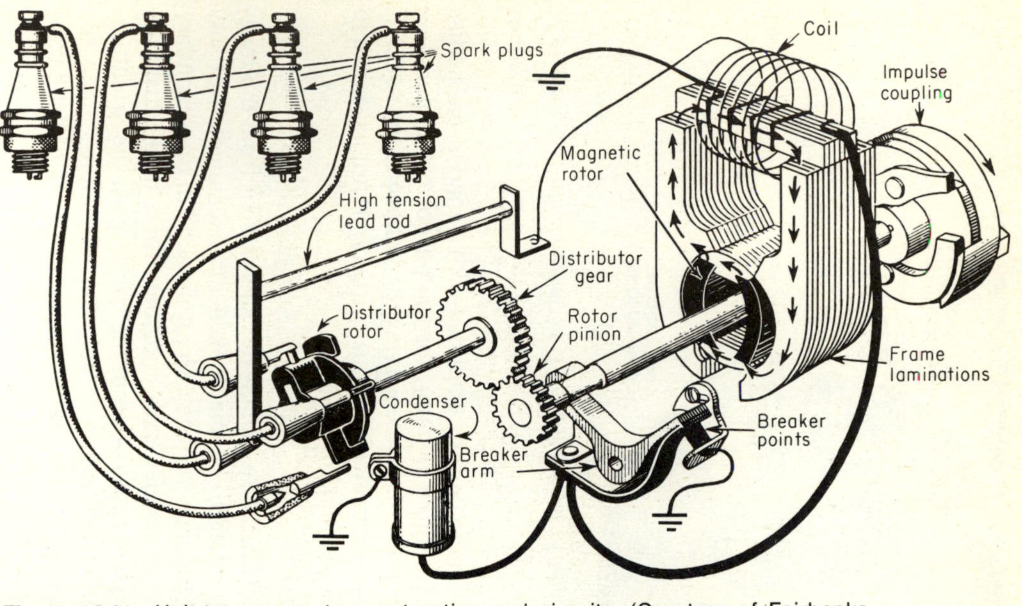

Figure 14-21 Unit-type magneto construction and circuits. (Courtesy of Fairbanks Morse and Company.)

ute the secondary circuit to the spark plugs in, for example, an engine with two or more cylinders.

Wires attached to the distributor terminals conduct the current to the spark plugs in the correct order. From these plugs the current returns through the engine to the base of the magneto, and thence to the grounded end of the secondary circuit. In other respects, the magneto is exactly the same as that used for a single-cylinder engine. The different makes of magnetos vary somewhat in detailed construction and operation, but the same general principles are followed in all makes. The greatest improvement in these has been the complete enclosure of the working parts, thus providing greater protection from dust, dirt, and moisture.

Flywheel Magnetos

Many small high-speed engines used on lawn mowers, small garden tractors, and motorboats have the magneto built into the flywheel. In a typical flywheel magneto (Fig. 14-22), the magnet is located in the outer rim of the flywheel, which revolves around the stationary coil, condenser, and breaker-point assembly. As the ends of the magnet pass by the polepieces, an alternating magnetic flux is created around the coil and a current is generated in the primary winding as long as the breaker points remain closed. At the instant the primary current is strongest, a cam on the crankshaft opens the points and instantly breaks the circuit and causes a collapse of the magnetic field. This induces a very high-voltage current in the secondary winding, which in turn results in a spark at the plug. This

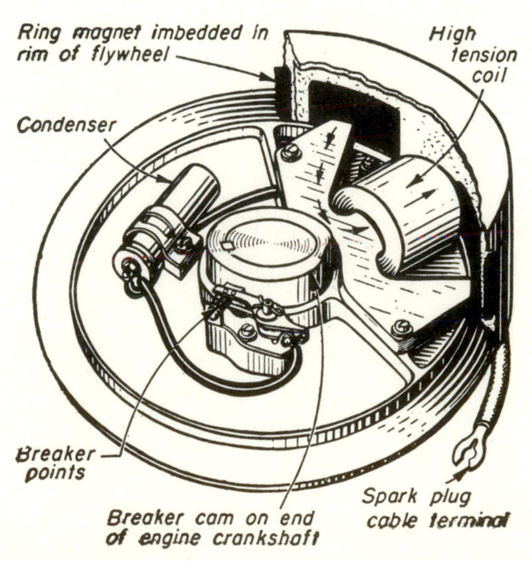

Figure 14-22 Construction of a flywheel-type magneto. (Courtesy of Fairbanks Morse and Company.)

type of magneto is very simple, has few moving parts, and requires very little attention or servicing.

Solid-State Magnetos

Electronics may be used to assist the breaker points, replace the condenser, cam, and breaker points, or increase the strength of the electric spark at the spark plug in magneto ignition systems. This is similar to the way it is used in battery ignition systems. Electronics that may be used in magneto ignition includes: (1) transistorized-assist breaker-point ignition system, (2) breakless transistorized ignition system, and (3) capacitor-discharged ignition system.

Solid-State Flywheel Magneto A capacitor-discharged electronic magneto ignition system is used on some small engines (Fig. 14-23). It is similar in one respect to flywheel magnetos (Fig. 14-22) in that the initial electric current is generated in a stationary coil by a moving magnet mounted on the flywheel.

Alternating current from the generating coil (Fig. 14-23) flows through a diode rectifier, where it is changed to direct current. Then the direct current flows to a capacitor where it is stored momentarily. The current remains stored in the capacitor until the flywheel rotates on half revolution. At this time the magnet on the flywheel passes the triggering coil and a small amount of current is generated in the coil. This current flows through a resistor to the transistor, which completes the circuit, allowing the charge on the capacitor to flow through the ignition coil. Electric

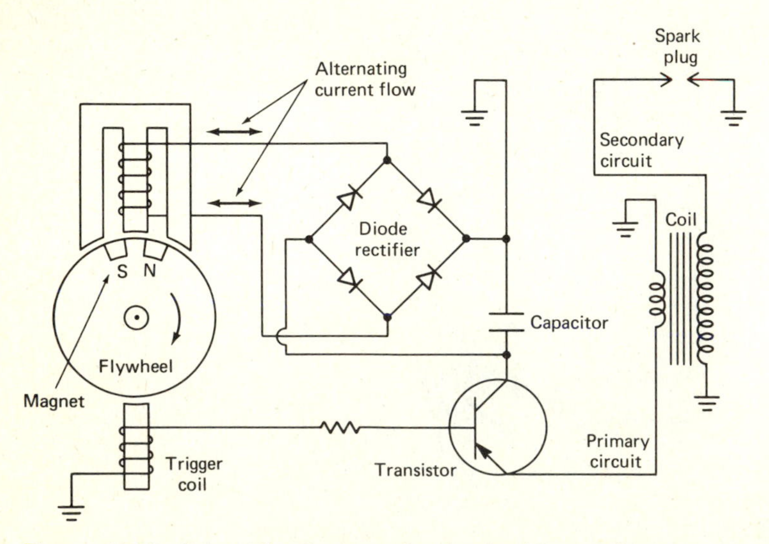

Figure 14-23 Schematic diagram of a flywheel-type solid-state magneto for one-cylinder engines.

magnetic induction in the ignition coil increases the voltages enough to jump the gap at the spark plug. This is known as capacitor-discharged type of solid-state magneto ignition.

Solid-State Flange-Mounted Magneto Multiple-cylinder solid-state magnetos with flange mounting and impulse coupling are available for many engines. A capacitor-discharge unit with breakless transistorized triggering is shown in Fig. 14-24. It utilizes a permanent steel magnet to generate the electrical energy for the primary and triggering circuits. Units of this type can produce a spark with 38,000 V at 3,600 rpm.

Impulse Coupling

Since most multiple-cylinder engines are of such a size that in cranking they can only be turned very slowly at best, a special provision of some kind is necessary to permit the magneto to supply a strong spark at this cranking speed. The device used for this purpose is known as an impulse-coupling, and it serves as a part of the magneto-drive coupling. Impulse coupling usually engages automatically whenever the speed of the armature drops below 150 rpm. Such a device is shown in Figs. 14-25 and 14-26. The action is dependent upon centrifugal force, which causes certain weighted members to move outward sufficiently to disengage the coupling when the proper speed is attained.

Impulse couplings are made up of certain small moving parts whose free movement is essential to the proper functioning of the device. If

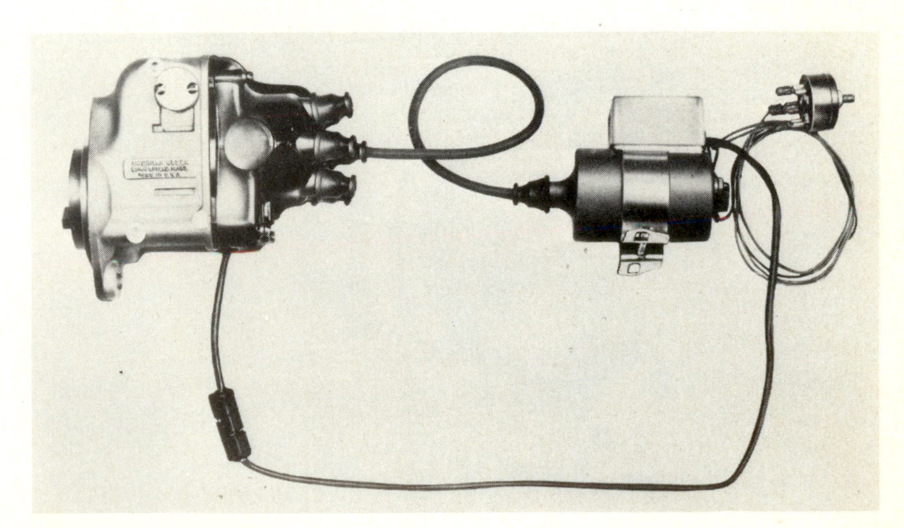

Figure 14-24 Solid-state magneto with base transformer, electronic module, and safety-switch adapter. (Courtesy of American Bosch Electric Products Division, AMBAC Industries.)

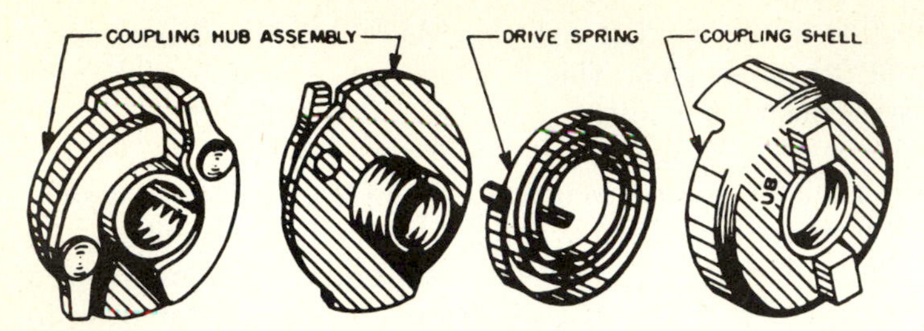

Figure 14-25 Automatic impulse coupling. (Courtesy of Fairbanks Morse and Company.)

lubrication is neglected, or if dust and dirt gum up the parts, poor action is likely. Most of the wear takes place in the catch block, catch notch, and catch release. The catch block, when released, is under considerable pressure. Consequently, the end of the catch and the notch holding it are subject to a certain amount of wear. Eventually the catch may fail to

Figure 14-26 Flange-mounted unit-type magneto with impulse coupling. (Courtesy of American Bosch Electric Products Division, AMBAC Industries.)

engage or may slip out, of its own accord. All impulse starters have a distinct click when working properly. Therefore, if this click is not heard or is weak, the trouble is likely to be due to the causes mentioned.

Since the spark made by an impulse starter is for starting the engine, it must always occur just as the piston reaches the end of the compression stroke. In other words, an impulse spark is always a retarded spark if the magneto is correctly timed.

Magneto Mountings

Magnetos must be mounted rigidly on multiple-cylinder engines to provide positive drive and accurate timing. The flange type of mounting (Fig. 14-26) is now used extensively on multiple-cylinder engines and some single-cylinder engines. The flange of the magneto fits directly to a mating flange on the governor housing or crankcase of the engine, thereby enclosing the entire magneto-drive assembly within a dustproof housing. A flexible coupling eliminates any difficulty from drive misalignment. Timing adjustment is provided by using slotted holes for the flange bolts. By loosening the two bolts, the entire magneto can be turned in either direction about the armature to give a timing adjustment range of about 20°.

Spark Advance and Retard The advancing and retarding of the spark on engines equipped with magnetos are accomplished by changing the time at which the breaker points are opened. Since all magnetos are now equipped with an impulse-starter coupling, a spark-advance and -retard device is unnecessary. The reason for this is if the magneto is correctly timed, the impulse or starting spark must occur at or very near dead center. As soon as the engine starts and speeds up, the impulse coupling is thrown out and the spark is not delayed but occurs earlier in the piston travel.

Timing Magneto with Engine When a magneto is disconnected and removed from an engine, it is essential that it be correctly timed when replaced so that the spark will come at the right instant in the piston travel. The following procedure for timing a high-tension magneto is applicable to any engine, regardless of the number of cylinders:

1 Place piston 1 on head dead center at the end of the compression stroke.

2 Select or determine the distributor terminal to be connected to spark plug 1. Unless the wires are already cut to an exact length and fitted, any terminal may be chosen for spark plug 1.

3 Place distributor rotor on terminal brush or segment 1. To do this it is usually necessary to remove the distributor cover.

4 See that breaker mechanism is in the retard position.

5 Slightly rotate or move armature in driven direction until breaker points are just opening. In doing this, the amount of rotation should not be sufficient to disturb the position of the distributor rotor.

6 Without moving the armature, place magneto in position on the engine and connect the drive coupling.

7 Check the timing by turning the engine over very slowly. See that the breaker points open just as the piston reaches head dead center at the end of the compression stroke. If the magneto is equipped with an impulse starter, it should snap just as piston 1 reaches compression dead center. If it snaps slightly before or after, carefully disconnect the coupling, rotate the armature a trifle forward or backward as the case may be, and again connect and recheck.

8 Connect the wires to the distributor and spark plugs in the proper manner according to the firing order of the engine.

Magneto Care and Adjustment Magnetos require very little attention and should seldom give any trouble. They are well enclosed and almost completely dustproof and moistureproof. This does not mean that they do not require some reasonable care. Occasional lubrication (according to the manufacturer's instructions), and prevention of excessive dirt accumulation around the magneto are important.

PROBLEMS AND QUESTIONS

1 Make a complete, detailed schematic sketch of the parts and circuits involved for a conventional ignition system for a six-cylinder automotive-type engine.
2 Explain the function and principle of operation of the following parts in a conventional spark-ignition system: 1) breaker mechanism; 2) condenser; 3) coil.
3 Make a complete schematic sketch of a magnetic-pulse ignition system and explain how it operates.
4 Explain the difference between a capacitor-discharged ignition system and the other types of spark ignition systems.
5 Name the factors affecting the spark advance needed for an engine and the effects of too late and too early spark settings.
6 Make a detailed sketch of the two circuits in a conventional magneto and explain how the magneto operates.
7 Explain the difference in the construction of so-called "hot" and "cold" spark plugs. State the engine and operating characteristics which determine the use of each.

Chapter 15

Alternators, Generators, and Starters

The various types of charging systems used on tractors keep the battery in a charged condition by replacing the electric energy taken from the battery by the starter motor when cranking a tractor engine. In addition, the charging system produces electricity for lights, heater, radio, ignition system, and other electrical loads when the tractor is running. If the electrical load, for short periods of time, is greater than the charging system can supply, the battery assists the charging system by supplying the additional electrical energy. The charging system includes either an alternator or generator, regulator, battery, ammeter, ignition switch, and wires for connecting these units.

TYPES OF CHARGING SYSTEMS

There are two types of charging systems that may be found on tractors. They include:

1. Alternating current (ac) charging circuit (alternators).
2. Direct current (dc) charging circuit (generators).

Both systems generate an alternating current (ac). The difference is in the way they rectify the ac current to direct current (dc).

The ac charging circuit has an alternator and regulator. The alternator is an ac generator that produces ac current but utilizes diodes to rectify it electronically to dc current. Generally, alternators are smaller than generators of equal electrical output and supply a larger electrical (current) output at slow engine speeds. The sole function of the regulator in ac charging circuits is to limit the alternator voltage to a safe, present value.

The ac charging circuit has an alternator and regulator. The alter-produces ac electrical power and rectifies it mechanically to dc by the use of commutators and brushes. The regulator has three functions: (1) opens and closes the charging circuit between the generator and battery, (2) limits the generators output to a safe rate; (3) prevents the generator from overcharging the battery.

ALTERNATOR CHARGING CIRCUIT

An alternator-charging circuit consists of an alternator, regulator, battery, ammeter, and connecting wires as shown in Fig. 15-12. This system is used on most modern tractors.

Alternator Principle Components

A typical tractor alternator consists of six basic components. They include the: (1) stator, (2) rotor, (3) rectifier, (4) brushes, (5) drive end; and (6) slip-ring head assembly (Fig. 15-1).

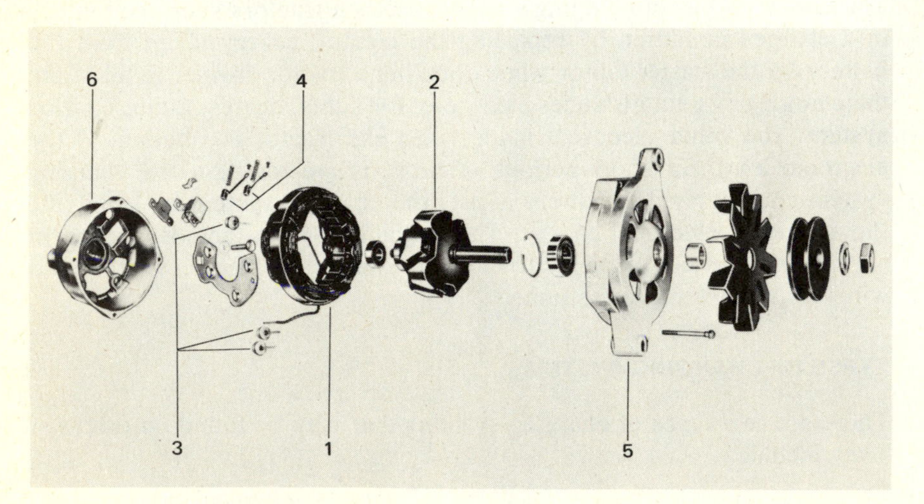

Figure 15-1 Principle components of an alternator. (Courtesy of The Prestolite Company.)

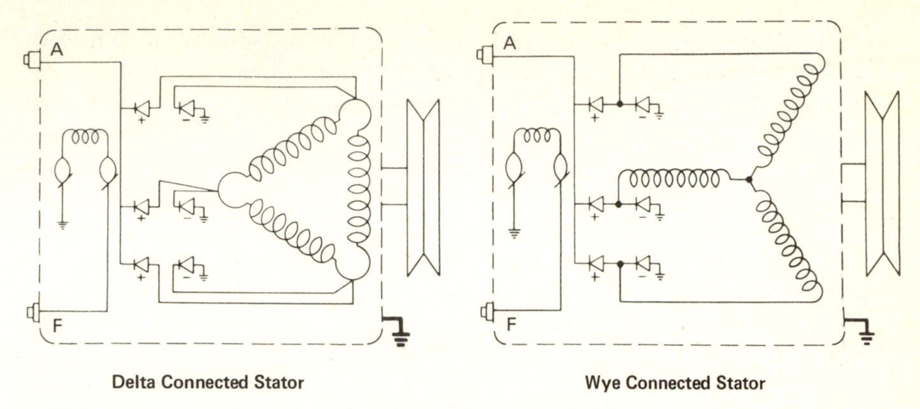

Figure 15-2 Delta and Wye connected stator for alternator. (Courtesy of The Prestolite Company.)

Stator The stator is fabricated from a number of iron stampings that are welded together to form a laminated soft-iron-ring core with slots inside the core to accommodate the three phase windings. The three windings are wound around the inside circumference of the core so that they are electrically 120° apart. The three stator circuits or windings can be connected electrically two different ways, with Delta or Wye connections (Fig. 15-2). Wye (“Y”)-connected alternators usually provide higher voltage and a moderate amperage. Delta-connected alternators produce higher amperage with lower voltage, and are used in heavy-duty operations. The Wye-connected stator circuit is used on most farm tractor alternators.

Rotor The rotor consists of a shaft, two rotor poles, soft-iron-center core, two slip rings, and a winding (Fig. 15-3). The two pole-pieces are assembled on the shaft so they will encase the winding and core. The two leads of the winding are connected to the slip rings, which are insulated from one another. Continuous sliding contact is maintained between the brushes and slip rings.

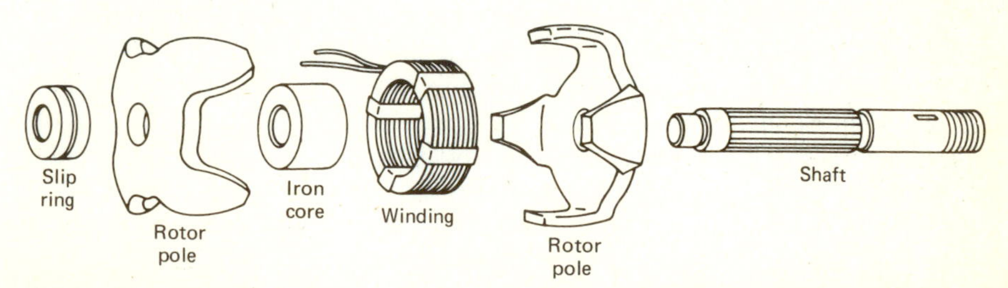

Figure 15-3 Alternator rotor construction. (Courtesy of The Prestolite Company.)

Rectifiers An alternator produces three-phase ac current, which is rectified to produce a relatively smooth flow of dc current. A set of six diodes rectifies the ac current. The leads, from each of the three stator circuits are connected to two diodes—one with negative polarity and the other with positive polarity as shown in Fig. 15-2. Most negative-grounded alternators have the negative diodes pressed into the end head, while the three positive diodes are pressed into a separate mounting plate that serves as a heat sink and is insulated from the ground.

In some units, all six diodes are mounted in heat sinks. When they are in operation, diodes produce lots of heat and become rather hot. The heat sink absorbs this heat and radiates it to air passing through the alternator.

Brushes Each brush contacts one slip ring and completes the electric circuit for current to pass through the rotor winding. This sets up the moving electric magnetic field through the coil of wire in the stator (Figs. 15-1, 15-3).

Drive- and Slip-Ring Head Assemblies The bearings on which the rotor shaft rotates are contained in the drive end and slip-ring assemblies (items 5 and 6, Fig. 15-1). The slip-ring end head also provides the insulated mountings for the rectifier and heat sinks, brushes, and other terminal studs.

Alternator Operating Principles

The purpose of the alternator is to convert the mechanical energy derived from the tractor engine into electric energy. The principle which alternators use to convert mechanical energy into electrical energy is known as electromagnetic induction. Three requirements for producing electrical energy by electromagnetic induction are: (1) a magnetic field; (2) a conductor that forms a complete circuit; and (3) motion between the magnetic field and conductor.

Basic Operation The alternator rotates a magnetic field so that the stationary conductor cuts the moving magnetic lines of force. A simple one-loop alternator (Fig. 15-4) has a rotating bar magnet furnishing the moving field. As the magnet rotates as indicated (Fig. 15-4) and with the S pole of the magnet directly under the top loop, the induced voltage, as determined by the right-hand rule, will cause current to flow in the circuit in the direction shown. Since current flow is from positive to negative, the end of the upper loop is (+) while the lower end is (−).

In the second half of the revolution as shown in the lower diagram, in Fig. 15-4, the bar magnet has reversed poles with the N pole under the top loop. The direction of current flow has reversed and the end of the loop

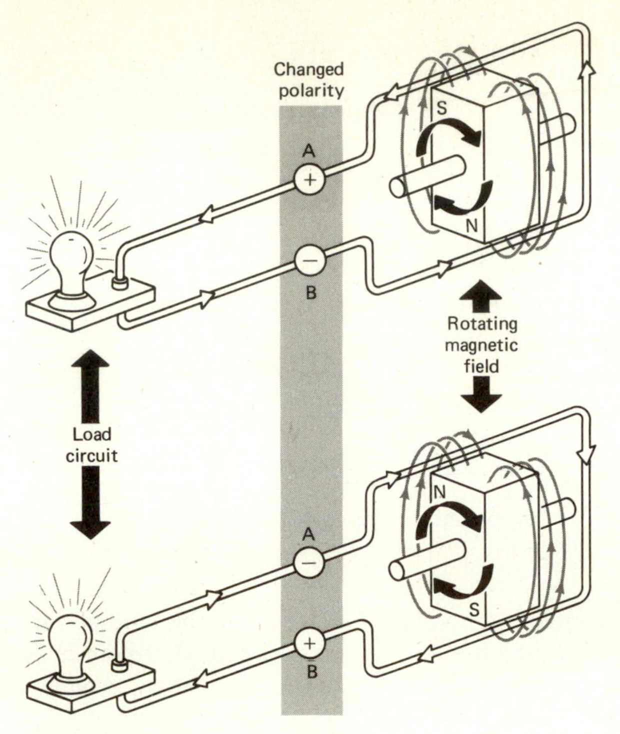

Figure 15-4 Operating principles of simple alternator. (Courtesy of Delco-Remy Division, General Motors Corporation.)

marked A has become negative (−) and the end marked B has become positive (+). As the magnet rotates and the two poles alternately pass the two legs of the loop, current will flow through the load or external circuit first in one direction and then in the other, thus producing alternating current.

The amount of electric energy induced in a circuit depends upon three factors, namely (1) the strength of the magnetic field about the conductor, (2) the number of turns making up the conductor involved, and (3) increasing the speed with which the magnetic field moves past the two legs of the loop.

The magnetic field is supplied by the rotor. When battery current passes through the rotor winding, an electric magnetic field is set up in the core, which causes one rotor pole to become a north pole and the other one a south pole (Fig. 15-5). This concentrates the magnetic field in the interlaced rotor-pole fingers to furnish alternate north and south poles around the outer circumference of the rotor.

Operation of Alternator Figure 15-6 shows different positions of the rotor as it rotates at constant speed. In the top portion of Fig. 15-6 is a

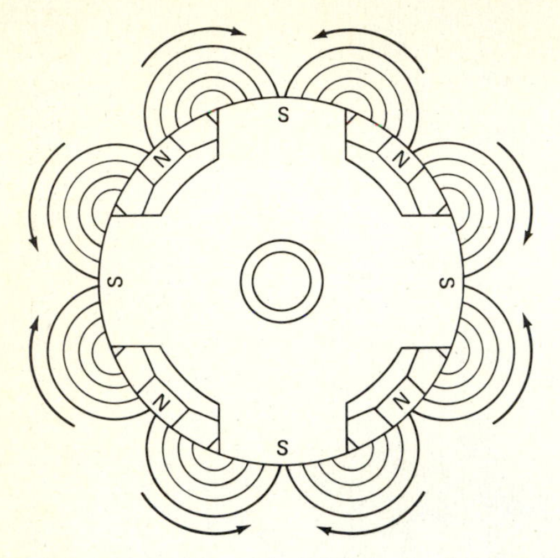

Figure 15-5 Magnetic fields in the interlaced rotor-pole fingers establish alternate north and south poles around the circumference of rotor.

curve showing the magnitude of the voltage generated in the loop of wire as the rotor revolves. With the rotor in the first position (A), there is no voltage being generated in the loop of wire because there are no magnetic lines of force cutting across the conductor. As the rotor turns and approaches the second position (B), the rather weak magnetic field at the leading edge of the rotor starts to cut across the conductor and the voltage increases. When the rotor reaches position B, the generated voltage has reached its maximum value, as shown above the horizontal line in the illustration. The maximum voltage occurs when the rotor poles are directly under the conductor. It is in this position that the conductor is being cut by the heaviest concentration of magnetic lines of force.

The magnitude of the voltage varies (Fig. 15-6) because the concentration of magnetic lines of force cutting across the loop of wire varies. The voltage curve shown is not the result of a change in rotor speed since the rotor is turning at a constant speed.

As the rotor (Fig. 15-6) turns from position B to position C, the voltage decreases until at position C it again becomes zero. As the rotor turns from position C to position D, the voltage increases again to a maximum negative voltage at D. The voltage again returns to zero when the rotor turns from position D to position E. Alignment of rotor poles, conductor cut by moving magnetic lines of force, and current flow in the conductor are illustrated in Fig. 15-7.

The sine wave pattern in Fig. 15-6 was produced by one phase of an alternator during one complete revolution or cycle of the rotor. When the winding for the other two phases is added, they are located electrically 120° apart. The rotating magnetic field and the stationary stator assembly

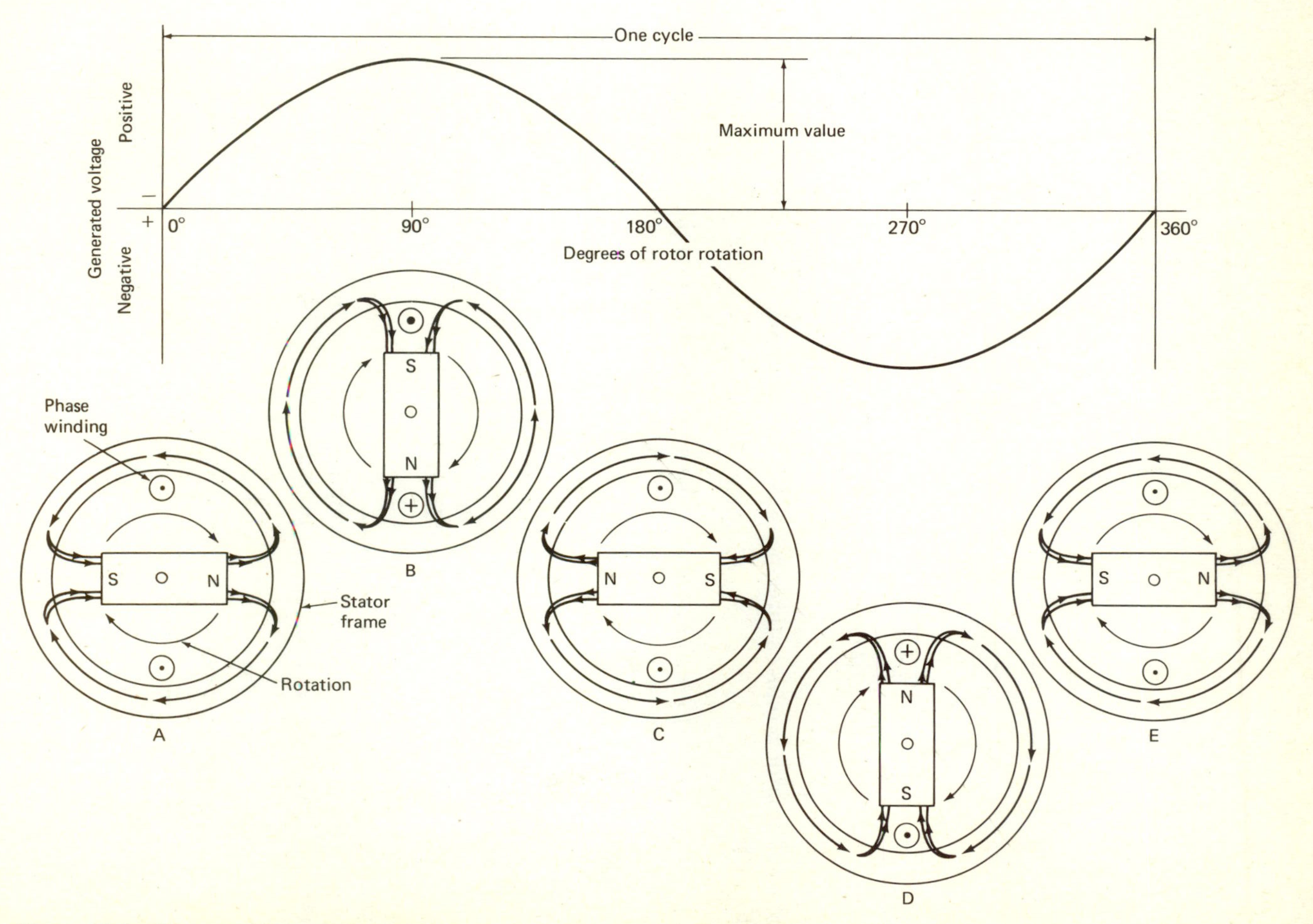

Figure 15-6 Sine-wave pattern showing the magnitude of the voltage generated in a loop of wire by a rotating magnet and stationary conductor.

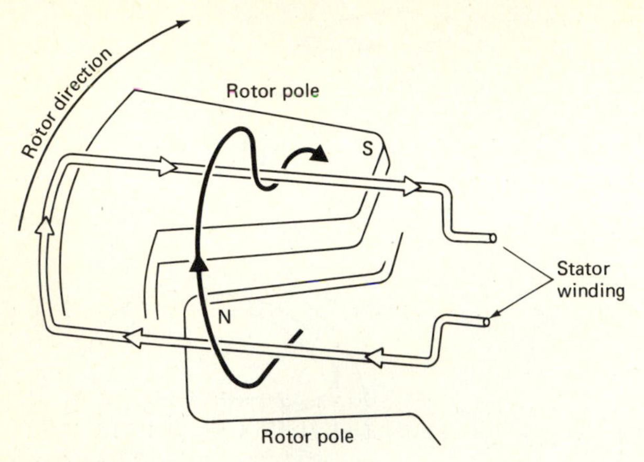

Figure 15-7 Rotor pole alignment and current flow in the conductor of an alternator.

of an alternator produces three-phase ac current as shown by sine wave in Fig. 15-8.

Converting Alternator ac to dc The alternating current (ac) produced by the rotor and stator assemblies must be changed into direct current (dc) in order to recharge the battery and supply other electrical loads. The alternating current is changed into direct current by diodes. As discussed earlier, a diode acts as a form of electrical check valve by allowing current to flow in only one direction. A positive diode will pass current traveling only in the positive direction while negative diode will pass current traveling only in the negative direction. Both positive and negative diodes will block current flow in the opposite direction.

Each phase winding of the alternator is connected to two diodes, one with positive polarity and one with negative polarity. When the phase winding of the alternator is producing positive current, the current flows through the positive diode to the output terminal (Fig. 15-9). When the

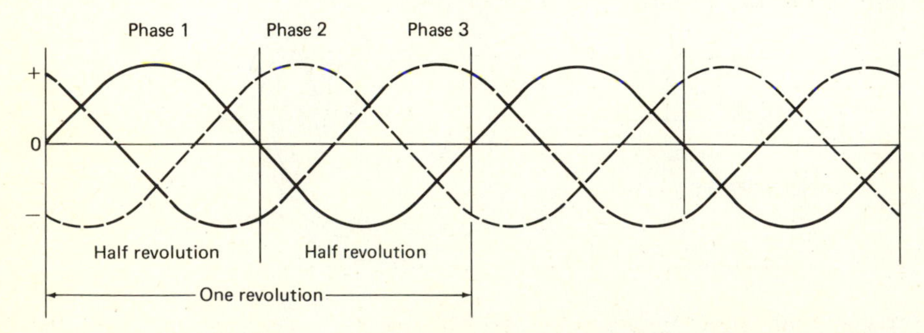

Figure 15-8 Three-phase ac-current sine waves produced by an alternator.

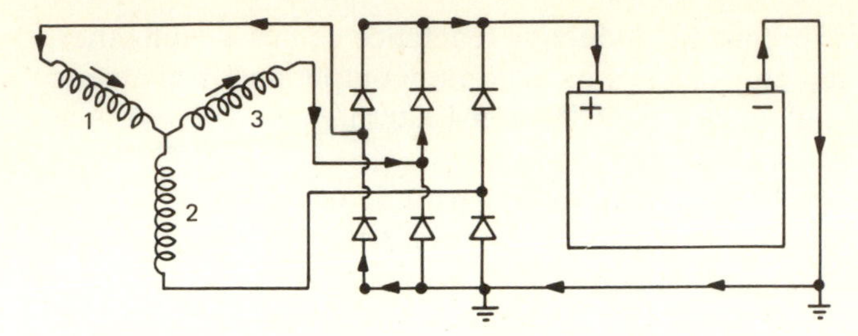

Figure 15-9 Alternator phase winding passing positive current: The current flows through the positive diode to the output terminal as direct current.

alternator phase winding is producing negative current, the negative diode allows the current returning through the ground circuit to pass through into the winding to complete the circuit for another phase winding (Fig. 15-10). The current flow from the alternator output terminal to the battery and accessories is rectified alternating current.

Regulators for Alternators

The output of the alternator is controlled by the number of revolutions per minute the rotor turns and the intensity of the magnetic field produced by the rotor. The rotor rpm is governed by the engine rpm, while the intensity of the magnetic field is controlled by the regulator.

Volts produced by an alternator drop when electrical load is heavy. The voltage regulator senses the low voltage condition and increases the magnetic field by allowing more current to flow through the rotor winding, thus increasing the output of the alternator. The reverse is true when electrical load is light and system voltage is near normal. The regulator senses the normal voltage and limits the amount of current passing through the rotor winding, thus decreasing the output of the alternator.

The maximum current output is self-limiting by the "inductive reactance" design of the alternator. Inductive reactance is a resistance to the current reversals in the stator windings. As rotor speed and current output are increased, the inductive reactance also increases until the alternator

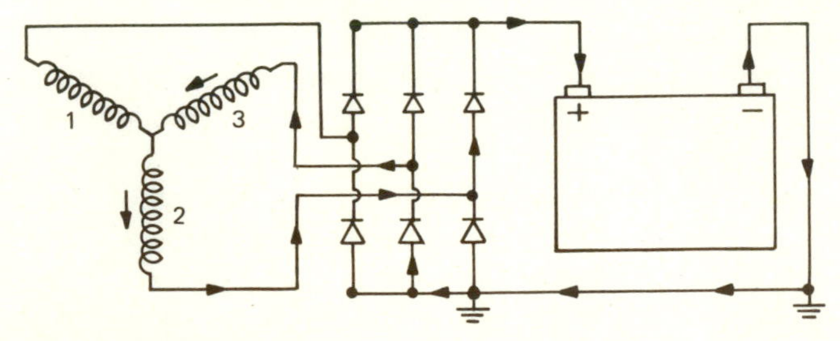

Figure 15-10 When the alternator phase winding is passing negative current, the negative diode allows the current to pass through the winding to complete the circuit.

reaches a point where output and inductive reactance equalize each other. At this point, the alternator reaches its maximum output and is unable to produce any increase in current flow. Thus the alternator is referred to as a self-current limiting machine.

No cutout relay is needed (as required with dc generators) because the diodes in the stator-to-battery circuit prevent the battery from discharging through the stator winding when the engine is not running. However, the field winding must be disconnected when the engine stops. This may be done by field relay or the field is connected to the battery directly (Fig. 15-12) through the ignition switch.

Many different designs are used by the manufacturers of regulators to control the field circuit in alternators. The designs can be broken down into four basic types (Fig. 15-11). They include:

1 Single-unit mechanical
2 Double-unit mechanical
3 Double-unit transistorized assisted
4 Solid-state or transistor type

Single-Unit Mechanical Type Battery voltage is applied to the ignition terminal of the regulator as shown in Fig. 15-12 when the ignition switch is turned to the "on" position. Current passes through the lower stationary point A and upper movable contact B through line E to the regulator field terminal F and on to the alternator. The field is now energized and the alternator will start producing current when the engine is started. The battery current is applied to the voltage coil D. As the alternator output increases with increased speed of the engine, the amount of current passing through the contact A-B and coil D also increases. The increased current through coil D produces an electromagnetic field strong enough to overcome the spring tension, opening circuit between contacts A and B. With contact B in the open or "floating" position, the field circuit is completed through the R-1 resistor, thereby reducing the current flow in the field circuit and output of the alternator. This reduces the magnetic pull of the voltage coil D and the spring tension closes movable contact B with contact A reenergizing the field circuit with full-battery voltage. The movable contact B vibrates between lower stationary contact A and the float position to control the voltage when the current demands on the electrical system are high.

When the electrical system current demands are low, the amount of current flowing through the R-1 resistor is sufficient to produce the low voltage required from the alternator, and the movable contact B remains in the floating position (no contact with A or C). As the system current demands are decreased, the system voltage tends to increase. The voltage regulator coil senses the increase in voltage, produces a stronger magnetic force, and pulls the movable contact B up against the stationary

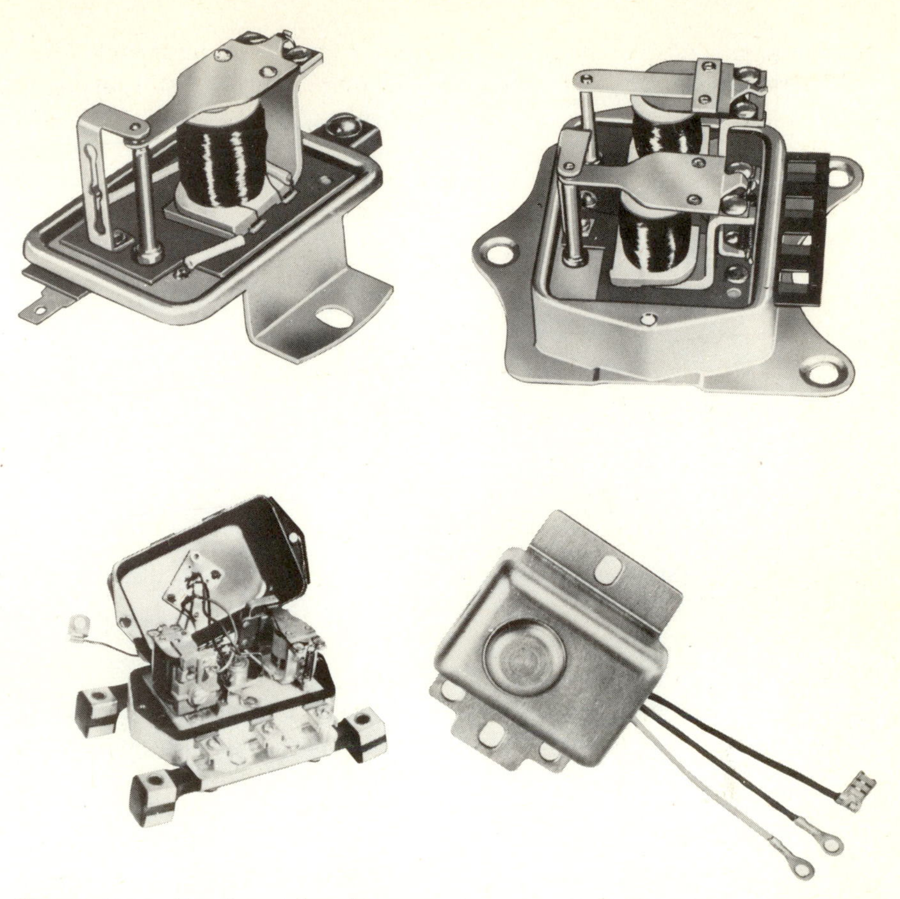

Figure 15-11 Regulators for alternators include: (A) single-unit mechanical, (B) double-unit mechanical, (C) double-unit transistorized, (D) solid state or transistor. (Courtesy of The Prestolite Company.)

contact C. Since the stationary contact C is connected to the ground, this momentarily turns off all current flow in the field circuit, and the alternator output drops. As the alternator output drops, the magnetic force of voltage coil D drops and movable contact C returns to the float position. By vibrating the movable contact B between the float position and contact with the lower stationary point C, the system voltage is controlled within very close limits, thus preventing battery overcharging during high engine speed with minimum load on the electrical system.

Double-Unit Mechanical Type The internal wiring of a typical double-unit mechanical type regulator is shown in Fig. 15-13. A voltage regulator D and a field relay F make up this unit. Voltage regulation is accomplished in the same manner as in the single-unit mechanical type, while the relay is used to operate a charge-indicator light.

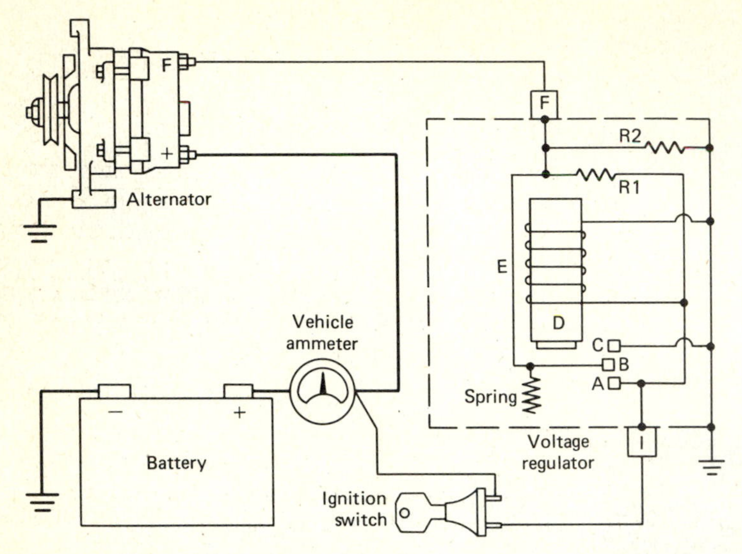

Figure 15-12 Typical alternator charging circuit with a schematic diagram of a single-unit mechanical regulator.

When the ignition switch is in the on position (Fig. 15-13), current flows through the charge-indicator light, which is connected to the number 4 or I terminal of the regulator, through the voltage regulator contacts A and B to the field terminal of the alternator, and through the rotor to the ground. This causes the indicator light to burn. When the alternator starts producing current, the current from the auxiliary terminal on the alternator causes the field relay points G to close, thereby connecting both sides of the indicator light to the battery's positive terminal, and the light goes out. If the alternator does not produce electric current, the indicator light remains burning.

Double-Unit Transistorized Assist This type of regulator for alternators is basically the same as the double-unit mechanical type, with one major exception. The field current is carried through a transistor rather than through the voltage-regulator contacts. This allows the use of only one set of voltage-regulator contacts, their function being to trigger the transistor. The relatively small amount of current required to turn the transistor on and off causes very little contact erosion, thereby giving greater contact life.

Solid-State or Transistor Regulator This type of regulator uses solid-state components such as transistors, thermistors, and zener diodes (see Chap. 13) to achieve alternator field current control. Current flows through R-1 (Fig. 15-14), the emitter-base junction of TR-1 and through

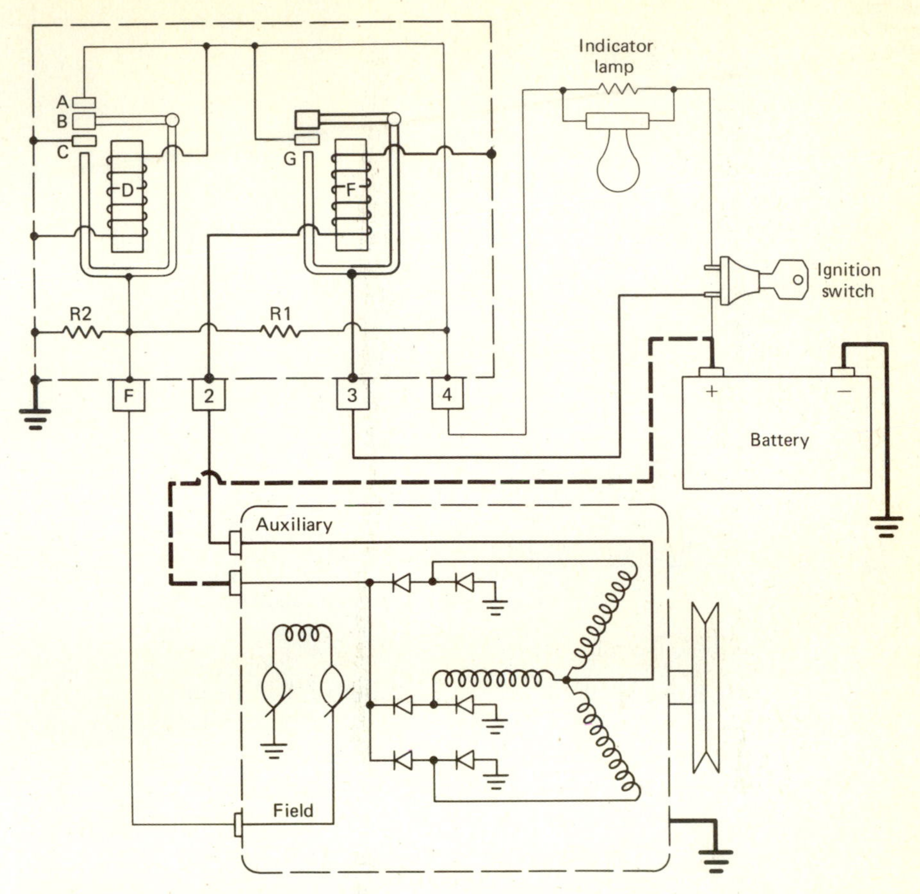

Figure 15-13 Schematic of a two-unit mechanical regulator for alternators with an indicator light.

R-2 to ground when the ignition is turned to the on position. This turns on TR-1, and current flows through the emitter-collector junction to the alternator field.

As the alternator output voltage increases, the voltage drop across R-6 and R-7 increases and triggers the zener diode D-2 into reverse conduction. This connects the emitter-base junction of TR-2 to ground through the thermistor NTC and R-5, thus turning on TR-2. With TR-2 turned on, current flows through the emitter-collector junction and R-2 to ground. This connects the base of TR-1 to essentially the same voltage that appears at the emitter of TR-2, causing TR-1 to switch off, interrupting the current flow to the alternator field. As the voltage produced by the alternator drops, the voltage impressed across the zener diode decreases, causing it to cease reverse conduction. This disconnects the base of TR-2 from ground causing it to turn off. With TR-2 turned off, current again

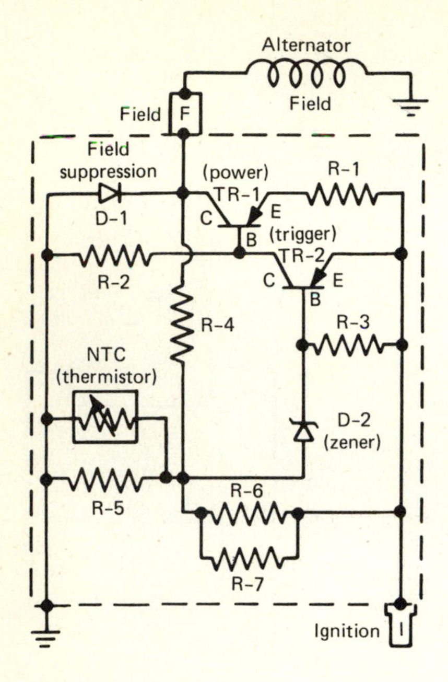

Figure 15-14 Wiring diagram of solid state alternator regulator.

flows through R-1, the emitter-base junction of R-2 to ground, turning TR-1 back on, allowing current to flow through the emitter-collector junction to the alternator field. This cycle of turning TR-1 on and off takes place many times per second during normal operation.

The network of resistors (Fig. 15-14) in the regulator is used to assure proper voltage levels throughout the solid-state circuitry.

The use of the solid-state regulator is increasing and is found on many tractors, trucks, and automobiles. Various locations are utilized for mounting regulators, but many of them are built into the alternator.

Types and Designs of Alternator

The alternator has only one function: to supply current to charge the battery and operate electrical accessories. Since each application may have its own special requirements on the alternator, there are many different types and designs of alternators. Some of the factors that determine alternator design are type of mounting; vibration; belt loading; minimum and maximum rotor speed; current output; service life required; and environmental factors such as dust, dirt, road splash, and the presence of explosive mixtures in the atmosphere.

Many late-model alternators have stationary field coils, thus eliminating slip rings and brushes. Some alternators are sealed and cooling is provided by a flow of engine oil through it. Other alternators are cooled by air using either external or internal fans.

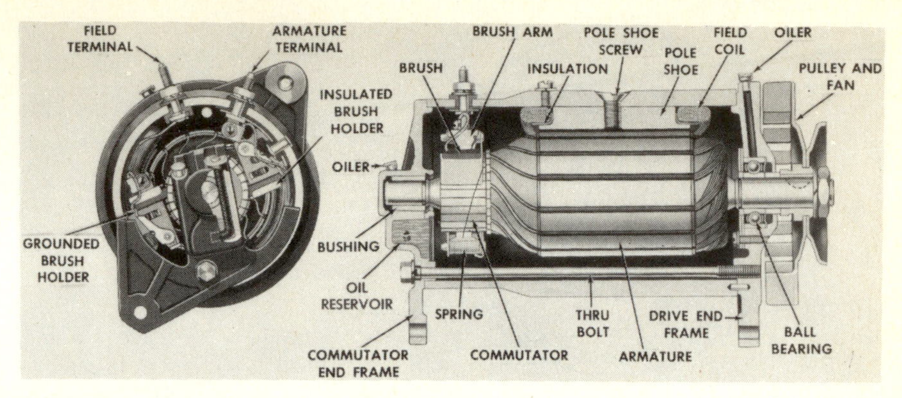

Figure 15-15 Cutaway and end view of a generator. (Courtesy of Delco-Remy Division, General Motors Corporation.)

GENERATOR CHARGING CIRCUIT

A generator charging circuit consists of a generator, regulator, cutout relay, ammeter, battery, and connecting wires as shown in Fig. 15-18. This system is found on many older tractors.

Generator Construction and Operation The generator (Fig. 15-15) produces a direct current and consists of a frame and soft-iron pole shoes, field winding, armature with commutator and winding, bearing, commutator brushes, brush holders and springs, field and armature terminals, and drive pulley. The theory of generation of an electric current by a generator is much the same as for a magneto, as described in Chap. 14. Figure 15-16 shows the construction of a simple direct-current generator and how it operates. The principal distinction is that the magnetic field is produced by a current passing through the winding of the pole shoes. These particular windings constitute the field circuit or winding as con-

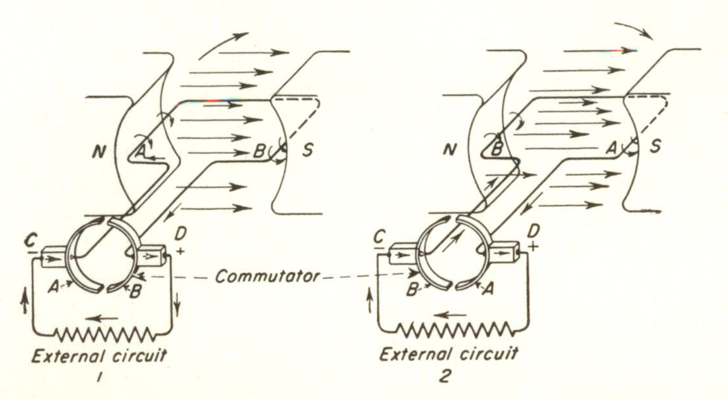

Figure 15-16 Operation of a simple direct-current generator.

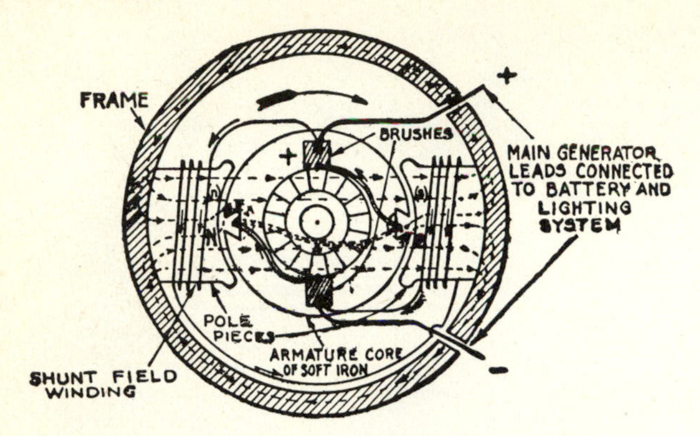

Figure 15-17 Principle of a shunt-wound generator.

trasted with the armature and main supply circuit. The field current is self-produced by the generator; hence, the field strength will vary according to the armature speed and other factors. In order for the field current to be self-produced, a small amount of magnetism must be present in the pole shoes at all times in order to create an initial current buildup. This limited permanent flux is known as residual magnetism. Generators are rated according to voltages as 6, 12, 24, and so on, in accordance with the voltage of the battery and the voltage specifications of the other electrical equipment. The actual generated voltage at normal speeds must slightly exceed the battery voltage in order to force current into the battery. This explains the noticeably brighter lights when the generator is running.

The most common type of generator is the shunt-wound type. It uses two brushes (Fig. 15-17), with both the field and external circuits coming from the same brushes. The field circuit is in parallel with the main supply circuit.

CONTROLS FOR TWO-BRUSH SHUNT GENERATOR

The current output of a shunt generator increases in direct proportion to its speed, therefore some external device must be used to limit the maximum output to a safe value. This is done by means of a combined current and voltage regulator with cutout relay (Fig. 15-18). The control unit consists of three distinct units: (1) the cutout relay, (2) voltage regulator, and (3) a current regulator.

Cutout Relay A generator-battery charging circuit must be equipped with a device that will automatically make and break the circuit, depending on whether the generator is operating or not, and on the voltage differential at low speed. This device as shown in Fig. 15-18 is known as a cutout relay. It automatically closes the circuit when the generator voltage

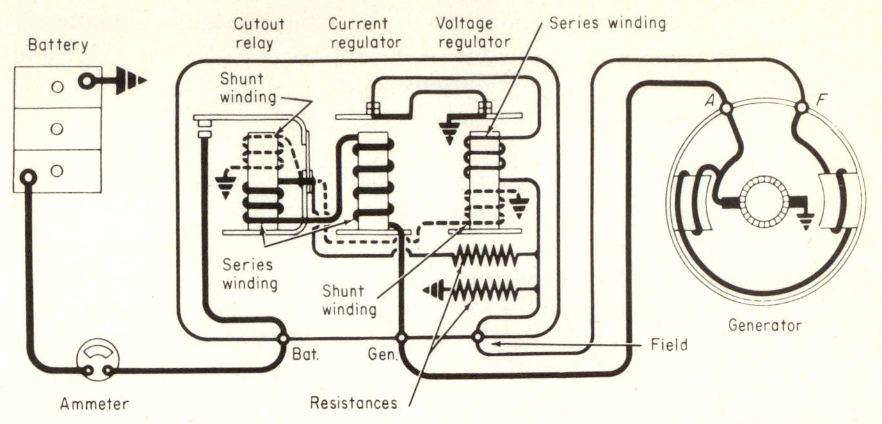

Figure 15-18 Shunt generator with combined current and voltage regulator with cut-out relay.

exceeds the battery voltage, and it opens or breaks the circuit when the generator stops running or its voltage is low. If the cutout relay were not used, the battery would discharge through the generator when the engine was idling or inoperative.

This relay consists essentially of a small iron core, the fine or voltage-shunt winding, the coarse or current-series winding, and the points. When the generator is not operating, the points are held open by a small spring. As the generator starts operating, a voltage is built up in the shunt-field winding. The fine or voltage-shunt winding of the relay, being a part of the field circuit, thereby carries a weak current that magnetizes the relay core and pulls the movable point into contact with the stationary point. The charging circuit is thus completed, and the stronger current now flowing through the heavy series winding on the relay core holds the points firmly in contact. If the generator speed drops sufficiently to permit the generator voltage to fall below the battery voltage, there is a reverse flow of current, the core is demagnetized, and the points open again. The cutout relay may be connected as a separate unit but is usually a part of the generator-charging-rate control unit (Fig. 15-18).

Voltage Regulator The vibrating-type voltage control (Fig. 15-18) consists of an iron core with two windings. The shunt winding consists of many turns of fine wire, and the series winding consists of fewer turns of heavy wire connected in series with the field circuit when the regulator points are closed. When the generator voltage reaches a certain value, the magnetic field produced by the two windings overcomes the armature spring tension, pulls the armature down, and separates the contact points. This breaks the series circuit, inserts resistance into the generator-field circuit, and reduces its voltage. Thus, it is observed that a continuous making and breaking action goes on because the armature actually vibrates rapidly. This vibration maintains the voltage at a constant value and permits the generator to supply varying amounts of current to meet the varying states of battery charge and electrical load.

Current Regulator The current regulator has a series winding of a few turns of heavy wire which carries the entire generator output. With the relay points closed, the generator-field circuit is completed to the ground through the current-regulator points in series with the voltage-regulator relay points. When the load demand is heavy, as when electrical devices are turned on and the battery is in a discharged condition, the voltage may not increase sufficiently to cause the voltage regulator to operate. Consequently, the generator output will continue to increase until the generator reaches its rated maximum, which is the current value for which the regulator is set. At this point, the magnetism in the core will pull the regulator armature down and open the points. With the points open, resistance is inserted in the field circuit and the output is reduced. As the output drops, the spring tension on the armature pulls it away from the core, the points again make contact and the output increases. Thus, this cycle continues; that is, the armature vibrates rapidly as long as the output remains at a maximum. When the electrical load is reduced or the battery becomes charged, the voltage increases and causes the voltage regulator to operate. This tapers off the generator output and prevents the current regulator from operating. The two regulators never operate simultaneously.

It will be noted that there are two resistances in the regulator and that they become connected in parallel in the generator field circuit when the current regulator operates. When the voltage regulator operates, only one resistor is inserted in the field circuit. The reason for this is that a higher value of resistance is required to reduce the generator output, when the voltage regulator operates, than is required when the current regulator operates to prevent the generator from exceeding its rated maximum.

STARTER MOTORS

Practically all automobile, truck, and tractor engines, most stationary power units, and other types of engines are cranked by means of a battery and a special electric motor. These motors vary somewhat in size and type and are usually equipped with a convenient switch, according to the service requirements.

Starter-Motor Construction Starter motors are similar to generators, with the exception that they have a series-wound instead of a shunt-wound field circuit, as shown by Fig. 15-19. Also, both the field and

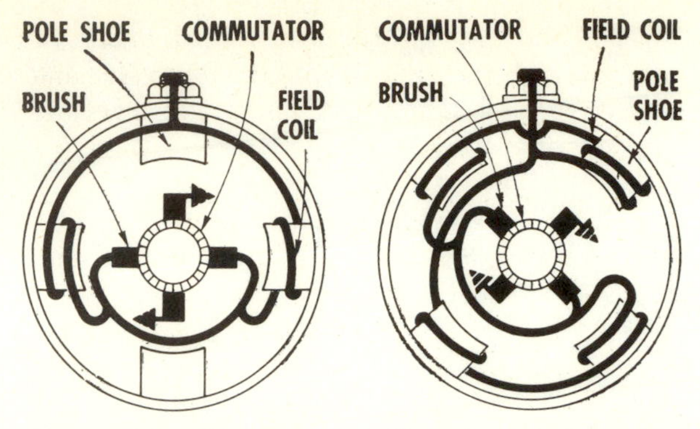

Figure 15-19 Circuits and windings for a starter motor.

armature windings (Fig. 15-20) are of very coarse low-resistance wire, which will allow a high-amperage current to flow. A starter motor must exert a strong turning effort or torque. To provide this, a heavy current must flow through the circuits and create a very strong magnetic field and consequent pronounced reaction between the armature and the field.

The principal variation in different starters is in the number of poles and brushes. For small engines, two-pole starters with two brushes are usually used. Larger engines use four- and six-pole starters with four or six brushes. Otherwise the construction and operation are the same.

Starter Drives The starter drive is the device that connects and transmits the power to the engine flywheel. A small spur pinion on the armature shaft meshing with teeth on the flywheel is the usual means of transmitting the power. However, the drive must embody some convenient means of disengaging as well as engaging the pinion, and the device must be one that will not give trouble or injure the starter motor when the engine begins to fire or if it kicks back.

Three common types of starter drives are the Bendix (Fig. 15-21), the overrunning-clutch type (Fig. 15-22), and the Dyer type (Fig. 15-23). In the Bendix, the construction is such that engagement of the pinion takes place as soon as the starter switch is closed and the armature begins to rotate. A threaded sleeve on the armature shaft carries this pinion and permits it to move along the shaft and into and out of engagement. If the armature

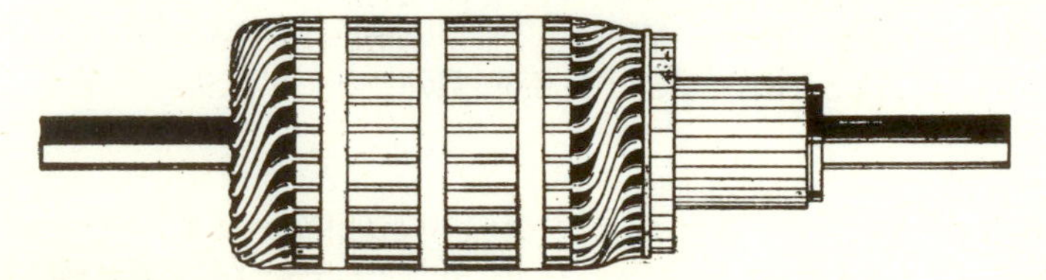

Figure 15-20 Starter-motor armature.

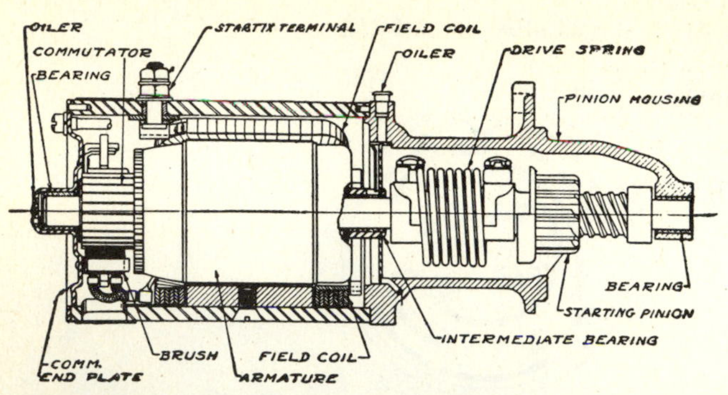

Figure 15-21 Sectional view of a starting motor using a Bendix-type starter-drive mechanism.

starts turning, the thread carries the pinion into engagement with the flywheel gear. As soon as the engine fires and turns the pinion faster than the armature shaft is turning it, the pinion is carried out of engagement by the threads. The device is equipped with a heavy spring to absorb the shock as the pinion comes into mesh with the flywheel or when the engine kicks back.

In the overrunning-clutch starter drive, the pinion meshes with the flywheel gear by a manually or solenoid-operated shift lever (Fig. 15-22). Further movement of the shift lever closes the electric switch, and the starter armature rotates and cranks the engine. To prevent damage to the starter motor when the engine begins to fire and turns the pinion at a higher speed than the armature is turning it, an overrunning clutch is built into the pinion and connects it to the armature shaft in such a manner that the pinion is automatically disconnected from the armature even though the shift lever has not been released.

The Dyer drive (Fig. 15-23) is a special type of drive combining some of the features of both the Bendix and overrunning-clutch drives. Its drive pinion is moved into mesh with the ring gear by a shift lever that is either operated manually or by a solenoid. This type of drive is used on large diesel and other heavy-duty engines. It provides for positive engagement of the drive pinion with the ring gear before the cranking-motor switch is closed and thus before the armature starts to rotate. This eliminates clashing and the possibility of broken or burred teeth on either the pinion or the ring gear. As soon as the engine starts, the pinion is automatically demeshed by the reversal of torque so that the armature will not be subjected to excessive speeds.

Cranking-Motor Controls The cranking circuit of any electric-starting outfit must be equipped with a convenient, easily operated, heavy-duty switch. This switch may be operated manually through a foot-

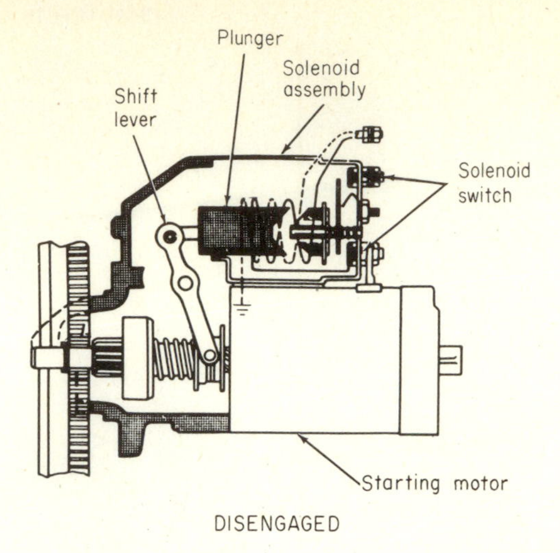

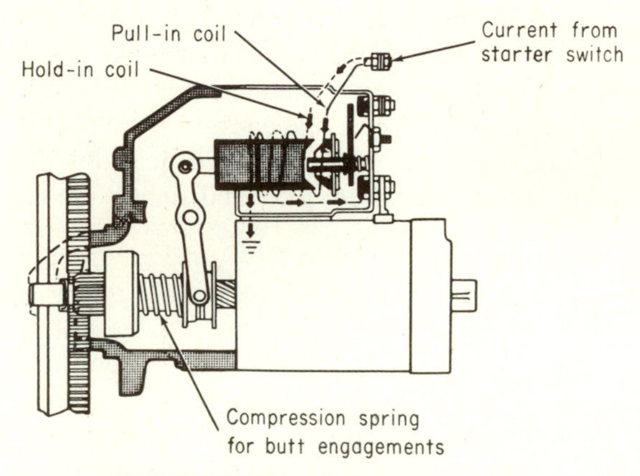

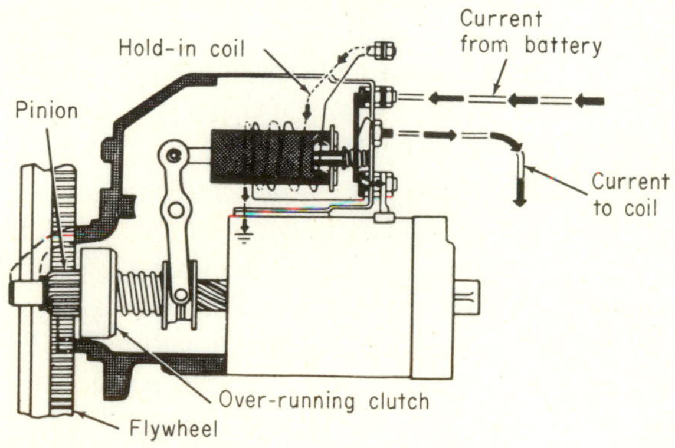

PINION FULLY ENGAGED AND
STARTING MOTOR CRANKING

Figure 15-22 Overrunning-clutch starter drive equipped with a solenoid-actuating mechanism.

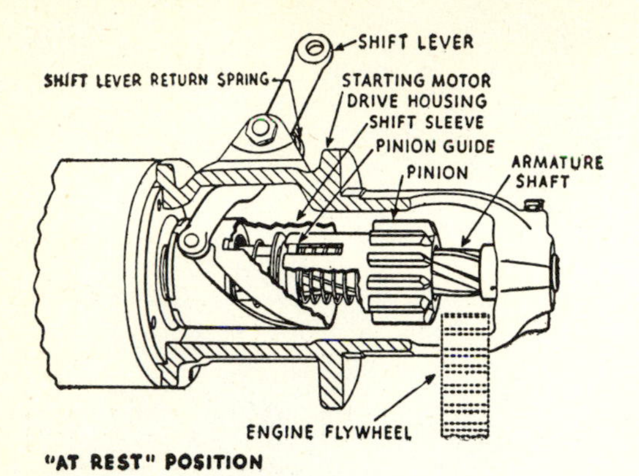

Figure 15-23 Dyer-type starter drive.

or hand-operated lever or semiautomatically by means of a solenoid switch or a magnetic switch. The usual method (Fig. 15-22) is to combine the ignition switch and starting switch in such a manner that the latter, when turned to a certain position, closes a circuit that energizes the pull-in coil of the solenoid. As the solenoid plunger moves inward, it slides the starter pinion into engagement with the flywheel gear and finally closes the main starter switch.

PROBLEMS AND QUESTIONS

1 Explain the major differences between the alternator and the dc generator.
2 Name the six basic components of a tractor alternator and explain the function of each.
3 Explain why alternators require diodes.
4 Explain how the generator converts the ac developed in the armature into dc.
5 The most common regulator used with dc generators is the three-unit type. Name the three units, state the function of each, and explain how they operate.
6 Why does the alternator not require a cutout relay.
7 Explain the fundamental differences between a generator and a starter motor with respect to construction and operation.

Chapter 16

Lubricants and Lubrication Systems

Lubrication plays an important part in the design and operation of any type of automotive machine. The life and service given by an automobile or a tractor are dependent largely upon the consideration and care given to its lubrication, both in the design of the system and during its use and operation. The fundamental purpose of lubrication of any mechanical device is to eliminate friction and the resulting wear and loss of power. Other important functions of lubrication of an internal-combustion engine are (1) to absorb and dissipate heat, (2) to serve as a piston seal, (3) to act as a cushion to deaden the noise of moving parts, and absorb shock, and (4) to assist in keeping the engine working parts clean and free of dirt, gum, corrosive acids, and other contaminants.

In the internal-combustion engine, lubrication is even more difficult than in other machines, for the reason that a certain amount of heat is present, particularly in and around the cylinder and piston, as a result of the combustion of the fuel. The lubrication of these heated parts is a somewhat more difficult problem than it would be if the heat were not present.

Friction Whenever two materials of any kind move against each other, a certain amount of a force known as *friction* tends to oppose the movement. For example, when a liquid or a gas is made to flow through a passage such as a pipe or a conduit, the flow is more or less retarded by the friction between the liquid or gas and the inner surface of the pipe. Such friction is undesirable. On the other hand, the transmission of power by belts and pulleys is entirely dependent upon the friction between the belt and the pulley surfaces. The greater the frictional contact, the more effective the arrangement and the lower the power loss.

The friction due to metal parts of machines moving in or against each other is usually highly undesirable, not only because of the power required to overcome it, but also because of rapid wear. Likewise, it often results in the generation of a certain amount of heat, which may damage the parts or produce a fire hazard. This friction cannot be completely eliminated, but it can be reduced to such an extent by the use of a suitable lubricant that the operation of the machine is greatly improved and its life lengthened.

LUBRICATING OILS AND GREASES

There are a large number and variety of lubricants available for the many different lubrication requirements. Each particular kind, quality, or type of lubricant has a comparatively limited use.

Automotive equipment such as automobiles, trucks, and tractors normally requires three general types of lubricants, namely, (1) a relatively thin, free-flowing oil for the engine, (2) a heavier high-viscosity oil for the transmission gears and bearings, and (3) various types of slow or nonflowing materials termed "greases" for wheel bearings, spring shackles, universal joints, and the like. As a result of the many changes and improvements being made in automotive machinery, oil refineries have found it necessary to develop many new and special types of lubricants having properties that better adapt them to operating conditions.

Source of Lubricants Lubricants in the past were derived from three general sources: (1) animal; (2) vegetable; and (3) mineral. Animal lubricants include lard, tallow, sperm oil, and fish oil. Vegetable oils are castor oil, cottonseed oil, olive oil, and linseed oil. Mineral lubricants are obtained by refining crude petroleum. Mineral lubricants are universally used on automotive equipment such as trucks, tractors, and automobiles.

Research has been carried on for many years to develop synthetic lubricants. Within the past few years, several brands of synthetic oil have been marketed for use in the crankcase of automotive-type engines.

Mineral Lubricants Crude petroleum, obtained from the earth and therefore usually classed as a mineral, serves as the greatest source of lubricating materials as well as fuels for internal-combustion engines. Petroleum lubricants retain their lubricating properties when subjected to abnormal temperatures and other conditions better than animal and vegetable oils.

Crude petroleum varies in quality, character, and chemical make-up, according to its geographical source. For example, some crude oils contain a high percentage of lighter products, such as gasoline and kerosene, and are therefore more valuable as a source of fuel. Other crudes, usually heavier in gravity, are low in these lighter products but contain a higher percentage of products that, when refined, produce lubricating oils and greases. The so-called paraffin-base crudes are usually lower in specific gravity and yield a high percentage of fuel products when distilled. The asphalt-base crudes are usually higher in specific gravity, darker in color, and yield a high percent of lubrication products.

Considerable argument has been put forth in the past concerning the relative merits of lubricants produced from different crude oils. Owing to the great development and advance made in the refining process, however, it is doubtful if any conclusive and consistent proof can be shown in favor of any particular crude-oil base as a base for a lubricant for a particular engine or purpose.

Manufacture of Lubricants Crude oil consists of a mixture of many hydrocarbons having different boiling points. Therefore, the first process in the conversion of any crude oil into its various products consists in heating it and distilling off the lighter fractions, such as gasoline, kerosene, gas oil, and so on. When a temperature of 350 to 400°C is reached, the heavier fractions, suitable for lubricants, distill over. If it is desired to secure a large quantity of lubricating oil from the crude, steam is introduced into the stills and they may be operated under a partial vacuum in order to prevent these heavy fractions from cracking or breaking up into lighter hydrocarbons. The lubricating-oil distillates may be redistilled with steam under a vacuum and thereby separated into heavy and light oils. Following distillation, the oils are subjected to a further rather complex refining process in order to obtain products which are adapted to every possible operating condition. Two processes have been used, the acid process and the solvent-extraction process. The former involves treatment with sulfuric acid, followed by washing with water and neutralization with an alkali. The solvent-extraction process is now used very extensively and involves six steps, as follows: (1) deasphalting, (2) dewaxing, (3) solvent extraction, (4) redistillation, (5) filtration, and (6) blending. Additional processes and treatments must be used in the preparation of certain special lubricants and greases.

ENGINE CRANKCASE OIL

The market is flooded with hundreds of brands and grades of oils of varying quality. Consequently, the owners and operators of gasoline and diesel engines, automobiles, trucks, tractors, and so on, realize the great importance of choosing the best possible oil for the engine. The price should not be more than a minor determining factor. During the early development and use of these machines, the selection of the correct lubricant was, in many cases, left with the operator. As time went on, the engine manufacturer realized the important part played by lubrication and lubricating oil in the satisfaction and service rendered by his product and, consequently, assumed the responsibility for determining and recommending the proper grade of oil to be used. As a rule, thorough tests are made either by the engine manufacturer or by the oil refiner, before any definite lubrication recommendations are offered. Such being the case, the user and operator of any type of internal-combustion engine should adhere closely to such advice. However, some knowledge of the fundamental factors involved in the choice of an engine oil is valuable and important.

Essentials of a Good Lubricating Oil As a general rule, the majority of the standard or better known brands of engine oil now on the market are reliable products and will give good results provided the correct grade is used. There are no hard and fast rules or specifications by which one may choose the proper oil to use in a given case. Likewise, there are no simple chemical or physical tests that the average individual may apply to an oil to determine its character. Such things as the color, feel, and general appearance mean nothing.

Thomsen[1] states that in order to satisfy certain fundamental lubricating requirements, a lubricant:

1 Must possess sufficient viscosity and lubricating power—oiliness—to suit the mechanical conditions and conditions of speed, pressure, and temperature. Too little oiliness means excessive wear and friction; too high a viscosity means loss of power in overcoming unnecessary fluid friction.

2 Must suit the lubrication system.

3 Must be of such a nature that it will not produce deposits during use when exposed to the influence of the air, gas, water, or impurities with which the oil may come into more or less intimate contact while performing its duty.

[1] T. C. Thomsen, *Practice of Lubrication,* McGraw-Hill Book Company, Inc., New York, 1937.

Tests of Lubricating Oils There are a large number of physical and chemical tests to which an oil may be subjected for the purpose of determining its quality and adaptability to a given purpose. Many of these, however, are of minor importance except to the oil refiner, the chemist, or the lubricating engineer. In a bulletin published by the American Society for Testing Materials and entitled "The Significance of Tests of Petroleum Products," the following statements are made:

> The rapid growth of the petroleum industry has been accompanied by the development of a variety of physical and chemical methods for the testing of petroleum products.
>
> Physical tests are more widely used than chemical tests. This is natural, in view of the fact that the utility of petroleum products depends to a large extent upon their physical characteristics. Some of these physical tests are of little value except as they serve the refiner in controlling manufacturing processes, while others are useful both to consumer and to manufacturer as an index of the value or fitness of products for particular uses.
>
> Such chemical tests as now exist serve principally to protect against impurities or undesirable constituents. This is because petroleum is an extremely complex raw material, varying greatly in composition in the various producing fields, with each new field bringing its own problems.

Viscosity The single most important property of lubricating oils is viscosity. The viscosity of a fluid is the measure of its resistance to flow. To be more specific, viscosity is the internal resistance exhibited as one portion or layer of a liquid moves in relation to another, and is due to the internal friction of liquid molecules moving past each other. The movement of molecules is affected by temperature; therefore, the most important variable affecting the viscosity of lubricating oil is temperature. In most commercial work, an expression involving the time in seconds required for a measured volume of oil to flow, under specified conditions, through a standardized tube, is known as the viscosity of the oil.

The significance of viscosity depends upon the purpose for which the oil is to be used. In a properly operating bearing, with a fluid film separating the surfaces, the viscosity of the oil at the bearing operating temperature is the property that determines the bearing friction, heat generation, and the rate of flow under given conditions of load, speed, and bearing design. The oil should be viscous enough to maintain a fluid film between the bearing surface in spite of the pressure tending to squeeze it out. While a reasonable factor of safety is essential, excessive viscosity means unnecessary friction and heat generation.

Viscosimeters for Lubricating Oils The Saybolt Universal Viscosimeter (Fig. 16-1) is now used almost universally in the United States for determination of the viscosity of lubricating oils. The apparatus con-

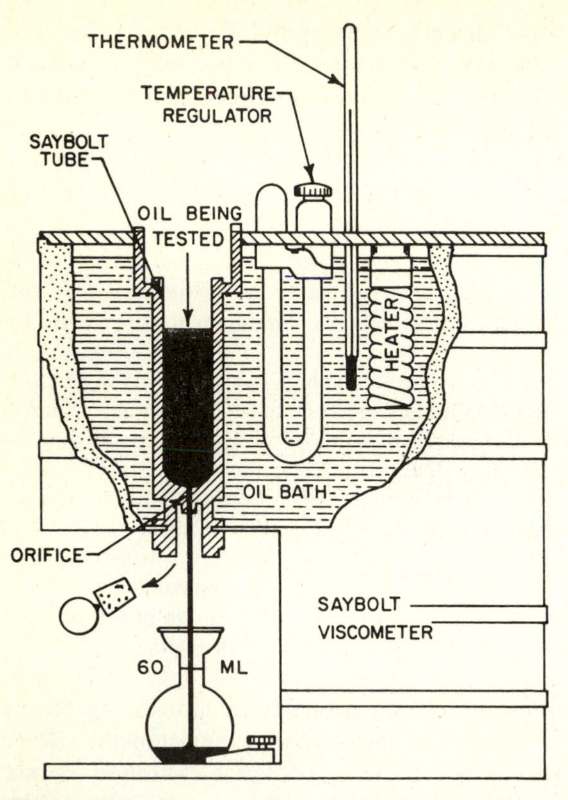

Figure 16-1 Cross section of the Saybolt Universal Viscosimeter. (Courtesy of Texaco's magazine *Lubrication*.)

sists of a central cylindrical reservoir in which the oil to be tested is placed. This reservoir is surrounded by a water bath to control the temperature. The viscosity of the oil sample is determined by taking the time in seconds required for 60 cm^3 to pass through a small orifice of specified size in the bottom of the reservoir. The viscosity is generally measured at a specified temperature such as 70, 100, 130, or 210°F. (21.1, 37.8, 54.4, or 98.9°C). Viscosities are quoted in terms of Saybolt Universal Seconds (SUS). Since viscosity changes rapidly with temperature, a numerical value of viscosity has no significance unless both the temperature and the instrument are specified.

Viscosity Index As previously explained, temperature is a vital factor in oil viscosity and the viscosity of any oil varies inversely with temperature. Furthermore, even though two oil samples may have the same viscosity at a certain temperature, say 130°F (54.4°C), yet there may be a considerable variation in their viscosities at temperatures lower or higher than 130°F (54.4°C). An ideal oil would be one which would have a desirable viscosity at some optimum temperature and whose viscosity would show little change at appreciably lower or higher temperatures. Unfortunately, very few oils have such a quality and the Saybolt Universal viscosity test of any oil at one temperature is not a true indicator of its viscosity at all temperatures. For this reason, a method was devised whereby this rate of change of viscosity of an oil with respect to temperature could be designated and the viscosity temperature characteristics of different oils compared. This designation is called Viscosity Index and is based upon the general observation that paraffin base oils show a relatively limited change in viscosity and are given an index of 100 while asphaltic base oils show a relatively high change in viscosity and are given an index of 0. Figure 16-2 shows an ASTM chart which can be used to determine the viscosity characteristics of engine oils. If the oil sample is tested for viscosity at two temperatures, such as 70 and 210°F (21.1 and 98.9°C), and the values are plotted on the chart and connected by a straight line, the slope of the line indicates the relative viscosity index of the oil and viscosities at other temperatures can be determined.

SAE Viscosity Classification In order to provide some standard means of designation of motor-oil grades, the Society of Automotive Engineers, cooperating with the oil refiners and automotive manufacturers, has worked out an oil-grading system based upon viscosity numbers as indicated by Table 16-1. As noted, the measured viscosity in seconds by the Saybolt Universal Viscosimeter (Fig. 16-1) is taken at 0 and 210°F (−17.8 and 98.9°C) to provide a knowledge of the relative change in fluid-

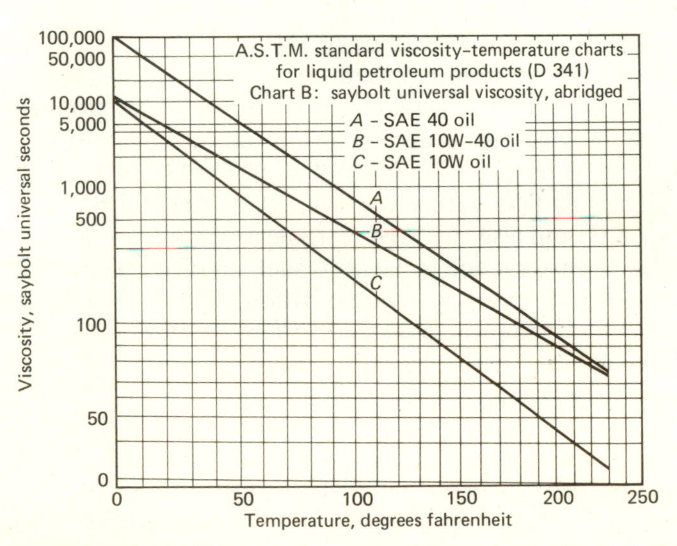

Figure 16-2 ASTM Standard viscosity—temperature chart. (Chart B, Saybolt universal viscosity, abridged, courtesy of American Society for Testing Materials.)

Table 16-1 SAE Crankcase-Oil Classification System

SAE viscosity number	Viscosity range, SUS			
	At 0°F		At 210°F	
	Min.	Max.	Min.	Max.
5W		4,000	39	
10W	6,000[1]	Less than 12,000	40	
20W	12,000[2]	48,000	45	
20	—	—	45	Less than 58
30	—	—	58	Less than 70
40	—	—	70	Less than 85
50	—	—	85	110

[1] Minimum viscosity at 0°F can be waived, provided viscosity at 210°F is not below 40 SUS.
[2] Minimum viscosity at 0°F can be waived, provided viscosity at 210°F is not below 45 SUS.

ity between average operating crankcase temperatures. The oil manufacturer designates the various grades by means of numbers instead of in the usual manner, the number being stamped on the container. The user, knowing the viscosity number recommended by the engine manufacturer, selects the oil accordingly.

Saybolt Universal viscosity is usually determined at temperatures of 100 and 210°F (37.8 and 98.9°C). Viscosity at lower temperatures such as 0°F (−17.8°C) is determined by extrapolation on a standard viscosity-temperature chart (Fig. 16-2). Summer-grade motor oils such as SAE 20 or 30 are based on viscosities at 210°F (98.9°C), while winter-grade oils such as SAE 10W are based on viscosities of 0°F (−17.8°C). Since the SAE viscosity classification system is based on viscosities at these two different temperatures, it becomes possible for a single oil, if properly formulated, to meet the viscosity requirements of more than one SAE grade. For example, an oil having a viscosity of 10,000 s at 0°F (−17.8°C) and 59 s at 210°F (98.9°C) falls within the SAE 10W range of 6,000 to 12,000 s at 0°F (−17.8°C) and also the SAE 30 range of 58 to 70 s at 210°F (98.9°C). It has become the practice to identify such an oil as SAE 10W-30. Similarly, it is possible to make such combinations as 5W-20, 20W-20, and 20W-40, for example. The principal advantage of multigraded oils is that, other factors being equal, they allow the engine to be operated safely over a wider range of atmospheric temperatures.

API Engine-Oil Service Classification In order to provide a more definite means of specifying the quality of engine oils and their adaptability to different types of engines, the various conditions under which the engines are operated, and the kind and quality of fuel used, the American Petroleum Institute (API), the American Society for Testing Materials

(ASTM), and the Society of Automotive Engineers (SAE), cooperated in establishing an entirely new classification system for crankcase oil in 1971 (Table 16-2). This system enabled engine oils to be defined and selected on the basis of their performance characteristics and the type of service for

Table 16-2 API Service Classification of Engine Crankcase Oil[1]

API service classification	Remarks
"S"—Service (service stations, garages, new car dealers, etc.)	
SA for Utility Gasoline and Diesel Engine Service	Service typical of older engines operated under such mild conditions that the protection afforded by compounded oils is not required. This classification has no performance requirements and oils in this category should not be used in any engine unless specifically recommended by the equipment manufacturer.
SB for Minimum Duty Gasoline Engine Service	Service typical of older gasoline engines operated under such mild conditions that only the minimum protection afforded by compounding is desired. Oils designed for this service have been used since the 1930s and provide only antiscuff capability and resistance to oil oxidation and bearing corrosion. They should not be used in any engine unless specifically recommended by the equipment manufacturer.
SC for 1964 Gasoline Engine Warranty Maintenance Service	Service typical of gasoline engines in 1964 through 1967 models of passenger cars and some trucks operating under engine manufacturers' warranties in effect during those model years. Oils designed for this service provide control of high- and low-temperature deposits, wear, rust, and corrosion in gasoline engines.
SD for 1968 Gasoline Engine Warranty Maintenance Service	Service typical of gasoline engines in 1968 through 1970 models of passenger cars and some trucks operating under engine manufacturers' warranties in effect during those model years. Also may apply to certain 1971 and/or later models as specified (or recommended) in the owners' manuals. Oils designed for this service provide more protection against high- and low-temperature engine deposits, wear, rust, and corrosion in gasoline engines than oils which are satisfactory for API Engine Service Classification SC and may be used when API Engine Service Classification SC is recommended.
SE for 1972 Gasoline Engine Warranty Maintenance Service	Service typical of gasoline engines in passenger cars and some trucks beginning with 1972 and certain 1971 models operating under engine manufacturers' warranties. Oils designed for this service provide more protection against oil oxidation, high-temperature engine deposits, rust, and corrosion in gasoline

Table 16-2 (*Continued*)

API service classification	Remarks
	engines than oils which are satisfactory for API Engine Service Classifications SD or SC and may be used when either of these classifications is recommended.
"C"—Commercial (fleets, contractors, farmers, etc.)	
CA for Diesel Engine Service	Service typical of diesel engines operated in mild to moderate duty with high-quality fuels and occasionally has included gasoline engines in mild service. Oils designed for this service provide protection from bearing corrosion and from ring-belt deposits in some naturally aspirated diesel engines when using fuels of such quality that they impose no unusual requirements for wear and deposit protection. They were widely used in the late 1940s but should not be used in any engine unless specifically recommended by the equipment manufacturer.
CB for Diesel Engine Service	Service typical of diesel engines operated in mild to moderate duty, but with lower-quality fuels which necessitate more protection from wear and deposits. Occasionally has included gasoline engines in mild service. Oils designed for this service provide necessary protection from bearing corrosion and from ring-belt deposits in some naturally aspirated diesel engines with higher sulfur fuels. Oils designed for this service were introduced in 1949.
CC for Diesel Engine Service	Service typical of certain naturally aspirated, turbocharged, or supercharged diesel engines operated in moderate- to severe-duty service and certain heavy-duty gasoline engines. Oils designed for this service provide protection from high-temperature deposits and bearing corrosion in these diesel engines and also from rust, corrosion, and low-temperature deposits in gasoline engines. These oils were introduced in 1961.
CD for Diesel Engine Service	Service typical of certain naturally aspirated, turbocharged, or supercharged diesel engines where highly effective control of wear and deposits is vital, or when using fuels of a wide quality range including high-sulfur fuels. Oils designed for this service were introduced in 1955 and provide protection from bearing corrosion and from high-temperature deposits in these diesel engines.

[1] Revised, 1979.
Source: American Petroleum Institute.

which they are intended. The classification system is open-ended and addition or deletion can be made at any time as needed. The classification system was published as SAE Recommended Practice J183 and J183a, and performance requirements are technically described in ASTM Research Report RR D2: 1002, revised January, 1971. The new API Engine Service Classification System does not affect the SAE Crankcase Oil Viscosity Classification System; therefore, it is used as before to indicate the SAE viscosity.

Comparison of present and previous API service classification of engine crankcase oil with related designation for industry and military is given in Table 16-3.

Engine Performance or Sequence Tests The fact has been recognized for a number of years that the performance of oils in engines cannot be predicted from the results of certain chemical and physical tests; the only real measure of oil quality is its performance in service. Some engine manufacturers and certain allied agencies have developed exhaustive engine test procedures to determine the specific performance characteristics

Table 16-3 Comparison of Present and Previous API Service Classification of Engine Crankcase Oils

API engine service classifications	Previous API engine service classifications	Related designations military and industry
	Service station engine services	
SA	ML	Straight mineral oil
SB	MM	Inhibited oil
SC	MS (1964)	1964 MS Warranty Approved, M2C101-A
SD	MS (1968)	1968 MS Warranty Approved, M2C101-B, 6041-M (Prior to July, 1970)
SE	None	1972 Warranty Approved, M2C101-C, 6136-M (Previously 6041-M Rev.), MIL-L-46152[1]
	Commercial and fleet engine services	
CA	DG	MIL-L-2104A
CB	DM	Supp. 1
CC	DM	MIL-L-2104B, MIL-L-46152[1]
CD	DS	MIL-L-45199B, Series 3, MIL-L-2104C[2]

[1] Oils meeting performance requirements of MIL-L-46152 meet the performance requirements of both API Service SE and CC.

[2] Oils meeting performance requirements of MIL-L-2104C meet the performance requirements of both API Service CD and SC.

Source: American Petroleum Institute.

of oils. These tests are able to measure the performance of oil under the most severe conditions to which they might be subjected.

Although not required, most oil companies identify those oils that have been tested and proved in sequence tests.

ENGINE-OIL ADDITIVES

Changes and improvements in the design of certain types of engines created new problems which were attributed largely to inadequate lubrication. Some of these were ring sticking, ring and piston scuffing, and excessive bearing corrosion and wear. These difficulties were encountered particularly in the high-speed diesel engines and other heavy-duty types used in tractors, trucks, and buses. As a result of the efforts of oil technologists and engine designers, it was found that certain chemicals, when added to a well-refined motor oil, would improve its lubricating properties and eliminate specific troubles. Most oil refiners are now producing special grades and qualities of motor oils containing chemicals to improve their lubricating properties, particularly under heavy-duty conditions. The term *additives* is usually applied to these chemicals.

Some lubricating-oil additives prevent or retard oil oxidation and thereby control carbon deposits and corrosion, while others improve such physical properties of the oil as its pour point, viscosity index, or film strength. The principal additives and their specific action are listed as follows:

1 *Oxidation inhibitor*. Excessive engine heat causes oil oxidation, which results in permanent thickening of the oil and can accelerate the formation of harmful deposits along with acids which attack engine bearing. Inhibitor slows down oxidation and protects bearing.

2 *Corrosion and rust inhibitor*. Combustion process produces water and corrosive acid. Certain compounds are added to form unbroken protective film on engine parts or completely encircle individual water and acid molecules.

3 *Detergents*. Certain compounds, when added to a motor oil, prevent the building up of carbon and gummy or carbonaceous deposits on the pistons and under the rings or on other engine parts by keeping the engine clean.

4 *Dispersants*. Certain chemicals, when added to an oil, cause any finely divided insoluble particles of carbon to remain in suspension in the oil rather than to separate out and form sludge deposits. The action of detergents and that of dispersants appear to be closely allied.

5 *Extreme-pressure agents*. Under certain conditions of high pressure and temperature, an ordinary lubricant, even with a high viscosity, will not provide sufficient oil film between the metal surfaces to control wear. The best example of this is the hypoid gear drive of the modern automobile. A special lubricant has been developed for this type of gear by adding certain chemical agents to an oil of proper viscosity. These

additives enable the lubricant to withstand the unusual high pressure and temperature conditions developed by hypoid gears and yet give satisfactory lubrication. Prevent metal-to-metal contact.

6 *Foam inhibitors.* Some types of motor oils have a tendency to absorb air when agitated vigorously, thus forming a foam. Certain chemicals, when added to such oils, accelerate the rate of breakdown of the foam but may not actually prevent foaming.

7 *Pour-point depressants.* Certain chemicals, when added to oils, lower the temperature at which oil will pour or flow, thus improve its pore-point characteristics, even though the lubricant was dewaxed in the refining process.

8 *Viscosity-index improvers.* Under certain conditions, it is desirable to use an oil the viscosity of which does not increase excessively at low temperatures. Chemicals have been found which, when added to these oils, aid in retarding this tendency to thicken as the temperature drops. They are called viscosity-index improvers.

9 *Antiscuff/antiwear agent.* Viscosity and natural friction-reducing properties of mineral oils may be insufficient to sustain the extreme pressure encountered in modern engines. Therefore additives are used to form lubricating films strong enough to carry the loads imposed on them.

SYNTHETIC LUBRICANTS

Since World War II, synthetic lubricants have been used in special applications where conventional mineral oil could not withstand the extreme temperatures involved. Applications included the low temperatures of internal combustion engines in the arctic and high temperatures in jet engines.

A synthetic lubricant is a nonnaturally occurring product made by the chemical reaction of two or more simpler chemical compounds which display lubricating properties. There are many types of synthetic lubricants. Boehringer[1] indicates that polyglycols, dialkylated benzene, polymerized alpha olefins, and diesters have been seriously considered for reciprocating-engine use.

There is a difference of opinion with regard to synthetic oils. Richman and Keller[2] state that synthetic oil provides performance superior to premium quality SAE 10W-40 mineral oils in engine cleanliness, fuel economy, oil economy, cold starting capability, and intake-system cleanliness. Rodgers and Kabel[3] indicate that synthetic oil shows better performance

[1] R. H. Boehringer, "Diester Synthetic Lubricants for Automotive and Diesel Application," paper 750686, Society of Automotive Engineers, Inc., Warrendale, PA 15096, 1975.

[2] W. H. Richman and J. A. Keller, "An Engine Oil Formulated for Optimized Engine Performance," paper 750376, Society of Automotive Engineers, Inc., Warrendale, PA 15096, 1975.

[3] J. J. Rodgers and R. H. Kabel, "Vehicle Evaluation of Synthetic and Conventional Engine Oils," paper 750827, Society of Automotive Engineers, Inc., Warrendale, PA 15096, 1975.

in two areas of deposit control, inferior performance with respect to wear protection, and equal performance of fuel and oil economies.

A service classification similar to the one for mineral oil has not been developed for synthetic oils.

GREASES

A grease is defined by the American Society for Testing Materials as "a semisolid or solid combination of a petroleum product and a soap or a mixture of soaps, with or without fillers, suitable for certain types of lubrication." Greases are used primarily for slow-moving parts where pressures are high, and for parts that are concealed or inaccessible—wheel bearings, spring shackles, universal joints, and water pumps. Advantages of grease as a lubricant include: (1) simplified seal design and may act as a seal, (2) shielded bearing can be prelubricated, and (3) generally less-frequent lubrication required. Many different kinds of greases have been developed to meet the specific requirements of automobiles, trucks, and tractors.

A grease is a mixture of metallic soap and mineral oil. Certain chemicals and additives such as graphite, talc, and asbestos may be incorporated in grease to provide stabilization, oxidation resistance, rust prevention, extreme-pressure characteristics, tackiness, and other desirable characteristics. The soap serves as a thickening agent, particularly at certain temperatures. Soap is produced by the chemical action between an alkali such as calcium hydroxide and a fat—usually animal or vegetable—or a fatty acid. The process is called saponification. Table 16-4 lists some of the most common thickeners and outlines their general characteristics.

Grades of Grease There are nine standardized grades of greases that have been adopted by the National Lubricating Grease Institute (NLGI). The grade numbers, which are an indication of firmness or penetration of the grease, are shown in Table 16-5.

The penetration test[1] determines the depth in tenths of a millimeter that a standard cone penetrates a sample of worked grease under prescribed conditions of weight, time, and temperature. The firmness or hardness of grease ranges from No. 000 (very soft) to No. 6 (very hard). Most manufacturers recommend either No. 1 or No. 2 grade.

GEAR AND TRANSMISSION LUBRICANTS

A gear or transmission lubricant is a heavy-bodied oil. It must have sufficient body to cushion and sustain the sudden high-pressure loads transmit-

[1] ASTM Method D217

Table 16-4 Characteristics of Greases

Class	Type	Kind of thickener	Approximate dropping point	Characteristics
Lime soap	Cup	Calcium-fat	190°F	Smooth, water-resistant, separates on loss of water content, limited consistency loss on working.
	Complex	Calcium-fat and calcium acetate	Over 500°F	Some or slight fiber, water-resistant, usually some E.P. properties, consistency change on aging.
Soda soap	Fiber or sponge	Soda-fat	375°F	Fibrous, not water-resistant, does not separate at elevated temperatures, variable consistency loss on working.
	Medium or short fiber, smooth	Soda-fat or fatty-acid	375°F	Semismooth to smooth, not water-resistant, does not separate at elevated temperatures, variable consistency loss on working
	Block or brick	Soda-fat, fatty-acid, and rosin	400°F	Smooth, hard, not water-resistant, does not separate at elevated temperatures.
Lithium soap	Multipurpose	Lithium-fat or fatty-acid	375°F	Smooth, water- and heat-resistant, does not separate at elevated temperatures, limited consistency loss on working.
Nonsoap	High temperature, multipurpose	Inorganic-clay or silica Organic-Arylurea	Usually 500°F or higher	Smooth, water- and heat-resistant, high temperature stability, good mechanical and chemical stability.
Residuum	Pinion or gear compound	Usually none, sometimes lead naphthenate	Rapid thinning when heated	Smooth, black, adhesive, excellent pressure and water resistance.

Source: National Lubricating Grease Institute.

Table 16-5 Grades of Greases Versus Penetration. National Lubrication Grease Institute (NLGI)

NLGI Grade, Number	ASTM Worked Penetration at 77°F ± 3
000	445–475
00	400–430
0	355–385
1	310–340
2	265–295
3	220–250
4	175–205
5	130–160
6	85–115

Source: National Lubrication Grease Institute.

ted to the gear-teeth surfaces and yet cling to these teeth. Also, it must not create undue resistance to motion and should flow in ample quantity to the shaft bearings. Gears such as the hypoid-type require a lubricant having the highest possible load-carrying capacity. Such oils are referred to as extreme-pressure (EP) lubricants and contain certain additives.

The most important single property of gear lubricants is viscosity. Table 16-6 gives the SAE viscosity recommendations for transmission lubricants. The viscosity of gear oils is not as heavy as the assigned numbers may indicate. SAE 80 gear oil has about the same viscosity as SAE 20 or 30 engine oil. The larger number is assigned to gear oil to separate the two types of oils.

Table 16-6 SAE Transmission and Axle Lubricants Classification System

SAE viscosity number	Viscosity range, SUS				Consistency, must not channel in service at
	0°F		210°F		
	Min.	Max.	Min.	Max.	
80	15,000	100,000	—	—	−20°F
90	—	—	75	120[2]	−10°F
140	—	—	120	200	20°F
250	—	—	200		

[1] The minimum viscosity at 0°F may be waived if the viscosity is not less than 48 SUS at 210°F.

[2] The maximum viscosity at 210°F may be waived if the viscosity is not greater than 750,000 SUS at 0°F (extrapolated).

There are some types of transmissions and differentials that require special oils that do not fall into the API service classifications for gear and transmission oil. Included in this group are tractors that use the same oil for transmission, differential, hydraulic systems and the hydrostatic-type transmission. In these cases, the manufacturer's operator's manual should be consulted regarding the correct type of oil to be used.

ENGINE LUBRICATION SYSTEM

The lubrication of an engine may be considered under two distinct heads, namely, (1) the choice and use of the correct kind and grade of lubricant and (2) the choice, design, construction, and operation of the lubrication system with which the engine is equipped. A good lubricating system must be efficient in operation, reliable, troubleproof, and simple. Yet, even though it possesses all these important features, if a poor-quality lubricant or one of incorrect grade is used, unsatisfactory service is likely to result.

As previously stated, in selecting the oil to be used, satisfactory results are more likely to be obtained if the advice and recommendations of the engine manufacturer are followed. Then, having selected a suitable oil, one should familiarize oneself with the construction and operation of the lubrication system of the engine itself and see that it functions properly at all times.

Parts Lubricated by Lubrication System The most important parts of an engine requiring lubrication are as follows:

1. Cylinder walls, pistons, and piston rings
2. Piston pin
3. Crankshaft and connecting-rod bearings
4. Camshaft bearings
5. Valves and valve-operating mechanism

In general, engine-lubrication systems may be classified as (1) splash; (2) pressure-feed and splash; and (3) full-pressure feed.

Splash System In the splash system, the lubrication of all the principal engine parts is dependent directly upon the splashing of the oil by a dipper on the bottom side of the connecting-rod cap that dips into the crankcase oil each time the piston reaches BDC. Some small engines also use oil slingers which are driven by the camshaft. These are gearlike parts that throw oil from the oil pan up into the moving engine parts. The splash system is used on some small four-cycle, one-cylinder engines for power lawn mowers.

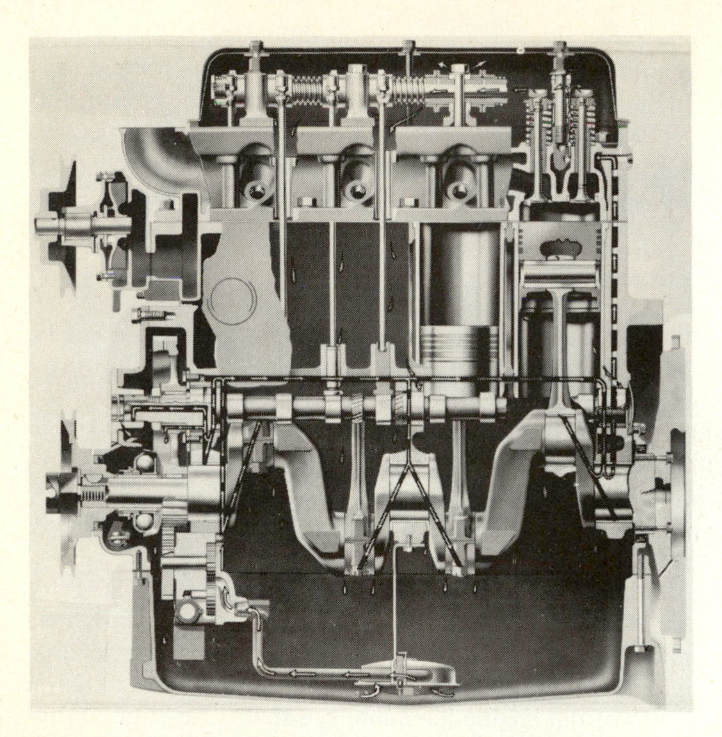

Figure 16-3 Pressure-feed and splash lubrication system. (Courtesy of Massey-Ferguson, Inc.)

Pressure-Feed and Splash System In this system (Figs. 16-3 and 16-4) the oil is forced directly to the main crankshaft, connecting rod, and camshaft bearings. Drilled passages in the crankshaft carry the oil from the main bearings to the connecting-rod bearings as shown. The oil oozing out of these bearings creates a spray that lubricates the cylinder walls, pistons, and piston pins. A pressure indicator shows whether the pump is working and pressure is being maintained. The valve mechanism is also oiled by pressure from the crankcase as shown.

Full-Pressure System This system goes a step further and forces the oil not only to the main crankshaft, connecting rod, and camshaft bearings, as previously described, but also to the piston-pin bearings through tubes or passages that lead from the connecting-rod bearings up the connecting rod to the piston pin, as shown in Fig. 16-5. The cylinders and pistons receive their oil from the piston pins and from the mist created by the oil issuing from the various bearings. The valve mechanism likewise is oiled by pressure feed as indicated.

Oil-Circulating Pumps—Oil Pressures—Relief Valves Two types of pumps, gear and rotor, are used for circulating the oil in engines equipped

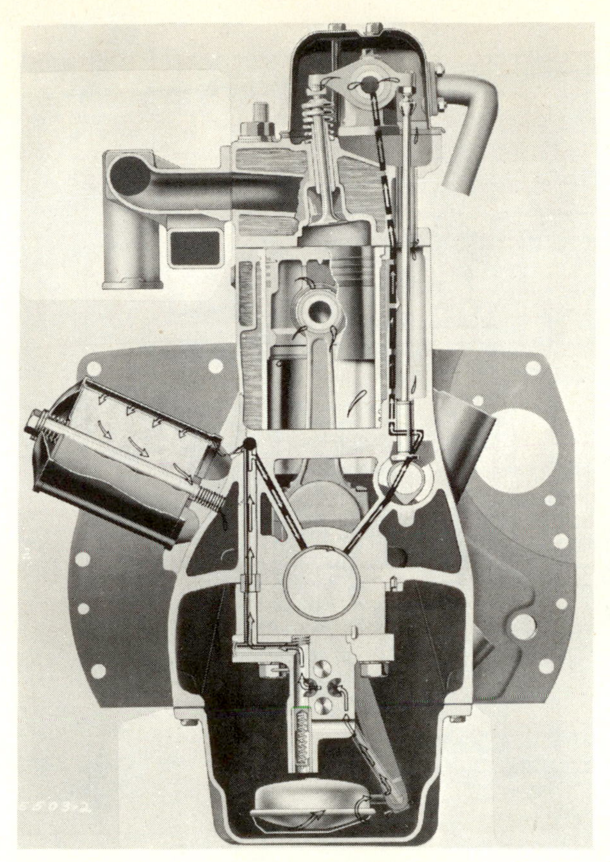

Figure 16-4 Pressure-feed and splash lubrication system showing full-flow screw-on oil filter. (Courtesy of Massey-Ferguson, Inc.)

with pressure-feed and splash, or the full-pressure feed system. The type most used is the gear pump (Fig. 16-6). It consists of two, small spur gears held in a horizontal position and enclosed in a close-fitting, oil-tight housing. A vertical shaft from the camshaft of the engine drives one pump gear; it, in turn, drives the second gear. The oil enters on that side of the housing on which the gear teeth are turning away from each other (going out of mesh) and is carried between the teeth and the inner surface of the housing around to the opposite side and discharged.

The force or pressure applied to the oil by the pump depends largely upon the type of lubrication system used and the clearance between engine parts. In different tractor engines, high pressure ranges from 30 to 70 lb/in^2 (2.1 to 4.9 Kg/cm^2).

In order to maintain the correct pressure, and control the quantity of oil circulated, a relief valve is built into the pump-discharge line. This valve consists, essentially, of a ball, held in place by a spring over an opening. The valve operates in such a way as to permit a certain amount

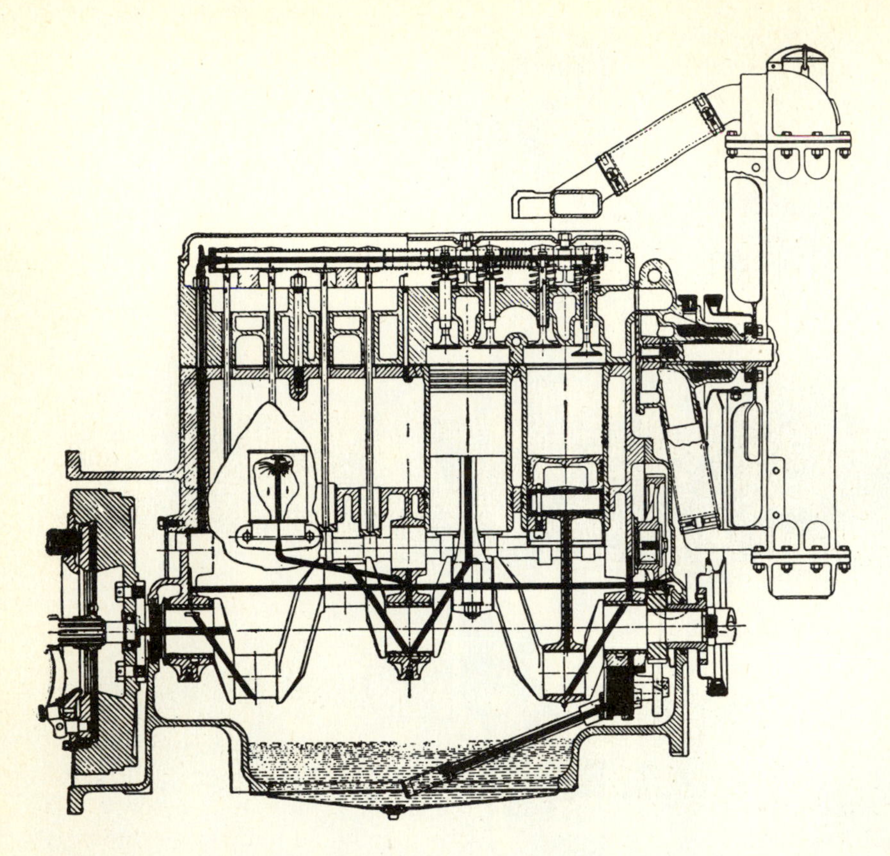

Figure 16-5 Full-pressure lubrication system.

of the oil to bypass back to the oil reservoir as it leaves the pump if pressure is too high, thus reducing the oil pressure to a preset maximum.

Oil-Pressure Indicators The oil-pressure indicator tells the tractor operator what the engine oil pressure is. This gives the operator a warning if something occurs in the lubrication system that prevents delivery of oil to vital parts. The two general types of oil-pressure indicators used on tractors are electric resistance or indicator light.

The electric resistance type utilizes changes in oil pressure to vary the resistance in the engine sending units. The change in resistance causes the pointer or needle on the dash unit to move, thus indicating the oil pressure.

The indicator-light system has a light that comes on when the oil pressure is low. When the engine switch is turned on, the oil pressure is

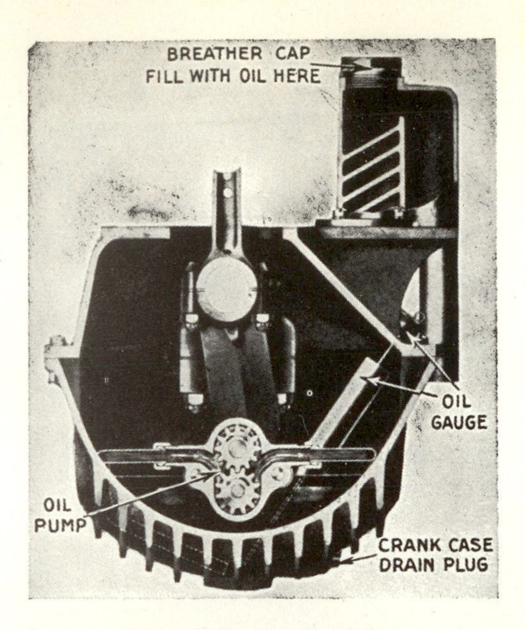

Figure 16-6 A gear-type oil pump.

low and the light comes on. Oil pressure builds up after the engine has started and turns off the oil-pressure light.

Crankcase Ventilation Air must circulate through the crankcase when an engine is running in order to remove water and fuel that collect in the crankcase when the engine is cold. The air also removes blow-by gases from the crankcase. Acids and sludge will form if the water, fuel, and blow-by gases are not removed. Sludge can clog oil lines and pump-screen, and acids may corrode metal parts and bearings, thus the engine may be ruined.

Tractor crankcases are ventilated by an opening at or near the front top of the engine and a vent tube near the back. The motion of the tractor and the rotation of the crankcase moves air through the crankcase, as shown in Fig. 16-7.

Crankcase ventilation (breather) is necessary because the pumping action of the pistons creates an uneven pressure, which might force oil upward past the pistons into the combustion chamber or out through the crankcase joints.

Oil Filters All tractors are regularly equipped with an oil filter, and it ranks with the air cleaner as a most important accessory for the purpose of reducing engine wear. The purpose of an oil filter is to remove sand, soil, metal particles, carbon, and all other undesirable foreign matter from the engine oil during operation.

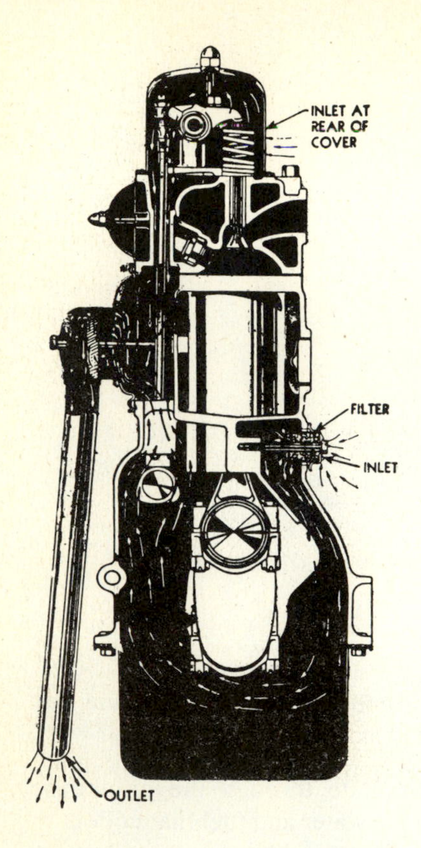

Figure 16-7 Tractor engine with crankcase ventilating system.

Oil filters are of two types: (1) the bypass filter, which filters only a part of the oil from the oil pump; and (2) full-flow filters, which filter all the oil from the pump before it is circulated through the system. Most modern tractors utilize the full-flow, screw-on-type filter shown in Figs. 16-4 and 16-8.

The full-flow filter includes a spring-loaded bypass system (Fig. 16-8). The valve protects the engine against oil starvation if the filter becomes clogged. Oil pressure increases when the filter becomes clogged. Thus, this increase in pressure causes the valve to open; the oil bypasses the filter and flows through the system. The filter element, to maintain filter efficiency, should be replaced according to instruction in the operator's manual.

Oil-Level Indicator The oil-level indicator shows the level of oil in the crankcase sump (oil pan) and indicates whether it is one quart low or full. The usual type of oil-level indicator for tractor engines is the bayonet-type or dipstick, as shown in Fig. 16-9. The dipstick is located on the engine so that its lower end is submerged in the oil. The oil level is

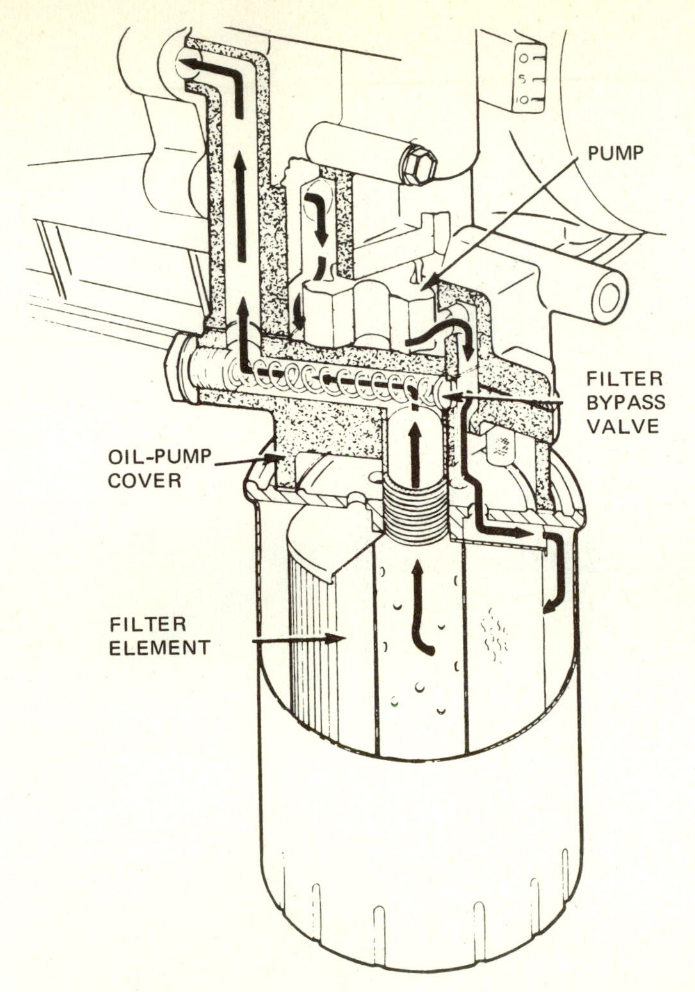

Figure 16-8 Full-flow screw-on oil filter cutaway to show bypass valve.

determined by withdrawing the dipstick and noting how high the oil mark appears on the dipstick.

Engine Lubrication Troubles As already stated, the primary requirements for the proper lubrication of an engine under all operating conditions are (1) the design and construction of a dependable system of oil circulation and distribution and (2) the selection of a lubricant of the correct grade and quality. However, the satisfactory operation of the engine from the standpoint of lubrication is also dependent upon certain other conditions. Some of these are:

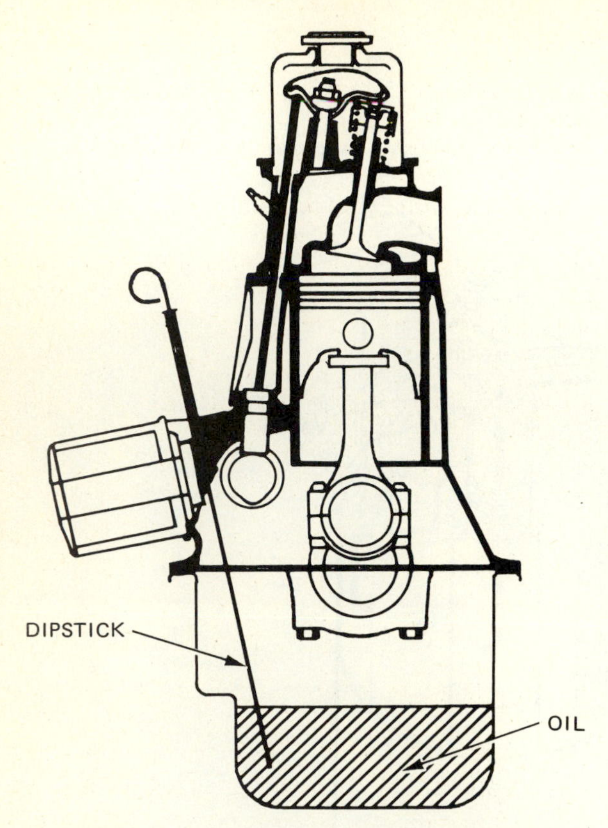

Figure 16-9 Dipstick used to indicate oil level in oil pan.

1 Periodic and regular changing of the oil.
2 Careful observation and regulation of the oil-reservoir supply and the quantity distributed or fed to the parts requiring lubrication.
3 Regular servicing of the oil filter according to instructions.
4 Prevention of pollution of the oil from any or all of the following sources:
   a Solid matter of foreign nature, such as dust, dirt, particles of vegetable matter, iron and steel particles, and so on, which may get into the reservoir through unprotected openings or by carelessness in handling the oil or working around the engine.
   b Water from leaks in the cooling system or through unprotected openings.
   c Liquid fuel which may be taken into the cylinders in an unvaporized condition and, remaining unburned, may get by the piston and rings, diluting the oil.
   d Particles of carbon due to excess carbon deposits in the combustion space and the piston-ring grooves.

PROBLEMS AND QUESTIONS

1 Discuss the refining process as it applies to the production of the various kinds of engine oils and other lubricants.
2 Explain the meaning of oil viscosity and viscosity index. Describe how each is measured or determined.
3 Explain the API engine-oil service classification system and the SAE system of designating viscosity.
4 Explain the use of lubricating-oil additives. Name the most important additives and give their specific action.
5 What is meant by synthetic lubricants? What are their advantages?
6 Define grease and explain what determines its characteristics.
7 Name and describe the systems used to lubricate tractor engines.
8 Describe the type of oil filter most commonly used and explain how it should be serviced.
9 Discuss engine-oil consumption, its causes and control.

Chapter 17

Diesel Engine Construction and Operation

Diesel Engines and Tractor Development For a number of years in the early development of the diesel engine its use was confined largely to very heavy-duty stationary applications such as electric power generation, irrigation water pumping, propelling large boats for both passenger and freight transportation, and for certain power needs in large factories and industrial plants. These engines were heavy, slow-speed, single- or multiple-cylinder units of either the two- or four-stroke-cycle type.

Lighter weight, higher speed diesel engines appeared about 1925, but development was slow. However, by 1930 reliable and well-designed multiple-cylinder, high-speed diesel engines were being used for some makes of heavy-duty farm and industrial tractors. This development was slow for a time until the end of World War II. Since that time, essential design and manufacturing processes have been developed that have resulted in the rapid growth and utilization of diesel power by all farm tractor and equipment manufacturers. Diesel tractors are available in all sizes from about 20 to 200 hp or more and have largely displaced gasoline and LP gas-burning tractors.

TRACTOR ENGINE CHARACTERISTICS AND DESIGN

In general, it can be said that the diesel tractor engine may be considered as a heavy-duty automotive type engine. The operating principles are identical with those of the auto engine, with the possible exception of the fuel handling and ignition process. In fact, diesel-powered automobiles have been in use on a limited basis for several years, and most buses and heavy-duty trucks are equipped with diesel engines. The principal difference is that all diesel engines are designed and built somewhat heavier and stronger and, in most cases, operate at lower speeds than the conventional automobile engine. In general, tractors use the in-line cylinder arrangement, with the number of cylinders varying from two to eight, depending on the tractor type and power requirements. Figure 17-1 shows the modern diesel tractor engine construction and parts.

Compared with the ordinary automobile engine for a given size and power rating, the diesel engine varies with respect to certain general design characteristics for the reason that its higher compression creates

Figure 17-1 Cutaway view of six-cylinder diesel engine. (Courtesy of Deere and Company.)

greater stresses, higher operating temperatures, and greater possible wear. In general, the problems have been solved largely by making certain engine parts of special metals or alloys, or just making them heavier. Some of these specific differences are:

1 Cylinder blocks are heavier.

2 Aluminum alloys are used for pistons because of lighter weight and better heat transfer characteristics.

3 The closed end of the piston or the combustion space in the cylinder head are specially shaped to provide better fuel mixture turbulence for maximum power and efficiency.

4 Valves and seat inserts are made of special steel alloys because of exposure to high temperatures and the hammering action during operation.

5 Crankshafts and connecting rods are made heavier to withstand greater loads and pressures.

6 Special alloy heavy-duty bearings are necessary.

7 The engine lubrication system must be designed to insure positive lubrication of all points exposed to high pressures and temperatures.

8 Heavy-duty air cleaners and oil filters are needed.

PRINCIPLES OF OPERATION

As explained in Chap. 7, the diesel engine differs from a carbureting-type engine primarily in two ways, namely, (1) only air is taken in on the intake stroke of the piston, the liquid fuel being injected directly into the combustion chamber at the end of the compression stroke and (2) the fuel mixture is ignited by high compression, and no special ignition device or mechanism is needed. On the other hand, all diesel engines operate on either the two- or the four-stroke-cycle principle like other internal-combustion engines.

Two-Stroke-Cycle Diesel The principles of operation and events involved for a two-stroke-cycle engine are shown by Fig. 17-2. The cycle begins with the upward movement of the piston from its CDC position. The intake and exhaust ports are closed and the charge of fresh air is compressed to approximately 500 lb/in^2. When the piston reaches HDC, a charge of fuel is injected into the combustion space. The high temperature existing ignites the mixture of atomized fuel and air, and combustion takes place in such a manner that a constant pressure equal to the compression pressure is maintained as the piston moves downward on the power stroke. Expansion continues until the exhaust port is opened and the burned gases are released. The intake port is likewise uncovered immediately after the exhaust port, and the cycle is repeated.

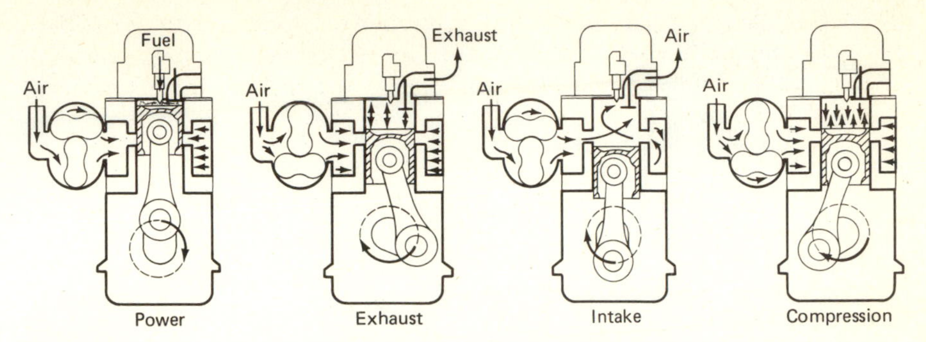

Figure 17-2 One working cycle of a two-cycle engine. (Courtesy McGraw-Hill Book Company.)

Four-Stroke-Cycle Diesel Figures 17-3 and 17-4 show the principles of operation of a four-stroke-cycle diesel engine. It will be observed that the events take place in the same manner as in the ordinary four-stroke-cycle carbureting-type engine (see Chap. 7), with the exception that air alone is drawn in on the intake stroke and the liquid fuel is injected into the cylinder at or near the end of the compression stroke. Obviously, two valves, intake and exhaust, operated by a gear and camshaft, and the usual mechanism, are necessary. Figure 17-5 shows the approximate valve- and fuel-injection timing for tractor- and truck-type diesel engines.

Fuel-Injection Systems The proper injection of the fuel into the combustion chamber against the high pressure is one of the most difficult problems encountered by the diesel-engine designer. Since the mechanism involved must supply a fuel charge sufficient only for a single explosion, it is obvious that it must be carefully designed to operate with the utmost precision. The principal requirements of a diesel fuel supply and

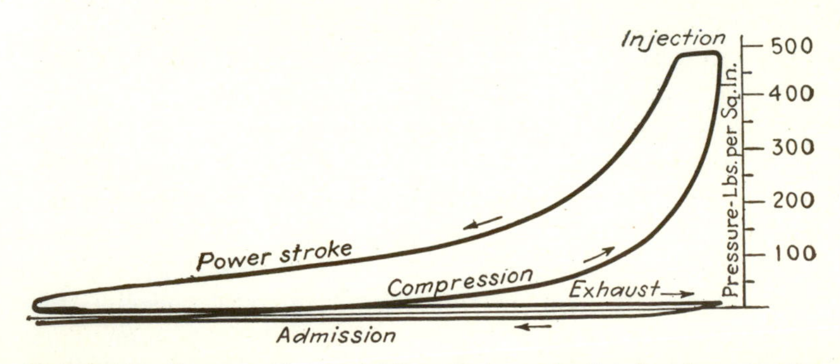

Figure 17-3 Pressure diagram of four-stroke-cycle diesel-engine operation.

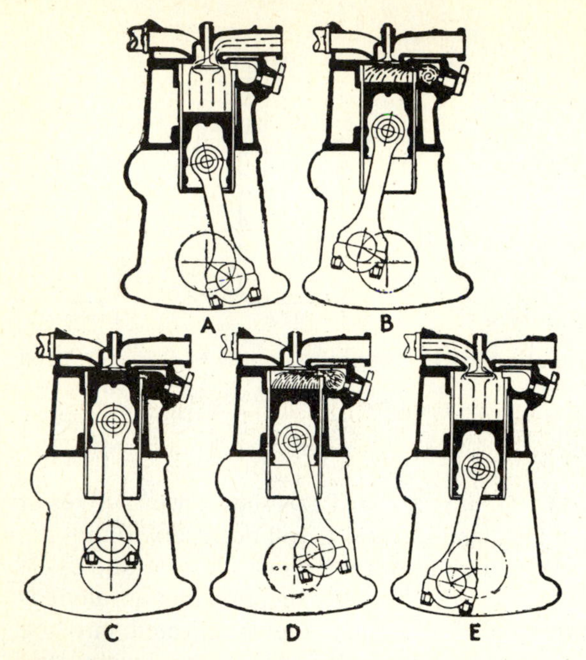

Figure 17-4 Cycle of events of a four-stroke-cycle diesel engine.

injection mechanism are (1) that it supply a correct fuel charge to each cylinder according to the engine load and speed, (2) that it inject the fuel at the correct time in the cycle, (3) that it facilitate efficient fuel utilization by atomizing the charge at the time of injection, and (4) that it not be subject to undue wear or require frequent adjustment or servicing. Considering these factors, it is obvious that for engines having limited piston displacement and high speed, the utilization of the diesel principle becomes increasingly difficult.

Although the earlier large, heavy-duty stationary diesel engines utilized compressed air to force the fuel charge into the cylinder, all modern multiple-cylinder, high-speed engines inject each fuel charge mechanically into the compressed hot air. This is known as direct or solid injection. The timing of the charge is controlled by timing the stroke of the injector pump with the crankshaft and piston position.

Referring to Figs. 17-6 and 17-7, a diesel fuel-injection system consists of certain major parts as follows: (1) fuel tank, (2) fuel transfer pump, (3) primary fuel filter, (4) two-stage secondary filter, (5) injection pump, (6) injection nozzles, and (7) governing mechanism.

The low-pressure transfer pump pushes the fuel through the filters to the high-pressure injection pumps which, in turn, force the necessary fuel charges to the injectors, and then into the respective combustion chambers. The primary filter removes coarser foreign particles, while the sec-

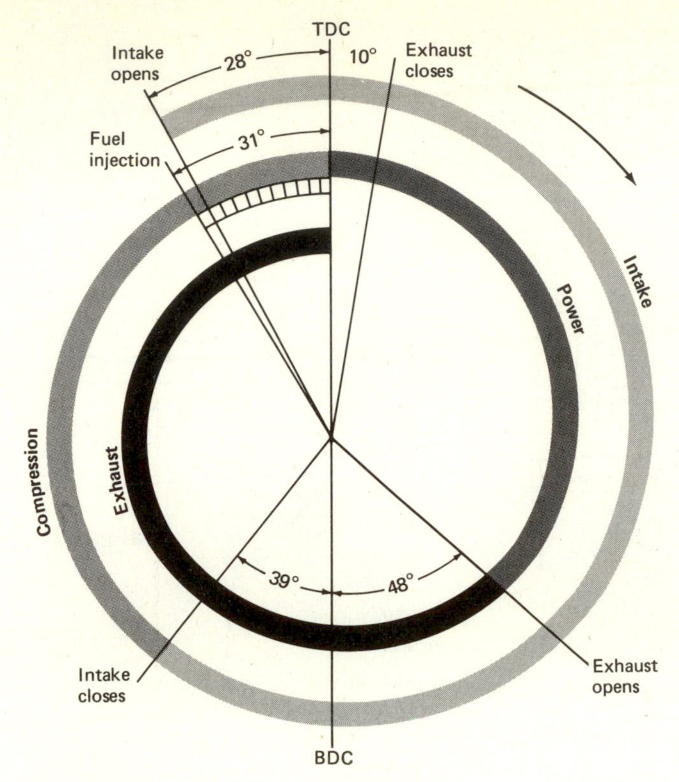

Figure 17-5 Valve and fuel-injection timing for a diesel engine.

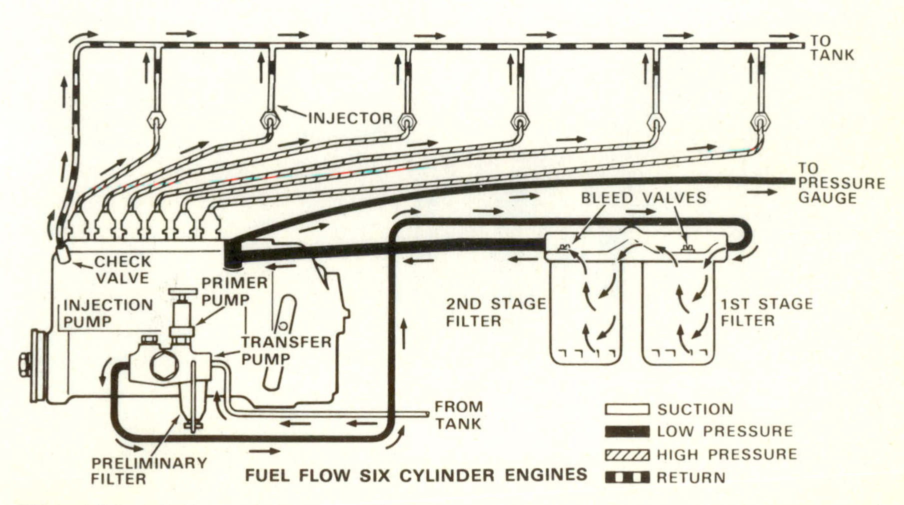

Figure 17-6 Fuel supply and injection system. (Courtesy of J I Case Company.)

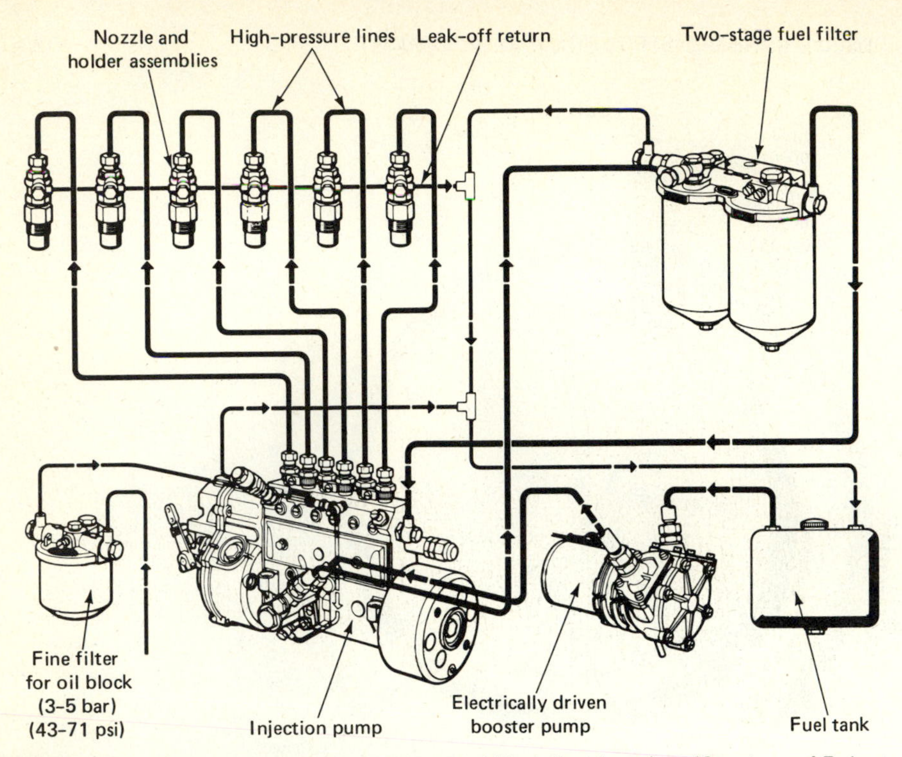

Figure 17-7 Injection system for dual- or multifuel diesel engine. (Courtesy of Robert Bosch Corporation.)

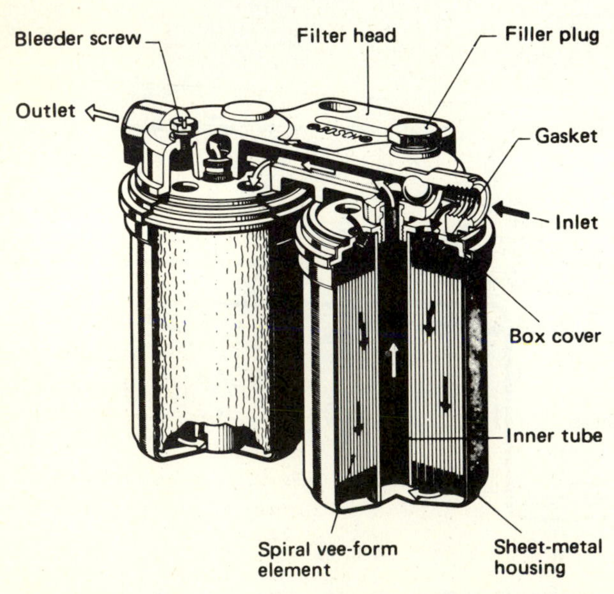

Figure 17-8 Sectional view of a two-stage box fuel filter. (Courtesy of Robert Bosch Corporation.)

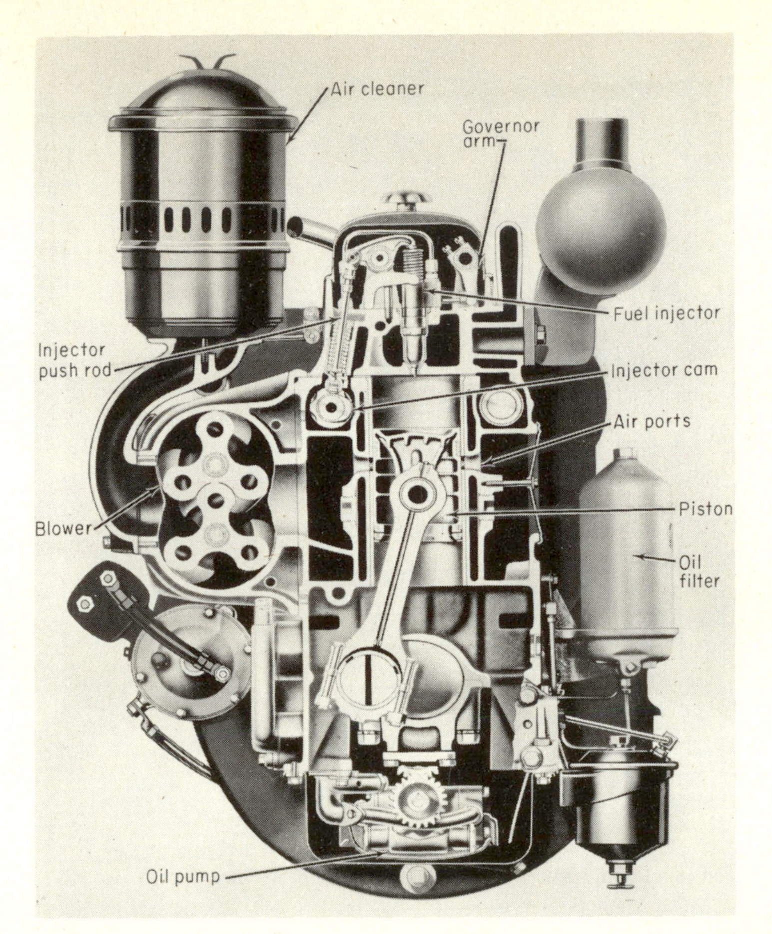

Figure 17-9 Cutaway view of a high-speed two-stroke-cycle diesel engine showing air blower and injection mechanism. (Courtesy of Detroit Diesel Engine Division, General Motors Corporation.)

ondary filters trap any very fine material. Figure 17-8 shows the parts and construction of a two-stage box fuel filter.

In general, direct-injection systems may be divided into three types, namely, (1) single-unit, with pump and injection nozzle built together in a single-unit mechanism; (2) two-unit with individual pumps and injectors for each cylinder; and (3) two-unit, with one master injector pump but with distribution valves and injectors for each cylinder.

Figure 17-9 shows the operating mechanism for the single-unit system, and Fig. 17-10 shows the detailed construction of the injector. This system is used in the GM two-stroke-cycle diesel engine.

Figure 17-11 shows the Caterpillar tractor two-unit injection system. In this case, a camshaft operates the plungers forcing the fuel charge to the injector valve and into the precombustion chamber. As the fuel is

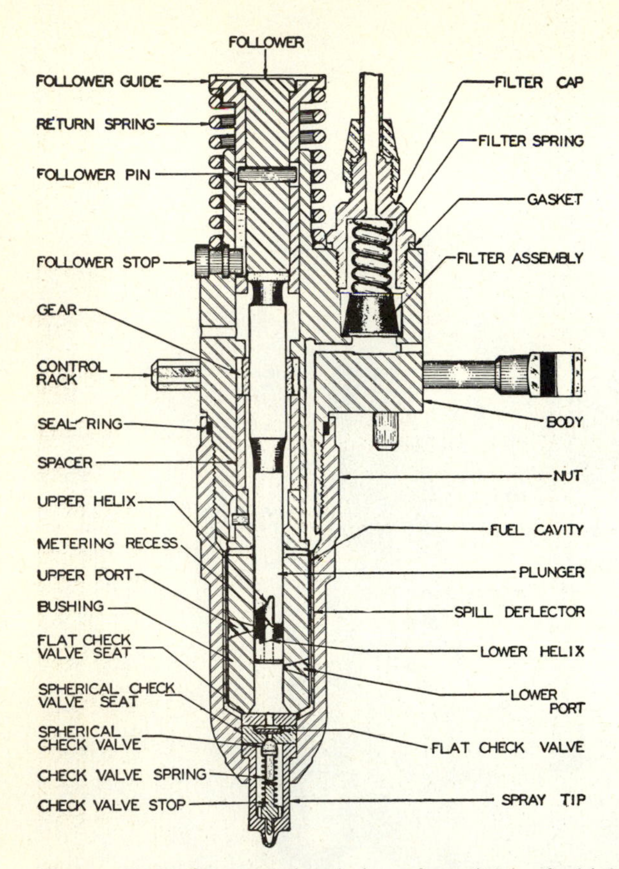

Figure 17-10 Cross-sectional view of a unit-type fuel injector.

sprayed into this chamber, the highly heated air causes a small portion to ignite. As injection continues, the fuel is enveloped by flame, becomes gaseous, and, as a result of the high pressure developed, rushes at high velocity into the main combustion chamber where complete combustion takes place. The use of a precombustion chamber rather than direct injection into the main combustion chamber does not require as high an injection pressure. Furthermore, it permits the use of an injector having a large single jet rather than several small jets, which would clog more readily.

Figure 17-12 shows the Robert Bosch two-unit injection system, with separate pumps for each cylinder.

Figure 17-13 shows the Roosa Master two-unit injector system, with a single master-injector pump and distributor with a fuel line for each nozzle. Its internal construction and operation are shown by Fig. 17-14.

Injection nozzles for different makes, types, and sizes of diesel engines vary in certain respects and particularly concerning the number, size, shape, and exact location of the jet orifices and the shapes and types of spray patterns. Several of these are shown in Fig. 17-15.

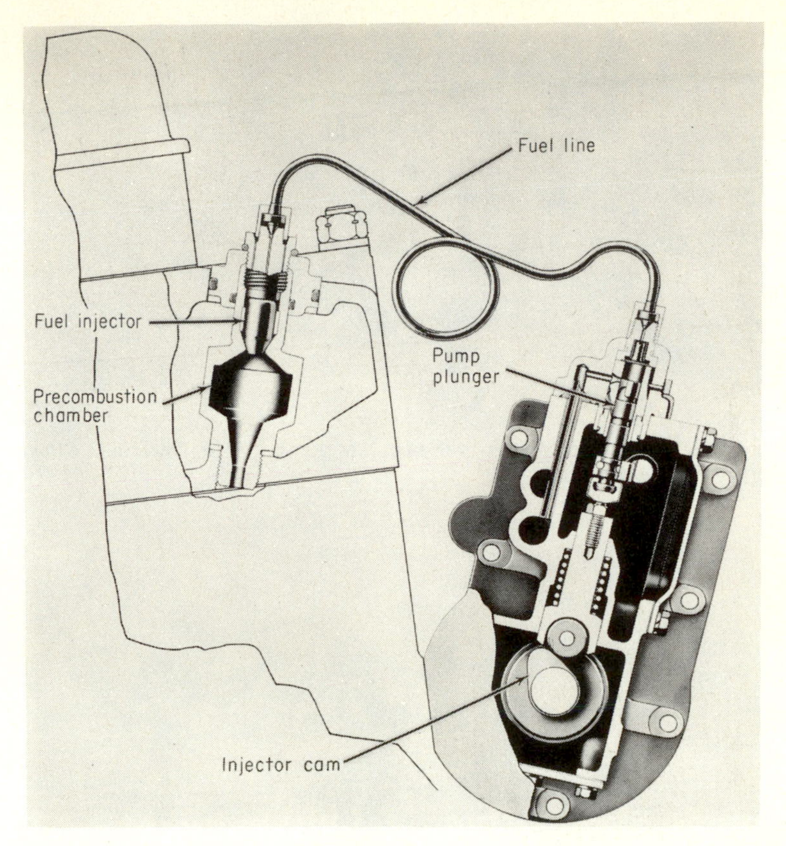

Figure 17-11 Two-unit fuel injection system showing precombustion chamber. (Courtesy of Caterpillar Tractor Company.)

Diesel Engine Governing The control of the power output and speed of a gas or a gasoline engine is relatively simple because the proper fuel mixture is created and in a gaseous condition before it enters the combustion chamber. Such is not possible in a diesel engine, for the proper air-fuel mixture cannot be produced in the same manner and fed into the cylinder under the existing high pressure and temperature conditions. Therefore, controlling the diesel engine speed and load must involve control of the charge of fuel injected. For this reason, the mechanism must be connected to and become a part of the injection pump. No attempt is made to vary the air charge, and it remains constant at all loads.

The control of the fuel charge with respect to varying speeds and loads in diesel tractor engines is largely the responsibility of the injector pump and its design and operation. As shown by Fig. 17-16, the plunger is equipped with scrolls and passages and the barrel with small ports. The plunger is moved vertically by a camshaft and turned horizontally by the rack and pinion mechanism. The two movements are correlated in such a

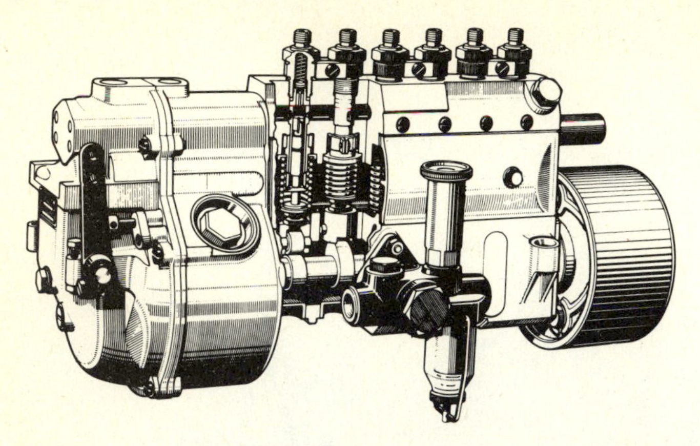

Figure 17-12 Two-unit injection system with separate pump for each cylinder. (Courtesy of Robert Bosch Corporation.)

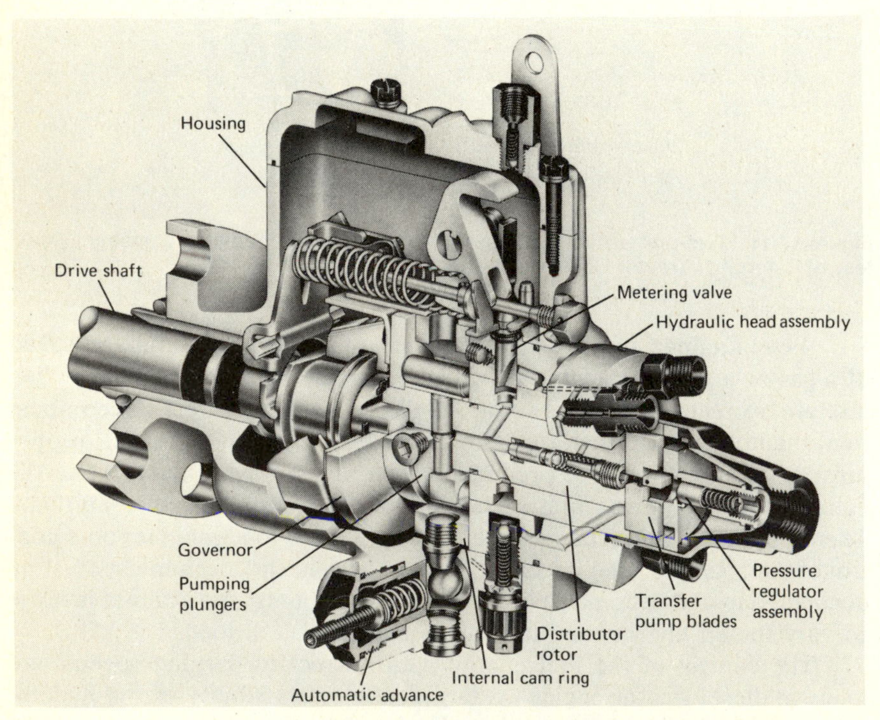

Figure 17-13 Two-unit Roosa Master injector system with single master-injector pump and distributor. (Courtesy of Stanadyne, Inc., Hartford Division.)

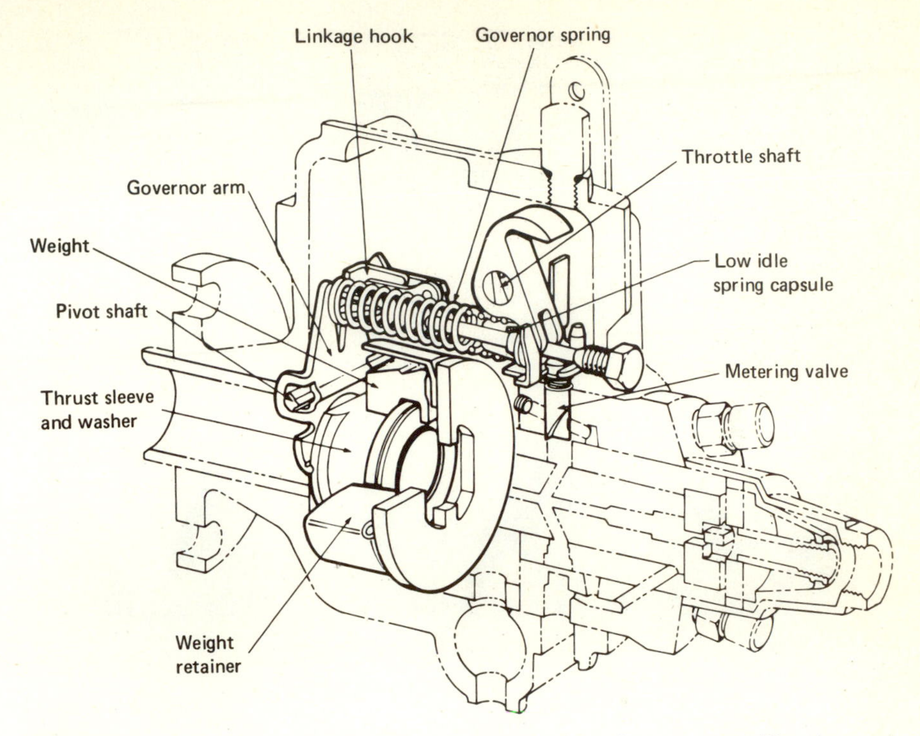

Figure 17-14 Skeleton view of Roosa Master injection system. (Courtesy of Stanadyne, Inc., Hartford Division.)

manner with respect to the scrolls and ports that a correctly metered fuel charge is injected into the combustion chamber at exactly the right time in the piston travel.

Various methods are being used to motivate tractor fuel-injection mechanisms. Probably the most common type utilizes the force of rotating weights; hence it is termed a centrifugal governor.

Figure 17-17 illustrates a complete fuel supply and injection system as used on Caterpillar tractors. Fuel from a low pressure gear transfer pump is fed to the fuel manifold (A). A separate plunger for each cylinder is driven by the camshaft (B). Plunger (C) turns on its vertical axis as the centrifugal governor changes rack (D) position. The scroll just beneath the top face of the plunger determines the amount of fuel fed to the injection valve and the precombustion chamber.

Figure 17-18 illustrates a pneumatically operated diesel engine governor, which is actuated by the air flow in the intake manifold. The velocity of the air flow varies with the changes in engine load and power output. Hence, a throttle valve located in the air manifold creates varying

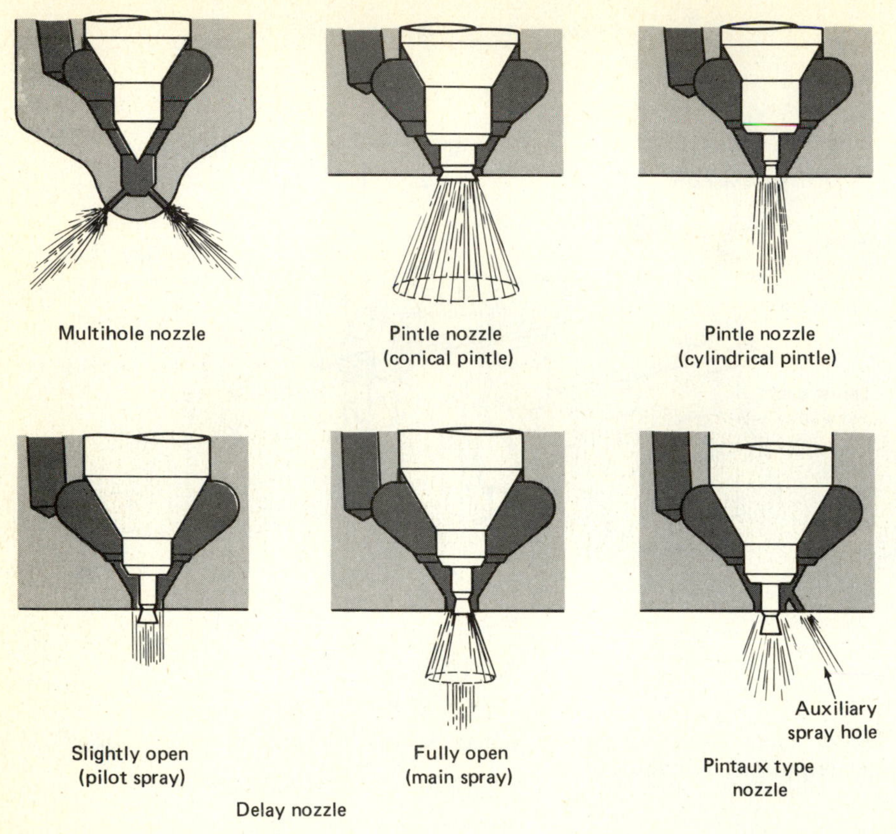

Figure 17-15 Spray patterns of various nozzle types. (Courtesy of Robert Bosch Corporation.)

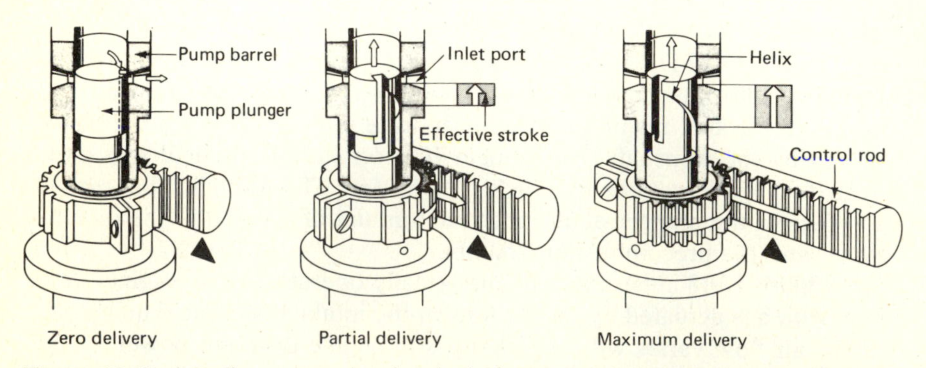

Figure 17-16 Injection pump showing fuel charge regulation. (Courtesy of Robert Bosch Corporation.)

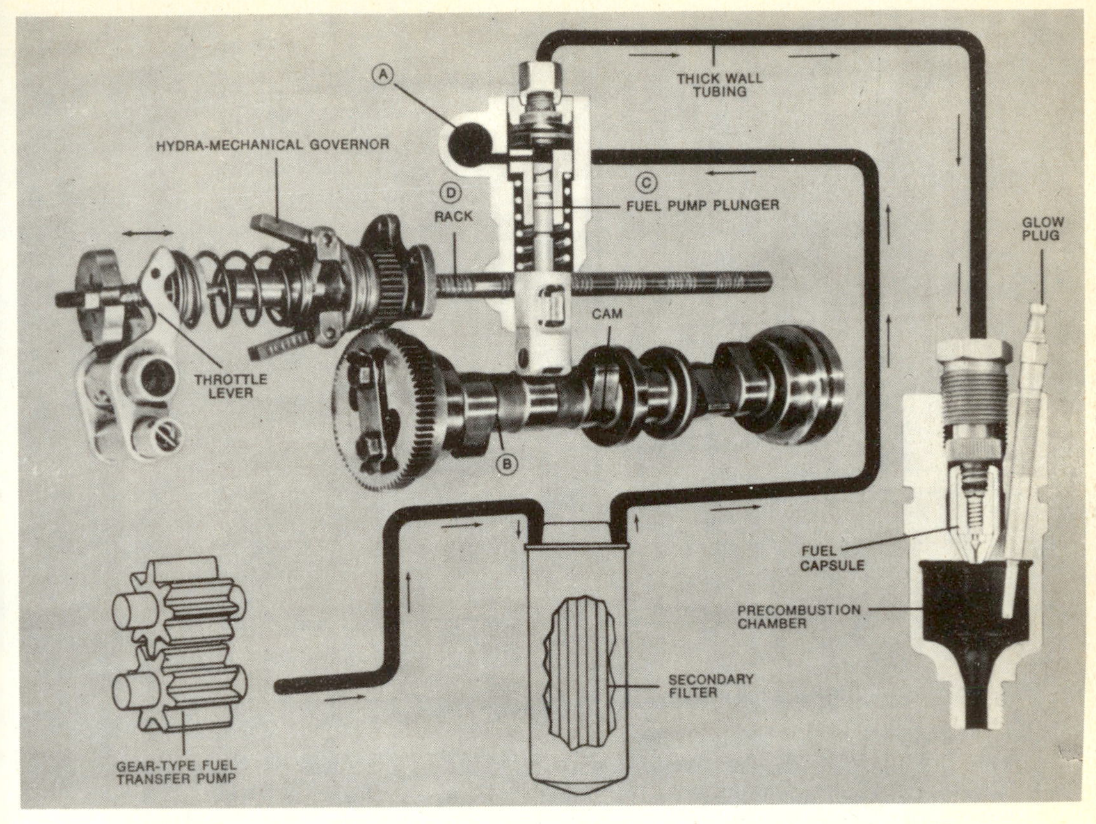

Figure 17-17 Caterpillar tractor fuel injection system. How the fuel system works: Fuel from a low-pressure gear transfer pump is fed to the fuel manifold at (A). A separate plunger for each system is driven by the fuel system's own camshaft (B). Plunger (C) turns on its vertical axis as the governor changes rack (D) position. The scroll just beneath the top face of the plunger determines the amount of fuel pumped to the injection valve. (Illustration and description of fuel system functioning courtesy of Caterpillar Tractor Company.)

pressure on a spring-loaded diaphragm whose movement in turn, depending upon the engine speed and power, controls the amount of fuel injected.

Turbochargers Chapter 7 explained the use of superchargers for gasoline burning engines for the purpose of forcing a greater amount of fuel mixture into the cylinder than would normally be drawn in and thereby increasing the power output. Diesel engines are frequently equipped with turbochargers that perform a similar function. However, since the fuel and air are not mixed in the diesel engine until they enter the cylinder, any similar increase in power must be dependent only upon increasing the amount of air combining with the fuel charge. Another difference between superchargers and turbochargers is that the former uses a gear-driven

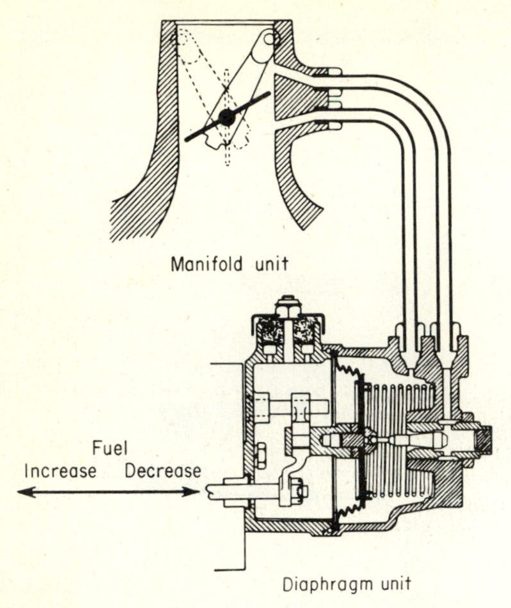

Figure 17-18 Pneumatically operated diesel governing mechanism.

blower, while the latter utilizes the exhaust-gas pressure to operate the blower.

Figure 17-19 shows the construction and operation of a complete turbocharger mechanism including air cleaner, exhaust-driven turbine, muffler, and exhaust outlet. In addition, it is equipped with an intercooler installed between the turbocharger and the engine intake manifold. The basic purpose of the intercooler is to reduce the temperature of the compressed air created by the turbocharger. The heated air expands and thereby reduces the power output. The air flows over a series of tubes through which the engine coolant is circulated and thereby reduces the temperature of the air and improves engine power.

Diesel Starting Methods One of the disadvantages of a diesel engine, regardless of size or number of cylinders, is the amount of energy required to crank it when starting because of the very high compression pressure. Hand cranking is out of the question and some type of mechanical starting arrangement is needed.

In general, four types have been or are being used. They are (1) compressed air system; (2) hydraulic system; (3) small gasoline engine; and (4) electric motor.

The compressed air system includes the air storage tank, which is connected to an air-cranking motor by suitable air lines and control valves. The air motor, in turn, is equipped with a special gear-drive mech-

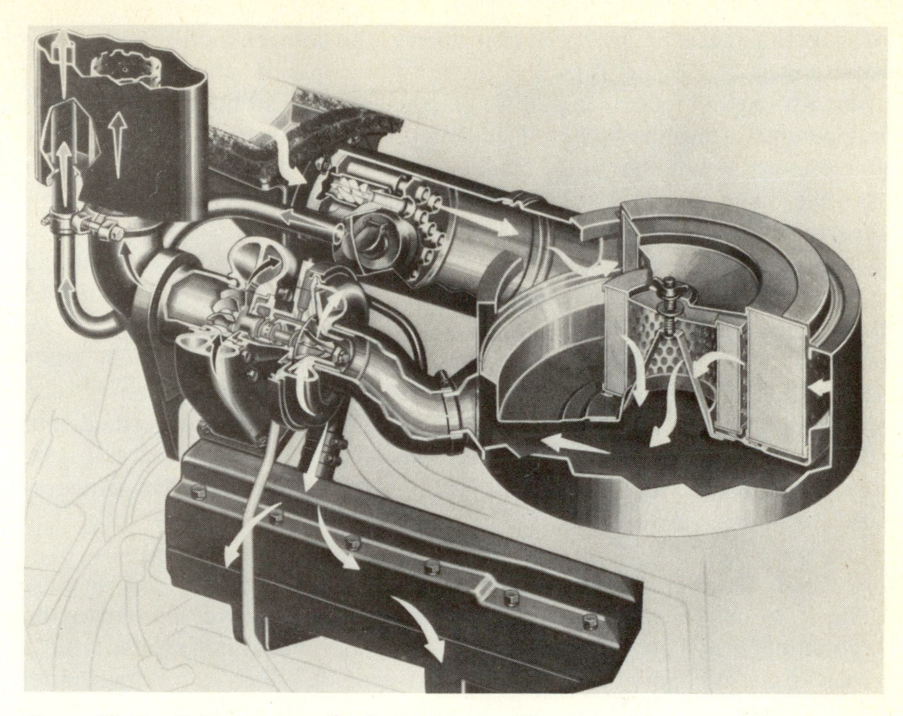

Figure 17-19 Cutaway view of a turbocharger. (Courtesy of Deere and Company.)

anism which engages with another gear on the engine flywheel and thereby cranks the engine.

A hydraulic starting system utilizes a fluid under high pressure to operate a hydraulic motor equipped with a gear, which meshes with a ring gear on the engine flywheel and thereby cranks it. The important parts of the system are (1) fluid reservoir, (2) engine-driven pump, (3) hand pump, (4) accumulator, and (5) hydraulic cranking motor plus the necessary connecting lines and filters. The hydraulic oil is first placed in a storage tank. Either a hand pump or an engine-driven pump transfers the fluid to the accumulator, which is made of a heavy-gauge steel to withstand pressures varying from 1,500 to 3,000 lb/in^2 as created by a charge of nitrogen. Assuming that the accumulator is under the required pressure, a control valve allows the oil to flow to the hydraulic cranking motor, the gears of which engage a gear on the engine and crank it.

Some large tractors equipped with diesel engines use a small, one- or two-cylinder gasoline engine for cranking. This engine is attached to the side of the main engine block. A special starter-drive coupling and gear engage a gear on the engine flywheel. In starting the tractor engine, the compression pressure is released by locking the exhaust valves open. As soon as the engine is turning over properly, the compression-release lever is disengaged, the fuel-injection pumps are engaged, and firing begins, after which the starter engine is disengaged and stopped. This small, independent, gasoline-powered cranking engine is particularly advantageous under low-temperature conditions in that it has sufficient power to spin the diesel engine's crankshaft as long as necessary under full compression.

Electric starting systems are now used almost exclusively for all types and sizes of tractor engines including diesels. They are similar to those used on automobiles and trucks and described in Chap. 15. The lead-acid type of battery is used with three different voltages available—12, 24, and 32, depending on the engine size. The electric system is relatively simpler and less expensive than the other systems described, but requires frequent servicing and attention. Also, it is important that it be kept fully charged at all times and particularly when exposed to temperatures of 40°F or lower.

Cold Weather Starting Aids Larger diesel engines are usually difficult to start at temperatures of 40° to 50°F or lower owing to the high-compression pressure plus the cold cylinder walls and resultant thickening of the oil. Some methods used to provide effective starting under such conditions are: (1) heating the coolant and lubricating oil, (2) using more volatile fuels, and (3) using glow plugs in the intake manifold or the precombustion chamber. Various types of heaters are available for heating the crankcase and engine block to thin the lubricant. Usually they are heated electrically. Placing an electric glow plug in the fuel injector heats the combustion space and aids in vaporizing the fuel charge. Another method used for tractor engines is the use of a volatile fluid, which is usually injected into the intake manifold. Ether is the fluid most commonly used, with various methods used to atomize and inject it into the manifold such as a special spray gun, aerosol cans and so on.

The outstanding operating characteristic of any diesel engine as compared with an electric-ignition engine is its lower fuel consumption/hph. Referring to Fig. 5-9, it will be observed that a diesel engine uses from 0.45 lb/hph at full load to approximately 0.90 lb at one-fourth load. The lower fuel consumption combined with the lower cost of the diesel fuel itself means a pronounced saving in operating costs. For example, the cost of the fuel per hour for a 50 hp diesel engine using fuel costing 70 cents per gallon and weighing 7 lb/gal would be:

$$\frac{50 \text{ hp} \times 0.45 \text{ lb/hph} \times \$0.70\text{/gal}}{7 \text{ lb/gal}} \approx \$2.25\text{/h}$$

For an electric-ignition engine of the same size, using gasoline costing 80 cents per gallon and weighing 6.15 lb/gal, the fuel cost per hour would be:

$$\frac{50\ \text{hp} \times 0.50\ \text{lb/hph} \times \$0.80/\text{gal}}{6.15\ \text{lb/gal}} \approx \$3.25/\text{h}$$

PROBLEMS AND QUESTIONS

1 Given two six-cylinder tractor engines—one a diesel engine and the other a gasoline engine—what specific differences would indicate their identity?
2 Name the three types of solid-injection systems and explain the general operation of each.
3 How is the timing of the fuel ignition and explosion controlled; of what importance is it to engine performance?
4 Explain the basic principles involved in diesel engine governing, and compare with spark-ignition-engine governor operation.
5 What is the difference in the starting procedure for a diesel engine under summer and winter conditions?
6 Enumerate the differences, if any, in the design and features of an automobile or a gasoline-burning tractor engine compared with a diesel engine of the same power rating.

Chapter 18

Clutches and Transmissions

Some means of disconnecting the power unit from the transmission gears and drive wheels of an automotive vehicle are necessary because (1) the internal-combustion engine must be cranked manually or by a special starting mechanism; (2) this type of engine must attain a certain speed before it will have any power; (3) shifting of the transmission gears must be permitted for the purpose of securing different traveling speeds; and (4) stopping the belt pulley must be permitted without having to stop the engine. All these can be taken care of by placing a clutch between the engine and the transmission gears and belt pulley.

Types of Clutches A number of different kinds of clutches used in tractors in former years are now obsolete. The most important of these are the contracting-band clutch, the cone clutch, and the expanding-shoe clutch. All present-day clutches are of the disk type, and the trend is toward a simple device such as the twin-disk and the single-plate clutch rather than the multiple-disk type.

A satisfactory tractor clutch must fulfill the following requirements: (1) it should not slip, grab, or drag and (2) it should be convenient, accessible, and easy to operate, adjust, and repair.

Multiple-Disk Clutch A multiple-disk clutch (Fig. 18-1) consists of a number of thin metal plates, arranged alternately as driving and driven disks. One set is attached to the flywheel and the other to the clutch shaft and transmission. If the plates are firmly pressed together, the clutch is said to be *engaged* and power is transmitted. This pressure is secured by means of a housing and a set of heavy springs as shown. The clutch throwout collar is attached to the rear part of the housing so that the depression of the operating lever slides the housing backward and compresses the springs. This enlarges the plate space and permits one set to rotate free and independent of the other set.

In the dry-type multiple-disk clutch (Fig. 18-1), the driven plates are faced on each side with friction material. The only part requiring lubrication is the throwout collar. The plates should be kept clean and dry. If slippage develops owing to wear, it can be overcome by increasing the tension of the three springs by tightening up on the nuts.

Single-Plate Clutch The single-plate clutch (Fig. 18-2) is a disk-type clutch in which a single, thick, iron plate, faced with friction material on both sides, serves as the driven member. It engages directly with the flywheel on one side and with an unlined iron driving plate on the other. The pressure produced by a number of springs, located between the driving plate and the housing, which is bolted to the flywheel, holds the friction surfaces firmly in contact. Three arms are hinged to the housing and have their outer ends connected to eyebolts that screw into the driving plate. When the inner ends of the arms are pushed toward the flywheel by the throwout collar, the driving plate is pulled away from the driven member and the clutch is thereby disengaged.

The driven plate (Fig. 18-3) used in one type of single-plate clutch, which consists of four lobes faced with special so-called cera-metallic pads which, it is claimed, wear longer than conventional clutch linings.

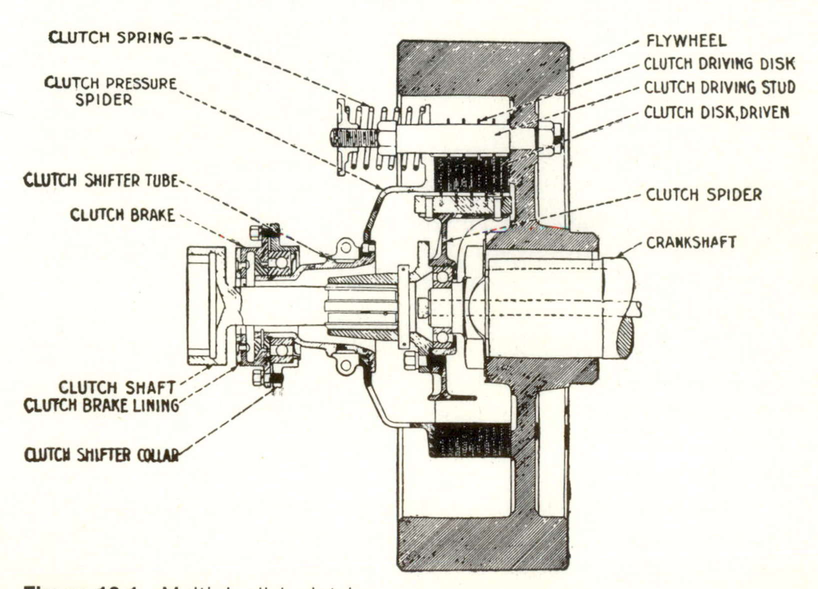

Figure 18-1 Multiple-disk clutch.

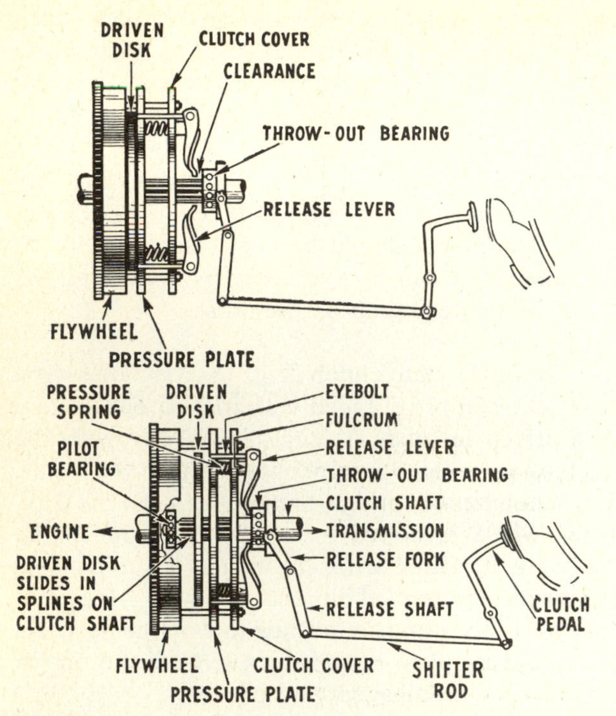

Figure 18-2 Construction and operation of a single-plage clutch (Courtesy of International Harvester Company.)

Hand-Lever-Operated Clutches The single-plate clutch shown in Fig. 18-4 differs in certain respects from those previously described. Instead of using spring pressure to produce positive plate contact, it utilizes short bell-crank levers that press the plates together. Instead of a foot pedal, a hand lever connected to the throwout collar actuates the bell cranks and thereby releases or engages the clutch. Unlike a foot-operated spring-compressed clutch, this clutch when disengaged will remain so without the operator's holding the lever. Another name applied to this clutch is "over center" because the pressure fingers are so linked to the lever and hinged that they snap over center when the clutch is fully engaged.

Figure 18-5 illustrates the construction and operation of a wet-type disk clutch, which is particularly adapted to constant and synchromesh, gear-type transmissions having a relatively large number of travel speeds. The clutch is actuated by oil pressure from the tractor hydraulic system. The clutch foot-lever merely opens and closes a valve to control oil movement and pressure.

Figure 18-6 shows a hand-lever-operated clutch, this one having one driving and two driven plates. The flywheel itself does not serve directly as a driving member but has the plate pinned to and rotating with it. The

Figure 18-3 Clutch plate with cera-metallic pad facings (Courtesy of International Harvester Company.)

two unlined driven plates carry the power to the transmission. The driving pressure in all clutches of this type is obtained by bell cranks and a sliding cone-shaped collar rather than by heavy springs. Some light springs placed between the driven disks are used merely to ensure the prompt release of the disks when disengagement is desired. The clutch fingers or bell cranks are attached to the threaded adjusting collar. Turning this collar to the right brings the fingers closer to the movable clamping disk and thus compensates for any wear of the lining. The spring-controlled lockpin fits into any one of a series of closely spaced holes and provides for any amount of adjustment desired.

Clutch Troubles and Adjustments The modern tractor clutch, if given ordinary care and attention, will seldom fail to perform satisfactorily. Any abnormal or unusual action is nearly always due to one of the following causes:

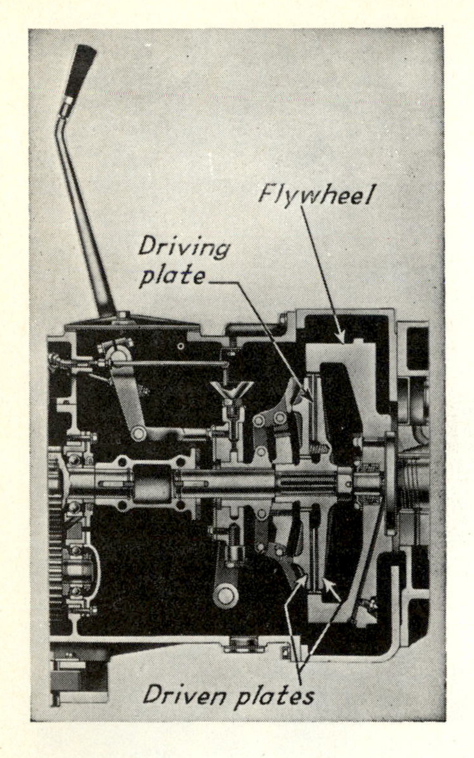

Figure 18-4 Hand-lever operated clutch.

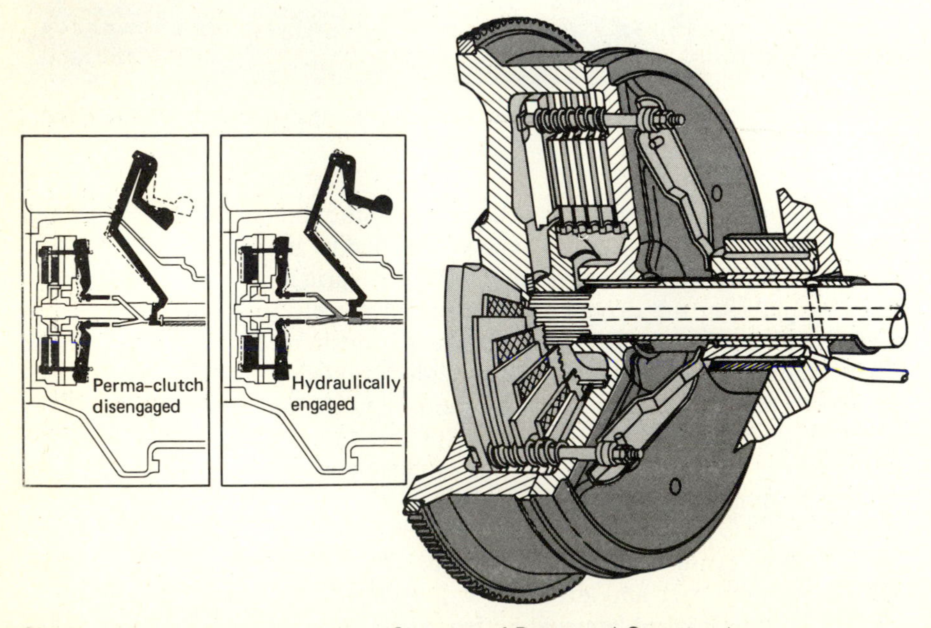

Figure 18-5 Wet-type disk clutch. (Courtesy of Deere and Company.)

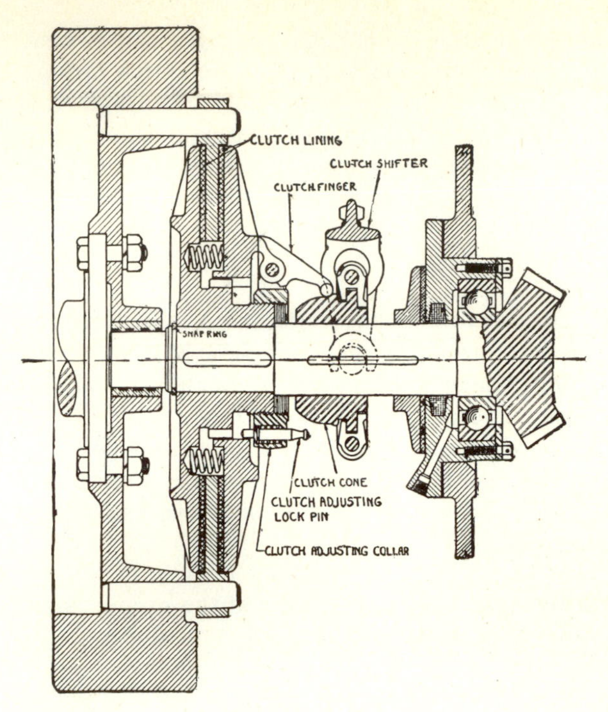

Figure 18-6 Twin-disk-clutch construction and operation.

1 Failure to keep the clutch fully engaged, due to riding the clutch pedal or permitting the hand lever to remain only partly in the engaged position.

2 Failure to keep the clutch adjusted to compensate for normal wear.

3 Permitting oil and grease to get on friction surfaces.

4 Failure to lubricate the throwout collar and bearings as directed.

Slippage is the most common clutch trouble and develops from one or more of the first three causes just named. It is immediately indicated and detected by the tendency of the engine to speed up, when the clutch is engaged and a load is applied, without apparently exerting any appreciable tractive or belt power. Naturally, if for any reason the clutch surfaces are not firmly pressed together, there will be some slippage and wear. If this condition exists for any length of time, it is apparent that undue wear, heat, and other effects will develop, which will produce slippage even with the lever fully engaged. The precaution, therefore, is always to keep a clutch either completely disengaged or completely engaged. If it is desired to let the engine run with the tractor standing still, the best practice is to put the gears in neutral and engage the clutch.

Leaving the gears engaged and holding the clutch disengaged causes heating and wearing of the throwout collar and may wear the clutch facings.

A dragging clutch is one that does not completely disengage. It is indicated by failure of the tractor or pulley to stop and by clashing of the gears when shifting. A dragging clutch is usually caused by incorrect adjustment of the clutch or clutch lever, or both.

TRACTOR TRANSMISSION FUNDAMENTALS

The mechanism involved in transmitting the engine power to the drive wheels or tracks of a tractor ordinarily includes four distinct parts, namely, (1) the clutch, (2) the change-speed gears, (3) the differential, and (4) the final-drive mechanism. There are some exceptions to this, as will be pointed out later. Unlike the automobile, there is considerable variation in the construction, arrangement, and operation of these four units in the different makes of tractors. In general, a tractor transmission must serve the following purposes:

1 It must provide a means of self-propulsion and the proper speed reduction between the engine crankshaft and the traction members to give the required travel speeds.

2 It must provide for an equalization of the power transmission to the traction members on both gradual and short, quick turns.

3 It must provide a means of reversing the direction of travel.

4 It must provide brakes for controlling the movement of the tractor in specific situations or holding it stationary for certain operations.

TRANSMISSION TYPES AND OPERATION

Travel-speed-changing systems found in tractors vary considerably particularly with respect to the number of speeds, kinds of gears, and methods of shifting. Based on these factors, transmission types can be classified as (1) sliding spur gear, (2) constant-mesh with shifting collars, (3) Synchromesh, and (4) hydrostatic. Also with respect to changing speeds, it can be done manually with a hand-lever or automatically by hydraulic pressure. These variations will be explained in connection with the descriptions of different makes of tractor transmissions. In general, the smaller sizes of tractors, 20 to 40 PTO hp use simpler transmissions with fewer travel speeds, while the larger sizes have speeds ranging from six or eight to twelve or sixteen. In such cases, provision is made for shifting from the "low-range" to the "high-range" speeds or vice versa while the tractor is in motion, and without disengaging the master clutch and coming to a dead stop. These advantages, in turn, permit a closer and more convenient adjustment and a quicker response of power, torque, and tractive ability of the tractor to the specific drawbar-load conditions and requirements.

Plain Spur-Gear Transmission Figure 18-7 illustrates the design and operation of the type of transmission originally used in earlier tractors and still used in some small, current models. As shown, the different speeds are obtained by sliding gears of a certain size into mesh with others of another size. Reverse requires an additional single gear on a separate short shaft. The entire mechanism operates in an oil bath in a special housing. The bearings may be plain or antifriction type. The different speeds are obtained by a manually operated lever attached to certain gears in such a manner as to slide them into mesh with others that are stationary. These changes can only be made with the main clutch disengaged and the tractor stationary. Figure 18-8 shows a similar transmission used in a high-clearance row-crop tractor.

Deere Horizontal Two-Cylinder Transmission Figure 18-9 shows a two-cylinder, horizontal engine equipped with a four-speed sliding gear transmission and spur-gear final drive for the axles and wheels.

Allis-Chalmers Dual-Range Transmission The Allis-Chalmers dual-range (Fig. 18-10) consists of an eight-speed, constant-mesh unit with helical gears and shifting collars. A special double-unit, hand-lever-operated disk clutch that runs in oil has three positions: "Low-range," "High-range," and "Neutral." The change from one range to the other can be made by means of this clutch and a hydraulic assist mechanism while the tractor is moving.

Ford Dual-Power Transmission Figure 18-11 illustrates a manual shift transmission, which normally has eight forward speeds and two reverse speeds. However, if the tractor must operate at slower speeds and under heavier draft conditions, this transmission can be equipped with an optional gear assembly that will provide eight additional forward and four additional reverse speeds. The shift is made by a single lever without stopping the tractor. The travel speed is reduced, but the pulling power is increased to permit the handling of unexpected or difficult situations.

Deere Hydraulic Hi-Lo Range Tractor Figure 18-12 is a six-cylinder, 80 hp tractor having a two-range transmission with twelve forward and three reverse speeds. Table 18-1 gives the travel speeds for each range.

International Harvester 60 hp Tractor Figure 18-13 shows the complete power transmission train for an International Harvester 60 hp tractor having two travel speed ranges with four forward speeds for each one. This is a constant-mesh synchronized transmission. A Hi-Lo-Reverse range lever permits shifting between forward and reverse for shuttle-type operation on back-and-forth operations. The entire power train is pressure lubricated. The wet-disk brakes are hydraulic actuated and self-adjusting.

Deere Four-Wheel-Drive Tractor Figure 18-14 illustrates the power train and transmission design for a four-wheel-drive tractor. Power is first transmitted to two sets of primary master gears. These gears, in turn, transmit power to two sets of change-speed gears by which the different

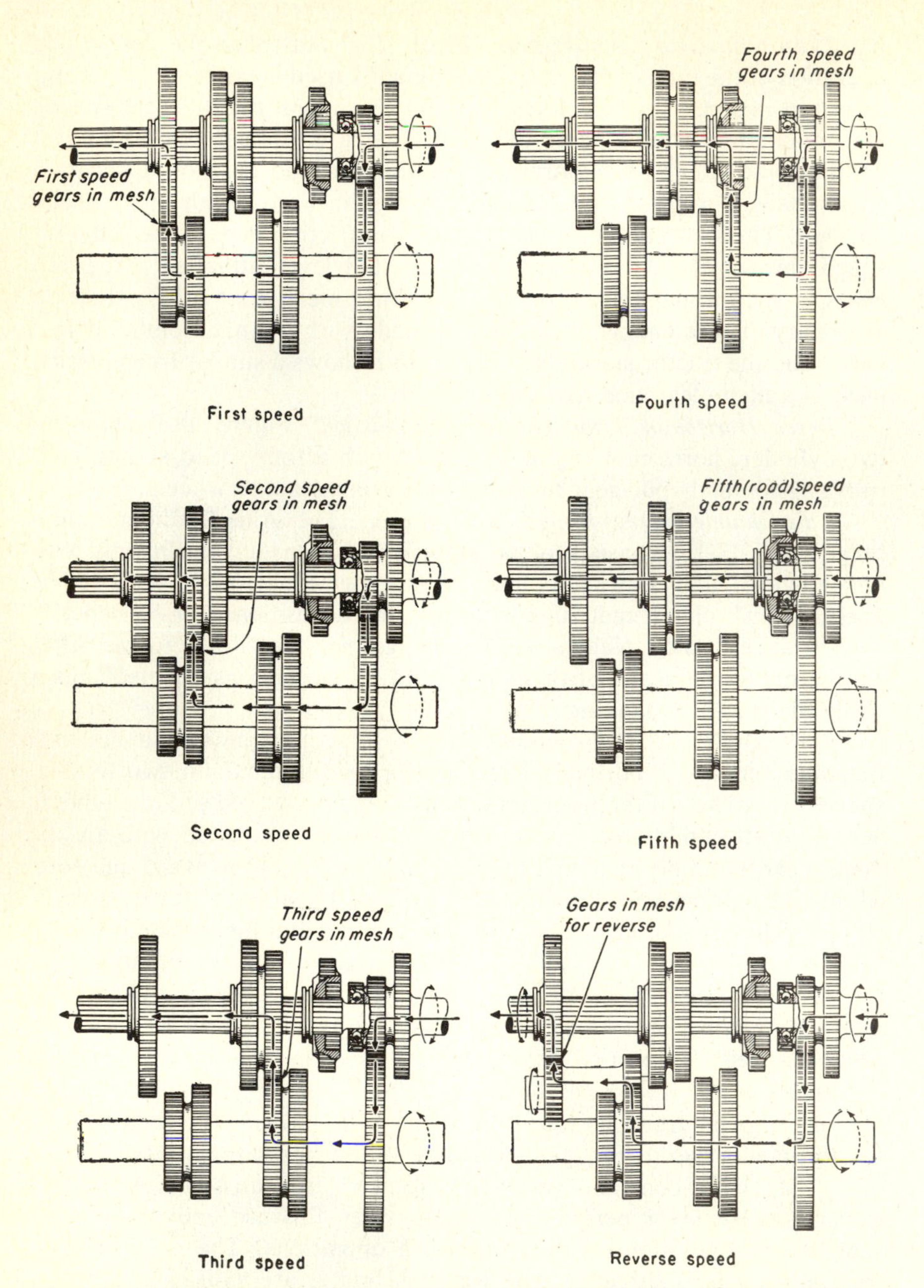

Figure 18-7 Operation of plain spur-gear five-speed transmission. (Courtesy of International Harvester Company.)

Table 18-1 Travel Speeds for Deere 2840 Tractor

	Operating Range, 1,300–2,500 engine rpm	
Gear	**Low range**	**High range**
1	1.4	1.8
2	2.4	3.0
3	3.8	4.9
4	5.2	6.6
5	8.6	10.9
6	13.8	17.6
R-1	1.7	2.2
R-2	2.9	3.6
R-3	4.7	5.8

travel speeds are obtained. These gears then transmit the power to the front and rear final-drive gears and axles. Hydraulically operated front and rear differential locks simultaneously lock both axles. Both locks are operated by a single foot pedal which can be engaged or disengaged on the go.

Track-Type Tractor Transmissions The transmissions for the track-type tractors are not unlike those in wheel machines except that the steering mechanism is incorporated in them. That is, the ordinary wheel tractor

Figure 18-8 Three-speed, all-purpose tractor transmission.

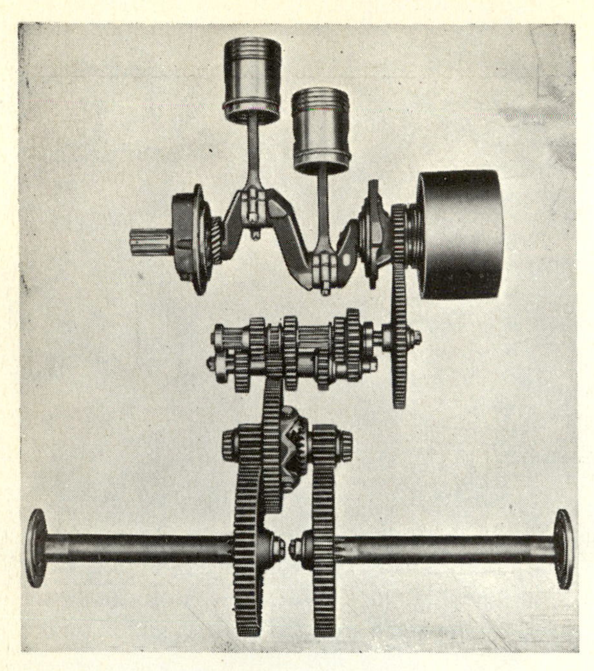

Figure 18-9 Horizontal two-cylinder engine with four-speed transmission. (Courtesy of Deere and Company.)

is propelled by the rear wheels and guided by means of the front wheels, whereas the conventional track-type tractor, having but two traction members, must be both propelled and guided through them.

Caterpillar Transmission Figure 18-15 shows the change-gear set for the Caterpillar tractor. The power is transmitted to a countershaft by means of bevel gears; then, through two steering clutches located on this countershaft on each side of the bevel gear, to the spur-type final-drive gears; and then to the sprocket. Steering is accomplished through the multiple-dry-disk steering clutches; that is, by means of hand levers, either clutch can be disengaged, which obviously cuts off the power to that particular track and causes the tractor to make a turn. Each clutch is equipped with a foot-operated brake that acts on the outside of the clutch drum carrying the driven plates. If a quick, short turn is desired, not only is the clutch released but the brake for that particular clutch is actuated and the track movement virtually stopped on the one side. With all the power going to the other track, the machine obviously will turn very short. By pressing on both brakes at the same time, the machine can be stopped almost instantly. It should be noted that a differential is unnecessary in the Caterpillar transmission.

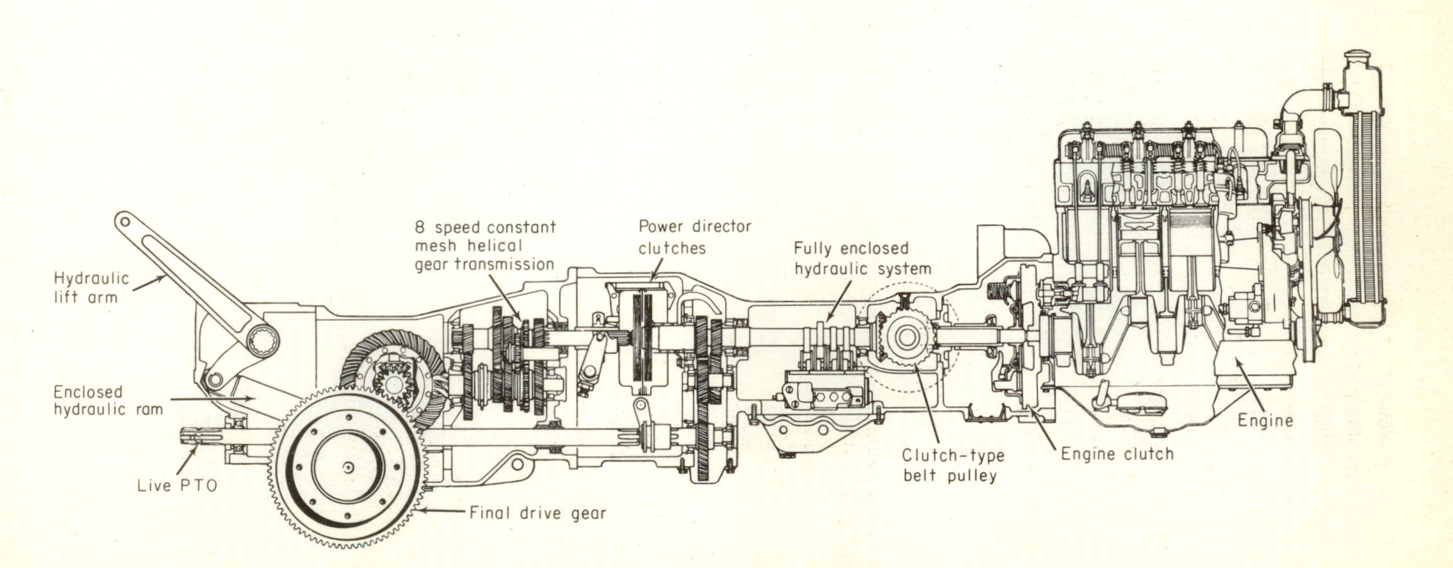

Figure 18-10 Allis-Chalmers dual-range eight-speed transmission. (Courtesy of Allis-Chalmers Manufacturing Company.)

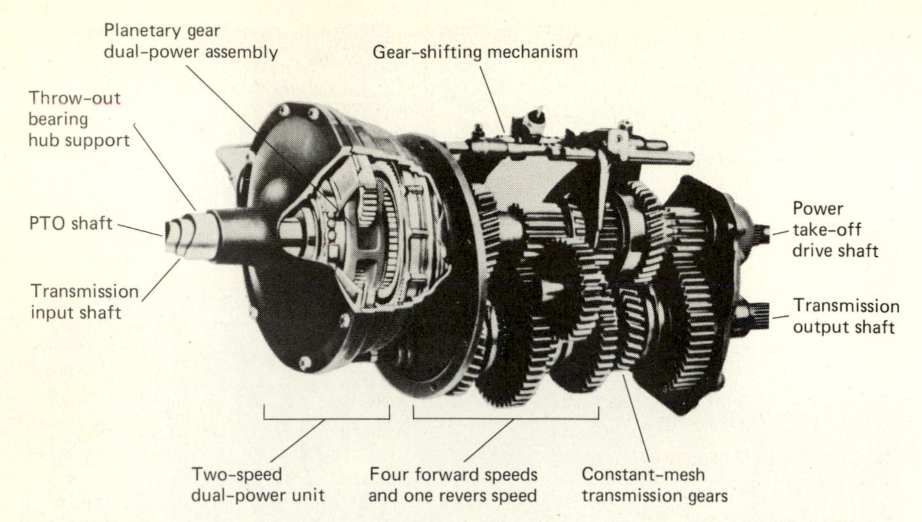

Figure 18-11 Ford dual-power transmission. (Courtesy of Tractor Division, Ford Motor Company.)

Figure 18-12 Eight-speed, all-purpose tractor transmission. (Courtesy of Deere and Company.)

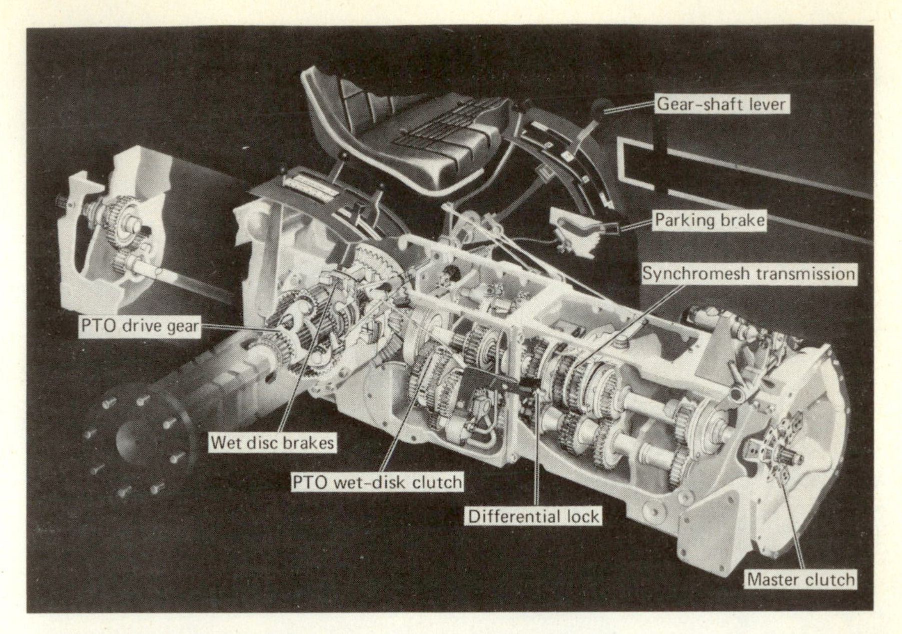

Figure 18-13 Operation of a five-speed transmission. (Courtesy of International Harvester Company.)

Caterpillar Planetary Transmission Figure 18-16 shows a planetary transmission and power train for a track-type tractor. Power from the engine is first transmitted to the torque converter by the crankshaft, then to the planetary power-shift transmission. This unit consists of a certain

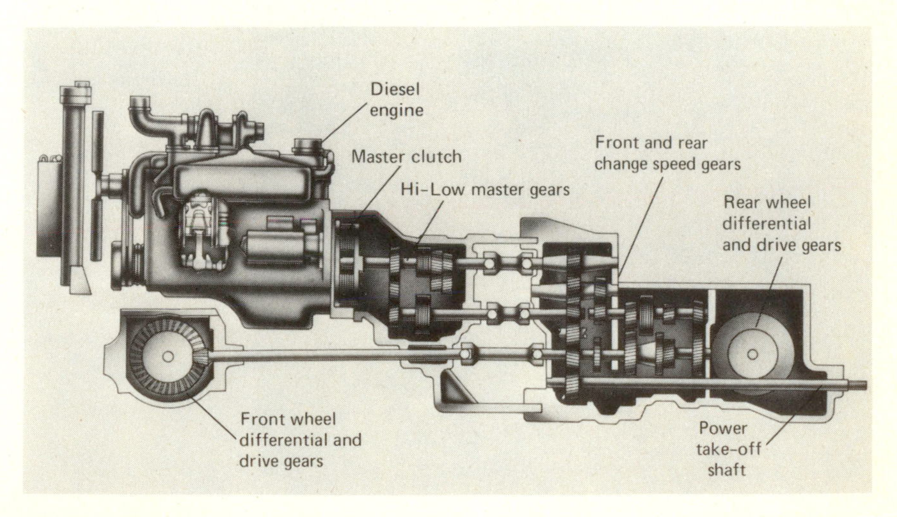

Figure 18-14 Power train and transmission design for a four-wheel drive tractor. (Courtesy of Deere and Company.)

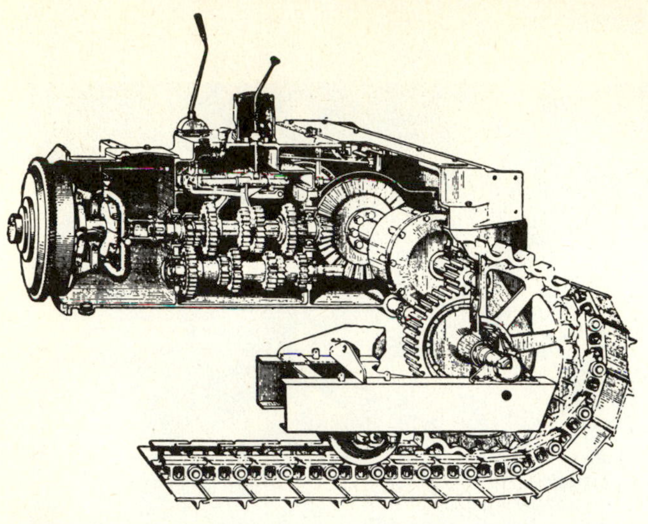

Figure 18-15 Track-type tractor transmission. (Courtesy of Caterpillar Tractor Company.)

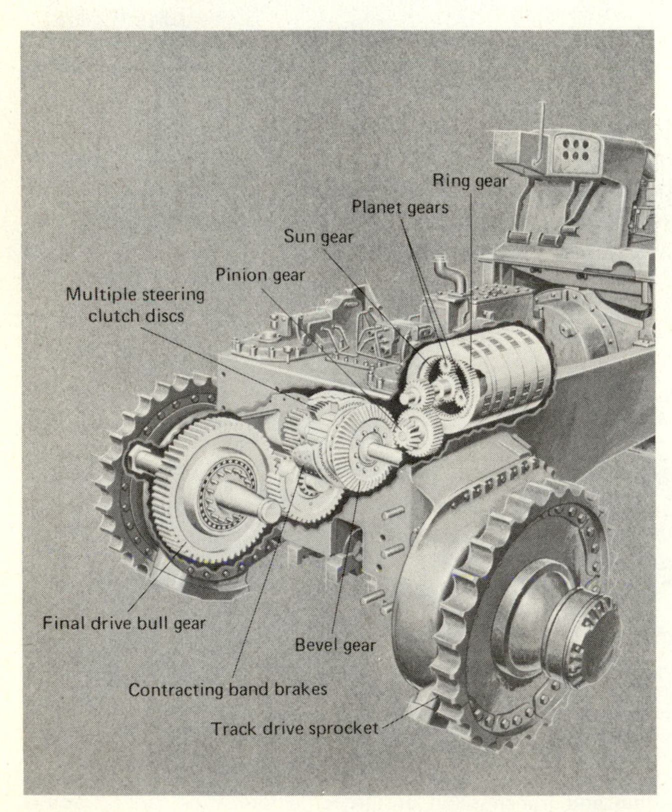

Figure 18-16 Caterpillar track-type tractor with planetary power-shift transmission. (Courtesy of Caterpillar Tractor Company.)

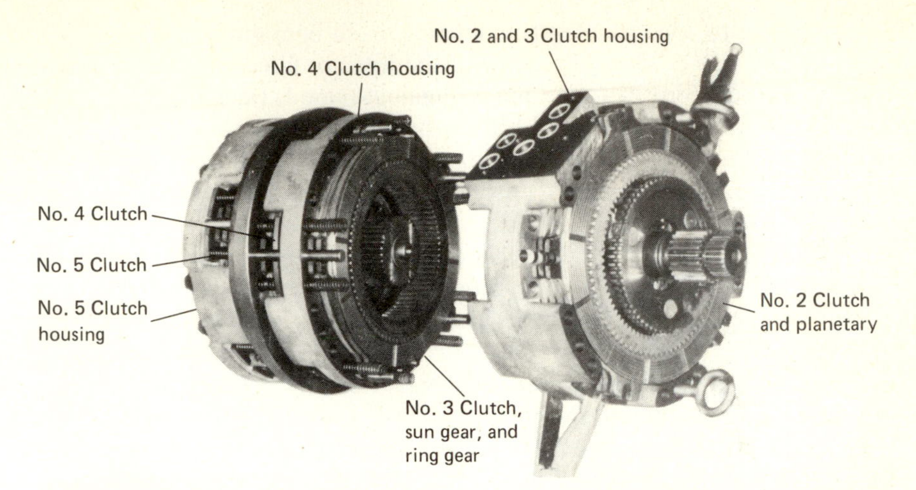

Figure 18-17 Track-type loader with power-shift transmission. (Courtesy of Caterpillar Tractor Company.)

number of planetary units that vary in size. Large clutch packs surround each gear set and engage with special hydraulic modulation for smooth, cushioned shifting. The clutch plates and gears are cooled by oil. For operations that do not require frequent direction changes this tractor can be equipped with a gear-type transmission.

Hydrostatic Drives and Power Transmission The basic principles involved in the operation of hydrostatic drives are explained in Chap. 20. A hydrostatic drive mechanism consists essentially of two major units, namely, a positive displacement pump driven by the original power unit and a positive displacement motor, which receives its energy from the pump. In other words, the tractor engine develops and transfers mechanical energy to the pump. The pump unit then converts this mechanical energy into hydraulic energy, which operates the motor unit and thus, in turn, produces mechanical energy to propel the tractor. Figures 18-18 and 18-19 illustrate the basic construction and operation of a simple hydrostatic drive.

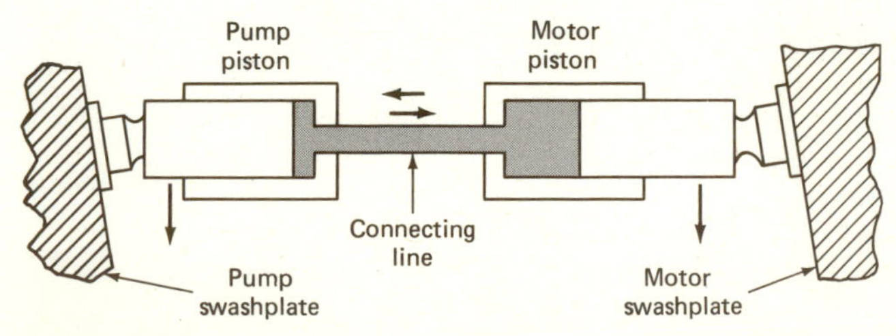

Figure 18-18 Two connected cylinders with swashplates. (Courtesy of Deere and Company.)

Referring to Fig. 18-18 two cylinders, each containing a piston, are connected by a line. The cylinders and the line are filled with oil. When a force is applied to one piston, it moves against the oil. Because oil will not compress, it acts as a solid and forces out the other piston.

In a hydrostatic drive, several pistons are used to transmit power—one group in the pump sends power to another group in the motor as shown by Fig. 18-19. The pistons are in a cylinder block and revolve around a shaft. The pistons also move in and out of the block parallel to the shaft. To provide a pumping action for the pistons, swashplates are located in the pump and in the motor. The pistons ride against the swashplates.

The angle of the swash plates (Fig. 18-20) may be varied to change the volume of oil pumped by the pistons, or to reverse the direction of oil flow. As the pump pistons revolve in the cylinder block, they move across the sloping face of the swashplate, sliding in and out of their cylinder bores to pump oil. The more the pump swashplate is tilted, the more oil it pumps with each piston stroke, and the faster it drives the motor. In this particular example, the motor swashplate is at a fixed angle, so the strokes of its pistons are always the same. The motor output speed depends entirely on the amount of oil supplied by the pump. Because the motor is linked to the drive wheels of the machine and the position of the swashplate, it determines the travel speed of the machine.

Three factors control the operation of a hydrostatic drive; they are: (1) rate of fluid flow which controls the speed of travel, (2) direction of fluid flow that gives the direction of travel, and (3) fluid pressure that controls the magnitude of the force applied to the output shaft and hence to the final drive gears and tractor wheels. The control of these factors is

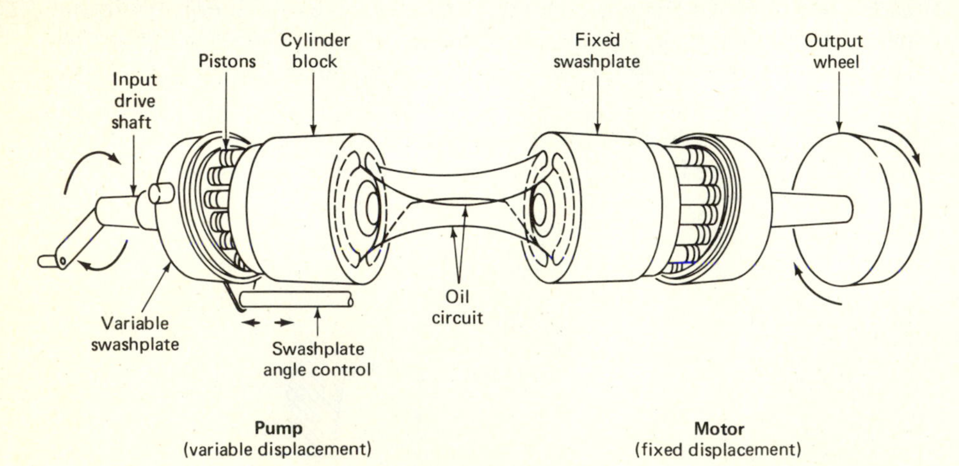

Figure 18-19 Variable-displacement pump driving fixed-displacement motor. (Courtesy of Deere and Company.)

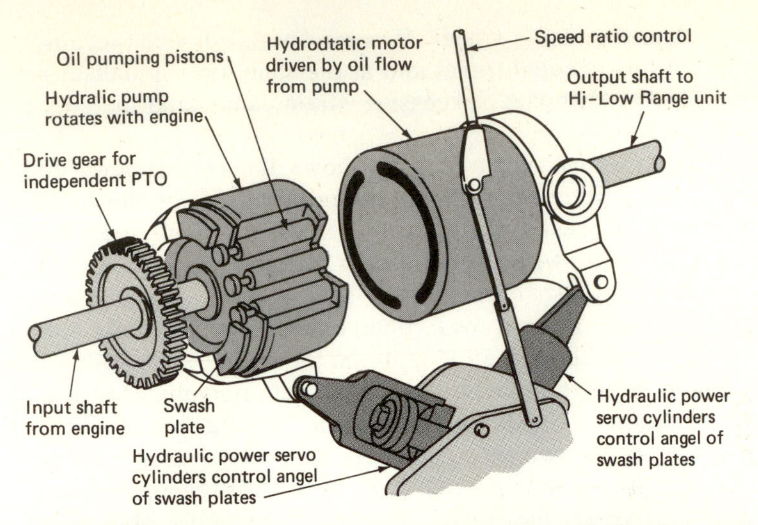

Figure 18-20 Hydrostatic unit showing power servo cylinders and speed-ratio control. (Courtesy of International Harvester Company.)

infinite, thus providing endless selections of speed and torque in a hydrostatic drive. Figure 18-20 further illustrates the appearance and construction of a hydrostatic drive, including the hydraulic power servo cylinders, by which the angle of the swashplates is operator-controlled by means of a hand lever that controls the speed and power of the tractor.

Differentials A differential is a special arrangement of gears so constructed and located in the transmission system of an automotive machine that it will permit one driving member to rotate slower or faster than the other and at the same time propel its share of the load. For example, referring to Fig. 18-21, it is quite evident that, in making a turn, the outside wheel of an automobile or tractor must travel farther and there-

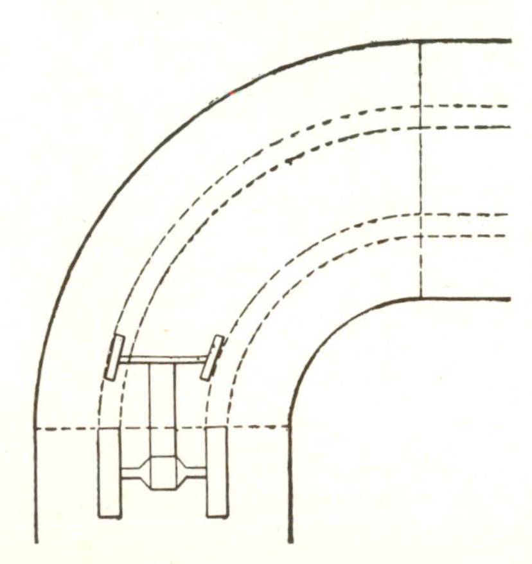

Figure 18-21 Sketch showing effect of turning on rear-wheel tread.

fore turn faster than the inside wheel. If some special device were not provided to permit this unequal travel and at the same time equalize the pull, it is obvious that slippage, excessive strain, and abnormal wear would result.

Differential Construction Figure 18-22 shows the construction and important parts of a differential. The operating principles of the bevel-gear-type differential are best understood by referring to Fig. 18-23. The main drive pinion B meshes with and drives the large gear C. A differential housing E is bolted rigidly to one side of gear C. This housing may be solid or split and carries one or more studs F, upon each of which is mounted a differential pinion G, which is free to rotate on the stud.

The two halves of the axle or shaft to be driven, namely, K and K_1, are inserted, one from each side as shown, and bevel gears H and H_1 placed on the ends. These two gears and their respective shaft ends are splined or keyed in such a manner that they must turn together. At the same time, gears H and H_1 mesh with differential pinion G. Main drive

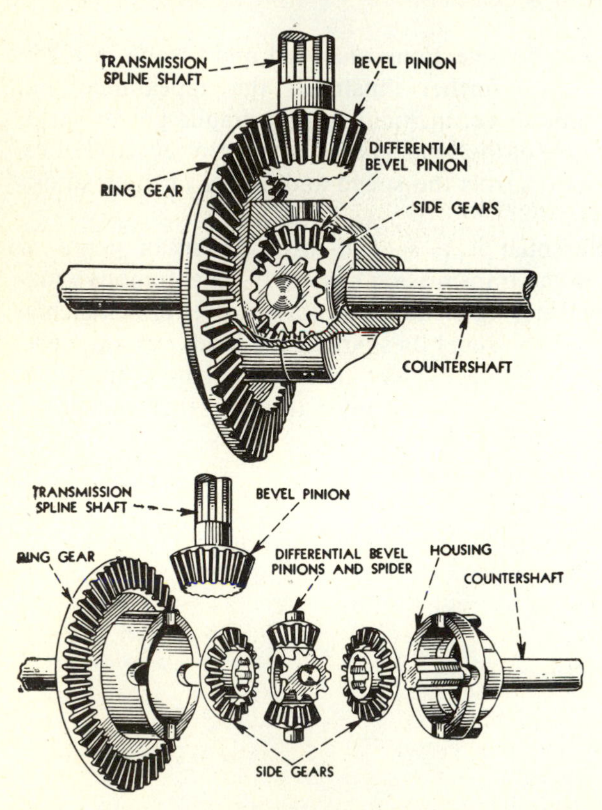

Figure 18-22 Construction and parts of a differential-gear assembly. (Courtesy of International Harvester Company.)

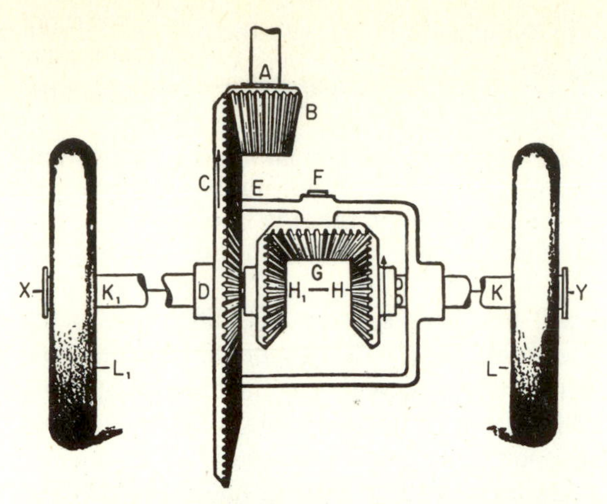

Figure 18-23 Differential construction and operation.

gear C and its attached housing E are free to rotate as an independent unit about the shafts K and K_1.

Differential Operation

1 *First condition,* with wheels L and L_1 fastened to outer ends of shafts K and K_1, respectively, and entire mechanism raised up so that the wheels are clear and free to rotate: Power is received by the drive pinion B and in turn is transmitted to gear C, turning it in the direction indicated. This likewise rotates the housing E in the same direction. Since the wheels L and L_1, shafts K and K_1, and gears H and H_1 are free to move, they rotate in the same direction also; that is, the entire mechanism rotates as one unit. The differential pinion G does not turn on its stud F, because the bevel gears H and H_1 are turning at the same rate and, therefore, lock pinion G. Consequently, with this condition existing, that is, with an equal resistance applied to both wheels, the latter will turn at the same rate and the tractor will move straight ahead.

2 *Second condition,* with one wheel locked and the other clear and free to rotate: Suppose wheel L is resting on a rough, firm surface and wheel L_1 is raised and is free to rotate. Power is again applied to large gear C from pinion B. This again rotates housing E as indicated and carries pinion G around with it. But wheel L and its shaft K are subjected to much greater resistance than wheel L_1 and its shaft K_1; therefore, L, K, and gear H remain stationary. Therefore pinion G, as it is carried around by the housing E, is also forced to turn on its stud F and, in so doing, causes gear H_1, shaft K_1, and wheel L_1 to rotate in the same direction as the drive gear and housing. Thus the pinion G, in making one revolution with the housing also makes one complete revolution on its stud by rolling on the stationary gear H. The axle gear H_1 is thus subjected to two rotative

actions—one revolution due to its being in mesh with the differential pinion G, which has been bodily revolved about the axis XY, and the other due to the rotation of the pinion G on its stud as it is rolled once around the stationary axle gear H. The free wheel thus makes two complete forward revolutions while the drive gear and housing are making one revolution. This is the condition that causes one wheel of an automobile or tractor to spin while the other remains stationary when there is unequal resistance applied to them, as when one is in soft mud and the other on firm footing.

In a similar manner it will be observed that any difference in the rotation of the wheels is compensated for by the rotation of the differential pinion G on its stud F while also revolving with the entire housing about the axis XY. Any rotation of this differential pinion on its stud means that it rolls on one of the axle gears, and the amount of motion in rolling on one gear is transmitted to the other as additional turning and driving effort. Any retarded motion of one wheel results in accelerated motion of the other. The power and driving force are thus transmitted to the wheels in proportion to the distance each must travel.

Differential Locks As previously explained, one objection to the common bevel-gear differential is that if the truck or tractor is moving forward in a straight line and either driving wheel encounters poor traction conditions such as a smooth wet surface or a mud hole, slippage occurs and that member turns faster than the wheel with good traction. Such a condition either reduces the travel rate of the vehicle or stops it entirely if resistance to the slipping wheel approaches zero. The basic solution for overcoming this difficulty is either to apply a braking action to the spinning wheel or to lock the differential in such a manner as to make both wheels turn together. Since tractors are equipped with independent drive-wheel brakes, it is possible to use them to lock partially and slow down the wheel which is slipping, thereby increasing the tractive effort of the other wheel and permitting movement of the entire machine. The use of the tractor brakes in this manner to overcome drive-wheel slippage depends on operator control, causes brake wear, wastes some engine power, and encourages engine stalling.

Differential locks, generally, are of two types, the first of which is built into the differential assembly (Fig. 18-24) and functions automatically with variations in wheel traction, slippage, and turning radii. Geiger[1] states:

> This type is sometimes called a torque bias or unequal torque differential. Driving torque causes locking by engagement of the friction clutch on each side through cam actuation. When one wheel spins, the cam action is re-

[1] M. L. Geiger, "Value of Differential Locks for Farm Tractors," *Agr. Eng.*, Vol. 42, No. 3, Mar. 1961, pp. 124–129.

versed which releases the friction clutch at the slipping wheel and thus the greater portion of available torque is directed to the better traction wheel. The action is similar in a turn in that the outer wheel overruns and the inner wheel drives.

This type of automotive locking differential was not adapted to the common two-wheel-drive tractors because tractors must have the outside wheels driving during turns, individual wheel braking, and short turning radii. With this automotive locking differential, the driving is done by the inner rear wheel in turns, while the outside tractor wheel must drive to pull heavy loads in turns. Individual wheel braking assists even a bare tractor in making sharp turns on loose soil surfaces. The desired turning radii of tractors for better maneuverability are very short compared to those of automobiles and trucks.

A second type of differential lock is the positive-locking type shown in Fig. 18-25 and described by Geiger[1] as follows:

One axle shaft is locked to the differential carrier through a sliding jaw clutch. This unit acts as a conventional differential when not engaged. Foot pedal actuation is normally used to engage the lock for emergency traction use only. Foot-pedal release returns the differential to conventional action. The foot-pedal engaging operation is somewhat equivalent to applying a brake, but no power is lost. Some differential locks are hand-operated and others have automatic control.

As a result of his investigations of differential locks for tractors Geiger[2] summarizes the situation as follows:

A differential lock may meet acceptance if the extra cost is reasonable. It is generally known to farmers that, in plowing, land wheels slip more than the furrow wheels unless they have added weight. However, most farmers do not

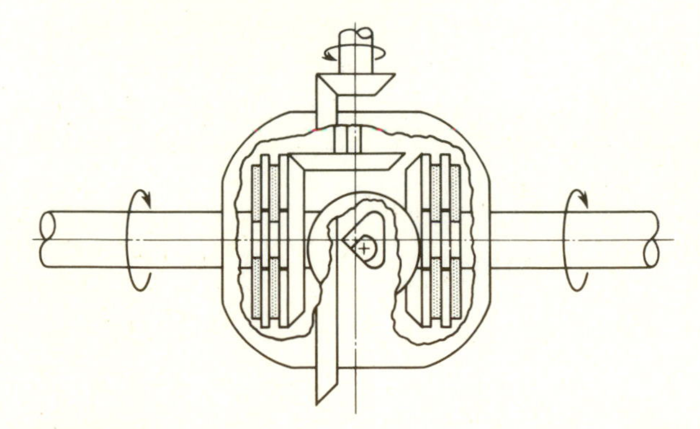

Figure 18-24 Automatic or unequal torque-type differential lock.

[1] Ibid.
[2] Ibid.

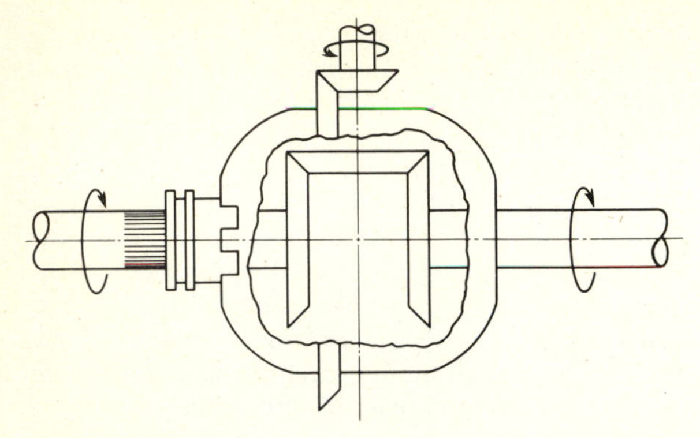

Figure 18-25 Positive-locking type of differential lock.

realize that the differences in slippage between the drive wheels are usually quite small.

One potential safety hazard with some differential-lock designs is the possibility of the operator failing to disengage the lock before reaching a turn.

Our results indicate that a differential lock is of little value in normal field operations, at least in most conditions in the United States. A little added weight on the land wheel is equivalent to the lock in equalizing wheel slippage in common field conditions. While compaction due to added weight should not be overlooked, it should be noted that the extra weight, in general, balances the weight distribution of the two drive wheels in plowing by bringing the land wheel weight up to the furrow wheel weight.

The other major deterrent to the differential lock is the essentially equivalent effect of braking for sustaining motion through intermittent areas of poor tractive conditions, providing sufficient engine power is available to offset the brake power loss. Brakes, of course, are already available as standard equipment for such use. The trend is to larger tractors which should have more reserve power available to meet this emergency.

Final Drives The illustrations accompanying this chapter show quite clearly the common types of final drives for tractors, that is, the means by which the power is transmitted finally to the rear axle and wheels. With the differential located on the rear axle, the final drive usually consists of a heavy bevel pinion on the countershaft meshing with a large bevel-differential gear.

Figure 18-26 shows a combined final drive gear and brake arrangement used on certain models of tractors. It consists of a planetary gear set with the disk-braking assembly built into it. A complete unit is used to drive and brake each rear wheel, the individual units being located on each side of the differential gear. The planetary gear set provides some speed reduction between the transmission and the rear axles.

The disk brakes, which are completely enclosed and operate in the

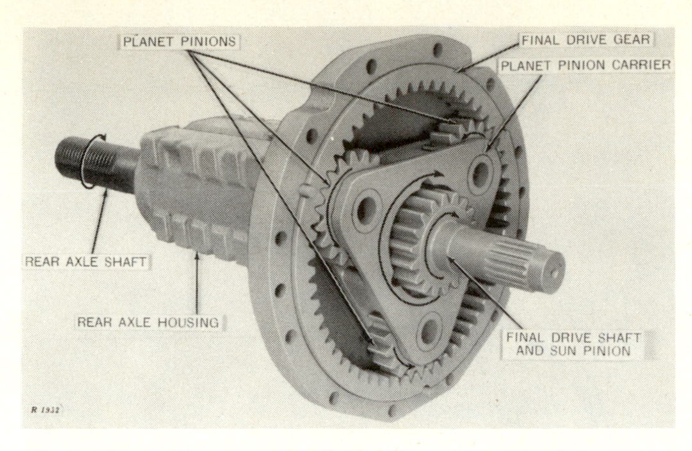

Figure 18-26 Planetary-type final-drive gear and brake assembly. (Courtesy of Deere and Company.)

transmission oil, are actuated by hydraulic pressure when the pedals are depressed.

Tractor Brakes Transmission brakes of some kind are essential on a tractor to control it on steep hills, to hold it stationary in doing belt work, and for making quick short turns. They are sometimes called differential brakes because they are located on each side of the differential gear and shaft assembly. They are operated by individual foot levers that can be locked independently or interlocked, if necessary, to hold the tractor stationary.

Three types of brakes are used, namely, (1) contracting band, (2) expanding shoe (Fig. 18-27), and (3) self-energizing disk (Fig. 18-28). In all cases they are well enclosed and protected and provision is made for periodic adjustment. Linings are made of wear- and heat-resistant material and are replaceable.

One of the essentials of a row-crop tractor is the ability to make a short, quick turn with ease in row-crop operations in order to save time at the ends. These tractors, therefore, are provided with two foot-operated differential brakes that permit holding one driver practically stationary and turning on it as a pivot.

Brake Linings and Clutch Facings The principal requirements for brake linings and clutch facings are (1) maintenance of the desired coefficient of friction over a wide range of operating conditions, (2) ability to withstand temperatures up to as high as 600 to 1000°F without breakdown or physical injury, (3) resistance to glazing and scoring, and (4) wear resistance and long life.

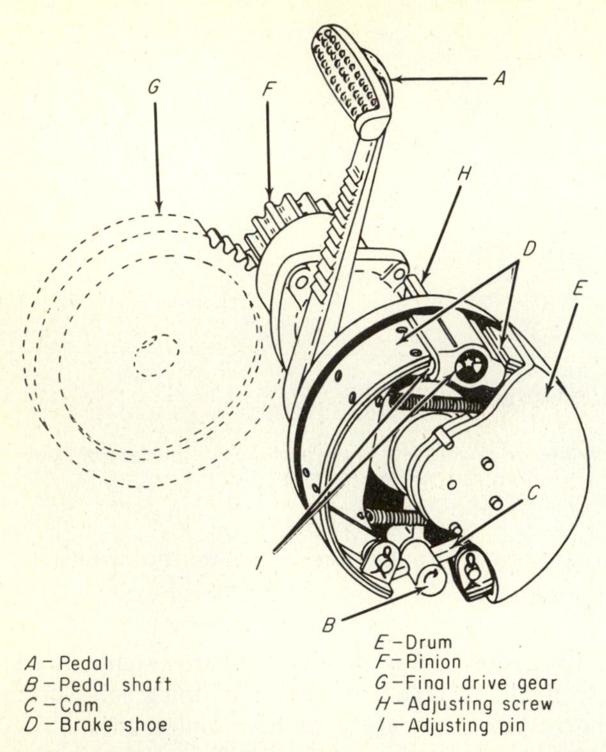

Figure 18-27 Expanding-shoe type of differential brake.

In general, brake and clutch materials are classified as (1) those having an asbestos base and (2) those using powdered or sintered metals as the friction material. Asbestos brake linings are of two types, namely, woven and molded. In the manufacture of the woven facing, long-fiber asbestos, together with a small percentage of cotton, is twisted around fine brass wires to form threads. These threads are then woven into a band of proper width and saturated with synthetic drying resins, drying oils, and asphalt to bind the fibers together when compressed.

Molded lining and facing materials contain 50 percent or more of short-fiber asbestos plus such wear-improving ingredients as particles of soft brass and powdered lead and zinc. The ingredients are molded under pressure by rolling and extrusion and impregnated with a binder such as graphite or ground soft rubber. Molded linings and facings are used more extensively than woven materials because of better resistance to high temperatures.

Powdered and sintered metal linings and facings are made in two steps. First, the friction material such as a mixture of powdered bronze, graphite, tin, and other metals and certain nonmetallic materials is compressed under a pressure of from 10 to 40 tons/in^2 into sheets and

Figure 18-28 Self-energizing disk brake. (Courtesy of International Harvester Company.)

wafers. This friction wafer is then bonded by heat-treatment in a special furnace to a copper-plated steel sheet or core. The core may be faced with the friction wafer on one or both sides. This type of brake material is more expensive, but it withstands relatively high temperatures, is not as much affected by water and oil, and requires less space than asbestos linings for the same amount of energy involved.

Some clutches and brakes operate in oil. This, in itself, reduces the coefficient of friction. Hence such devices require the best type of lining available such as a special grade of molded asbestos material or a powdered and sintered metal material. Also any oil-type clutch or brake requires a much higher pressure per unit of surface area.

The coefficients of sliding friction for dry brake linings and clutch facings are as follows:

Asbestos base, woven	0.30–0.55
Asbestos base, molded	0.20–0.47
Powdered metals	0.20–0.60

For the same materials operating in oil, the coefficient would be 0.07 to 0.15.

Clutch facings and brake linings are attached to their supporting metal members by rivets or by bonding the friction element directly to the metal with synthetic resin or special cement. The bonding process is used more extensively because it eliminates rivet holes and possible contact of rivet heads with the metal clutch or brake member.

Hydraulic Transmission Devices The utilization of the energy of a fluid in motion or under pressure for the purpose of transmitting power in automotive vehicles and tractors is now widely practiced. The principal advantages are: (1) manual gear shifting is either eliminated or simplified; (2) additional torque multiplication is automatically applied to any gear ratio to meet varying load demands; (3) the engine can run at an efficient speed under heavy load conditions; (4) engine stalling is prevented; and (5) smoother vehicle performance over a wide range of engine and travel speeds is obtained. The two general types of hydraulic drives are (1) the fluid coupling or drive and (2) the torque converter. The fluid coupling (also called *fluid drive, fluid flywheel,* or *fluid clutch*) consists of a driving member or impeller with radial vanes, a driven member or runner with similar vanes, and a cover that is welded to the driving member to form a housing. The entire assembly is mounted on the engine crankshaft in place of the conventional flywheel and is about three-fourths filled with a special oil. The driven shaft is made oiltight with a special spring-loaded sealing ring. When the crankshaft rotates, the oil is thrown by centrifugal force from the center to the outside edge of the impeller between the vanes. This increases the velocity of the oil and its energy. It then enters the runner vanes at the outside and flows toward the center, imparting rotation to the runner. The oil circulates as long as the impeller and runner rotate at different speeds, but when they rotate at the same speed, circulation stops. A fluid coupling does not increase the applied torque but merely transmits all torque exerted on it when the runner speed approximates the impeller speed. Since the amount of slip is determined by the torque required by the driven member, the slip is 100 percent when the vehicle is stationary but drops quickly as the vehicle gathers speed. Under certain conditions the slip may be 1 percent or less. Advantages claimed for a fluid coupling are that (1) it provides a cushion effect between the engine and transmission and thus eliminates sudden jerks and strains and (2) it helps to control and eliminate vibration. Figure 18-29 shows a fluid coupling with conventional plate clutch to permit shifting of gears.

The torque converter is a hydraulic device similar to the fluid coupling in that the power is transmitted entirely by hydraulic means. However, it differs in construction and operation from the fluid coupling in such a way that the engine torque is actually multiplied several times under certain conditions. The term *torque converter,* is, therefore, a misnomer, and *torque multiplier* would seem more appropriate.

The mechanism consists of four basic parts, namely, (1) the pump rotor driven by the engine, (2) the turbine or driven member connected to the transmission, (3) the stators or reactors, and (4) the housing or hydraulic chamber. Referring to Fig. 18-30, the pump, driven by the engine, throws the fluid outward and gives it kinetic energy. As the fluid strikes the turbine blades, it tends to induce rotation of the latter. If the load on

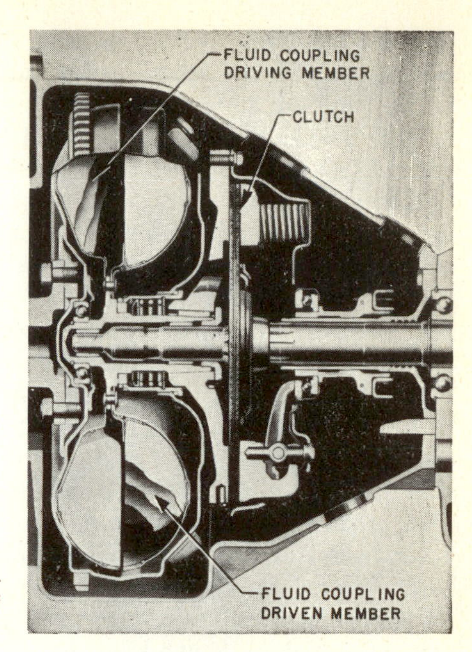

Figure 18-29 Fluid-coupling-type flywheel with plate clutch. (Courtesy of Chrysler Corporation.)

the output shaft is too great, little or no rotation will result; the fluid will rebound from these blades, and its direction of flow will be reversed. But only a small portion of the fluid energy is lost, and because of the shape of the hydraulic chamber, the fluid is forced to flow through a set of reactor blades attached to the inner wall of the chamber. This reverses the direction of motion again; that is, the first set of reactor blades causes the fluid to rebound again. This time, instead of being forced outward, it is forced downward through a second set of turbine blades, where it again attempts

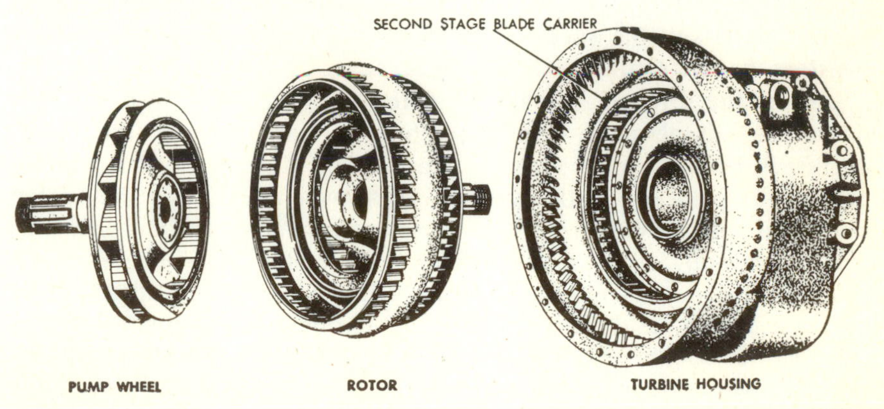

Figure 18-30 Principal parts of a hydrokinetic torque converter. (Courtesy of GMC Truck and Coach Division, General Motors Corporation.)

to rotate the output shaft in the same direction as the engine is turning. In passing through the second set of turbine blades, more of the energy is utilized and the direction is again reversed. Again some energy is spent, and the fluid is forced inward toward the center of the chamber through a second set of reactor blades and finally to the third and last set of turbine blades on the output shaft. Here the remaining energy in the fluid is spent, and it is conducted by this third set of turbine blades in toward the center of the centrifugal pump, where it started its cycle.

If the load on the output shaft is small, the blades will be rotated easily and the greater part of the energy in the fluid will be spent at the first set of blades with no reversal of the fluid direction taking place. As the fluid is forced to flow through the remainder of the blades, there would be no further push exerted on the output shaft. Multiplication of input torque depends upon the number of times the direction of oil flow is reversed by the driven rotor blades before its energy is spent and upon the completeness of reversal. Reversal of flow is complete only when the turbine or rotor blades are stationary. If the rotor blades move with the oil, the torque decreases accordingly. Therefore, the multiplication of torque is at its maximum when the driven rotor is stationary, or when starting a motionless vehicle. It is at a minimum when the rotating speed of the driven shaft is approximately the same as that of the driving rotor, or when the vehicle is fully in motion. Maximum multiplication of torque is approximately 4.8 : 1.

Throughout the cycle of operation it must be remembered that power or energy cannot be increased. Consequently, if the torque is increased, the speed is correspondingly reduced. The natural losses incurred in circulating the fluid, due to the internal friction of the fluid itself and the friction of the reactor blades, also contribute to the loss in efficiency of the unit. These losses, however, are more than offset by the fact that an automatic torque ratio between the engine and drive members is maintained throughout the speed range of the vehicle with the engine at a constant throttle setting.

From the foregoing, it will be noted that maximum torque at the output shaft is obtained when the vehicle is stationary and the torque decreases as the vehicle speed increases. The efficiency of the unit increases to a point of slightly over 80 percent, when the output shaft speed reaches a point of one-half to two-thirds of engine speed, and then efficiency drops off. Hence, if vehicle speeds of greater ranges are desired, a selective transmission must be incorporated or a sacrifice of efficiency and overheating will result. Even during normal operation, a means must be provided for cooling the fluid as well as an expansion tank to accommodate its volumetric increase due to expansion.

Heat generated within the converter is directly proportional to its loss in efficiency; since it is operated normally above 70 percent efficiency, cooling radiators capable of dissipating 30 percent of the maximum engine horsepower must be provided. Circulation of fluid through the radiator is obtained by utilizing the pressure differential across the converter pump. This unit does not operate well when cold, because the thickened fluid does not circulate readily. For this reason a very thin fluid is used, ranging from an SAE 10 engine oil to a straight diesel fuel.

Automatic Transmissions An automatic transmission is one which is so constructed that manual shifting of gears to obtain the different travel speeds is eliminated and the progressive gear changes are performed by mechanical means within the unit itself. A fluid coupling or torque converter unit is a part of the complete assembly. The speed changes are brought about by the hydraulic pressure of a special oil maintained by a pump. This pressure, by means of control valves, actuates planetary gear sets, which in turn create the necessary speed changes. A selector lever permits shifting to reverse, low-gear, and parking positions. The use of automatic transmissions is confined largely to automobiles.

Planetary Gear Unit The preceding discussion of special transmissions has shown that in most instances a planetary gear set is utilized in some manner to obtain certain speed and torque changes. For this reason, a knowledge of the basic construction and operation of the planetary gear set is important. Referring to Fig. 18-31, a planetary gear set consists of four basic parts: a central sun gear, two or more planet gears, a planet-gear carrier, and a large outer (internal) gear.

All gears are in constant mesh with each other and the planet gears turn free on their spindles. By incorporating a clutch and a brake in the assembly, it can serve as a clutch, a direct coupling, a reduction gear, or as a reversing gear. Referring to Fig. 18-31, these conditions and the method of producing them are as follows:

1 *Forward drive and major torque multiplication.* If the internal gear is held stationary while the sun gear is driven clockwise, the planet gears are rotated counterclockwise around their own axes. Since they are also meshed with the stationary internal gear, they must "walk" clockwise around the sun gear and thereby rotate their planet carrier in a clockwise direction at a much lower rotational speed than the sun gear and with correspondingly higher torque.

2 *Forward drive and minor torque multiplication.* If the sun gear is held stationary while the internal gear is driven clockwise, a similar action but of lesser effect occurs. In this case the internal gear drives the planets clockwise and forces them to "walk" clockwise around the stationary sun gear, carrying their planet carrier clockwise at a lower speed with a higher torque than the internal gear.

3 *Direct-drive coupling.* If any two of the three major assemblies of a

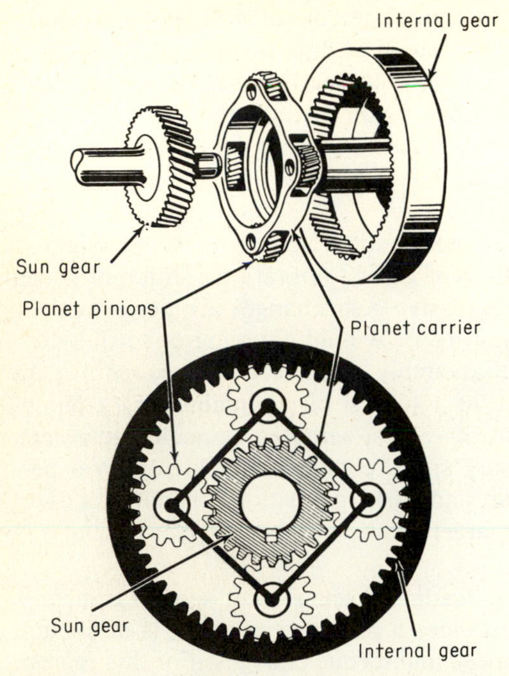

Figure 18-31 Construction of a planetary-gear assembly. (Courtesy of Texaco.)

planetary gear set are locked together in some manner, the entire set is locked and rotates as a solid unit or coupling which merely transmits speed and torque without changing either.

4 *Neutral or idling position.* If power is applied to any one of the three planetary gear major assemblies, and neither of the other two is restrained, no torque is transmitted and the elements merely rotate or "idle" in their neutral position.

5 *Reverse.* If power is applied to the sun gear in a clockwise direction and the planetary carrier is restrained, then the planets must rotate counterclockwise and drive the internal gear in a counterclockwise or reverse direction from the input sun gear.

TRACTOR POWER TAKE-OFF (PTO)

The change in the past four decades from animal to mechanical power and the tremendous growth in the utilization of tractors for all types of agricultural production has resulted in the design and use of many new types of tractor-operated machines and the adoption of improved crop-production practices. It was soon observed that if a shaft were attached to the clutch and transmission, the tractor could operate the mechanisms of various types of field machines as well as pull them. Hence, the power take-off was introduced by tractor manufacturers and in a few years machines such as mowers, combines, hay balers, corn pickers, and stalk shredders were being operated by tractors, thus expanding their utility and service life.

In 1955 the ASAE-SAE established certain recommendations and definitions relative to PTO design and use. In 1955, the following definitions were adopted.(The term *master clutch* is generally used to describe a clutch that transmits all power from the engine, and controls both travel and the power take-off. Likewise, when disengaged, both stop.)

Transmission-Driven Power Take-off Power to operate both the transmission and the power take-off is transmitted through a master clutch, which serves primarily as a traction clutch. The power take-off operates only when the master clutch is engaged. The transmission-driven power take-off ceases to operate at any time the master clutch is disengaged.

Continuous-Running Power Take-off Power to operate both the transmission and the power take-off is transmitted through a master clutch. Both operate only when the master clutch is engaged. Auxiliary means are provided for stopping the travel of the tractor without stopping the power take-off. The continuous-running power take-off ceases to operate any time the master clutch is disengaged.

Independent Power Take-off Power to operate the transmission and power take-off is transmitted through independent transmission and power take-off clutches. Travel of the tractor may be started or stopped by operation of the transmission clutch, without affecting operation of the independent power take-off. Likewise, the power take-off may be started or stopped by the power take-off clutch without affecting tractor travel.

In 1958, certain specifications were established that allowed any 1,000 rpm power take-off driven machine to be operated with any make of tractor having a 1,000 rpm power take-off drive.

Practically all types of farm tractors having a PTO horsepower rating of twenty or more are equipped with both 540 and 1,000 rpm PTO shafts. In most cases, they have special clutches that can be operated independently of the transmission.

PROBLEMS AND QUESTIONS

1 Explain the construction and operation of (a) multiple-disk clutch; (b) single-plate, foot-operated clutch; and (c) hand-lever-operated clutch.
2 What are the important characteristics of a suitable clutch lining material?

3 Explain the expression "over-center" clutch and just how this type is adjusted for wear.
4 What is meant by "riding" a clutch and why is this practice objectionable?
5 How is a slipping clutch detected, and what are the common causes?
6 Compute the crankshaft-rear-axle speed ratio required to give travel speeds of 2, 4.5, and 15 mph for a tractor having an engine speed of 2,000 rpm and rear wheels of 46-in diameter. Compute the same for the rear wheels of 60-in diameter.
7 What is meant by a constant-mesh transmission and how does it differ in construction and operation from a conventional tractor transmission?
8 Examine the transmissions shown or described in this chapter and list three distinct designs used, by which travel speed can be changed while the tractor is moving.
9 Explain why it is undesirable to make a short turn with a wheel tractor in a high gear with a high engine speed.
10 In what part of a tractor transmission would you expect to find the brakes?
11 What are the essential parts of a brake and what specific care and attention do brakes require?
12 Distinguish between the terms fluid flywheel, torque converter, and automatic transmission.

Chapter 19

Traction Devices—Steering Mechanisms—Transmission Accessories

Modern tractors are designed not only to perform efficiently but to supply power in a multiplicity of ways. Much attention has been given to design features that provide ease of operation and handling and simple and quick adjustment of the field machine or equipment being driven. A consideration of recent changes and developments with respect to traction devices, steering, power take-off operation, and hydraulic mechanisms is essential to a full understanding of farm tractor design and application.

TRACTION DEVICES AND EFFICIENCY

The proper working of a tractor under certain conditions often depends upon its means of securing traction and upon the use of the correct equipment. In general, two types of tractors are available according to the method of obtaining traction: (1) the rubber-tired wheel type and (2) the track type.

Considerable investigation has been made in connection with the factors involved in the effective and efficient operation of tractor wheels, wheel equipment, and other traction devices.

The power applied to any traction device, either wheel or track, is consumed largely in about four ways, namely, (1) by rolling resistance, (2) by wheel slippage, (3) by the action of the lugs on the soil, and (4) by the tractor-drawbar resistance. Obviously the most efficient device is one in which the first three factors cited are low, so that the net power available at the drawbar is as high as possible. In other words, the problem of securing efficient traction is dependent upon the reduction of these apparent power losses. Rolling resistance varies with the soil type and conditions and the weight upon the tractor. A certain amount of weight is essential for traction, but too much weight produces a high rolling resistance and therefore may reduce the net power output of the wheel. The second factor, wheel slippage, is likewise apt to prove excessive under certain conditions and result in low tractive efficiency.

The most important factors affecting the efficiency of a tractor wheel and its equipment are (1) wheel diameter, (2) wheel width, (3) weight on wheel, (4) type of tread surface, (5) speed of travel, (6) soil type and condition, (7) grade, and (8) height of hitch. These factors likewise apply to track-type tractors.

The great variation in soil types and conditions is perhaps the most outstanding handicap encountered in the solution of the problem of tractor traction. It seems quite impractical to attempt to provide a distinct type of equipment for every condition, but it would be desirable to have one type of equipment that is adapted to as many conditions as possible.

Rubber Tires for Tractors

The first tractors to be equipped with solid or pneumatic rubber tires were those used for industrial purposes around factories, airports, and for highway maintenance, since satisfactory traction was secured and jarring, vibration, and damage were reduced. In about 1931, investigations were begun concerning the possibility of using low-pressure pneumatic tires for agricultural tractors. Extremely favorable results were immediately observed, and a number of advantages in favor of rubber tires over steel wheels and lugs were disclosed.

Ply Rating of Tires Tire capacity is indicated by the ply rating molded in the side of the tire. In the early days, when all tires had cotton cord fabric of the same size and strength, the number of fabric plies was a meaningful measure of tire capacity. Presently, however, the improved properties of synthetic fabrics has made the number of actual plies in the tire less meaningful. The Tire and Rim Association and the Rubber Manufacturers' Association have defined ply rating as follows: "The term 'Ply Rating' is used to identify a given tire with its maximum recommended load, when used in a specific type of service. It is an index of tire strength and does not necessarily represent the actual number of cord plies in the tire.'' Ply rating for agricultural tires ranges from two to twelve, depending upon the type of service.

Tread Design Tread-bar angle or shape is probably the most publicized competitive aspect of tractor tires. It is difficult to design and specify a single tire-tread for traction wheels on all types of tractors and self-propelled machines that would be suitable for all conditions and situations. Tractors must operate on many different types of surfaces such as grass, weeds, bare soil, hard paved surfaces, loose plowed soil, and many different types of soils with shear strength that varies with type and content. Obviously, all these factors are subject to change. The tire designer must seek the best compromise that gives reasonably satisfactory performance in as many combinations of these situations as possible. Different types of treads found on power driven tires are shown in Fig. 19-1.

In the past, most tractor tires had straight tread bars molded at a 45° angle to the center line of the tire. This was necessary because of limitation on the physical strength of the tire material. If the tread angles were less than 45°, the forces imposed upon the tread bars were often greater than the materials could withstand. Tread bars can now be formed at angles less than 45° because of increased strength resulting from the use of new material in tire construction. As tread-bar angle decreases, traction capability increases in many conditions, but self-cleaning ability in muddy conditions decreases, thus reducing the effectiveness of the tire. Tires have less tread overlap as tread-bar angles decrease, thus giving a bumpier ride on hard surfaces. Research has indicated that tread-bar angle of 20 to 30° gives the best performance in soft soil conditions. Several manufacturers offer tires with a tread-bar angle that is less than 45°.

Inflation Pressure Proper inflation pressure is extremely important for tractor tires. Improper tire pressure is a major contributor to tire failure. Underinflation causes sidewalls to flex excessively and will result in breaks and separations in the cord body. Overinflation will make the tire body rigid and reduce its resistance to impact, making it more susceptible to sidewall breaks. Manufacturers' recommendations for tire pressure should be followed. Recommended air pressure for front tires on two-wheel-drive tractors range from 20 to 36 lb/in^2 (1.4 to 2.1 kg/cm^2) and power tires range from 12 to 26 lb/in^2 (0.8 to 1.8 kg/cm^2). The air pressure of in-furrow-wheel tires should be increased 4 lb/in^2 (0.2 kg/cm^2) when plowing.

Tests made by Sauve and McKibben[1] show that low inflation pressure gives somewhat better traction, especially in loose soils, than higher pressure. Results of these tests are shown in Figs. 19-2 and 19-3.

[1] E. C. Sauve and E. G. McKibben, "Studies on Use of Liquid in Tractor Tires," *Michigan State University, Agr. Expt. Quart. Bull.*, vol. 27, no. 1, Aug. 1944.

Figure 19-1 Types of tire treads for power-driven tractor wheels. (A) Standard type farm tread, code R-1; (B) cane and rice, code R-2; (C) industrial and sand, code R-3; (D) industrial lug type, code R-4. (Courtesy of Goodyear Tire and Rubber Company.)

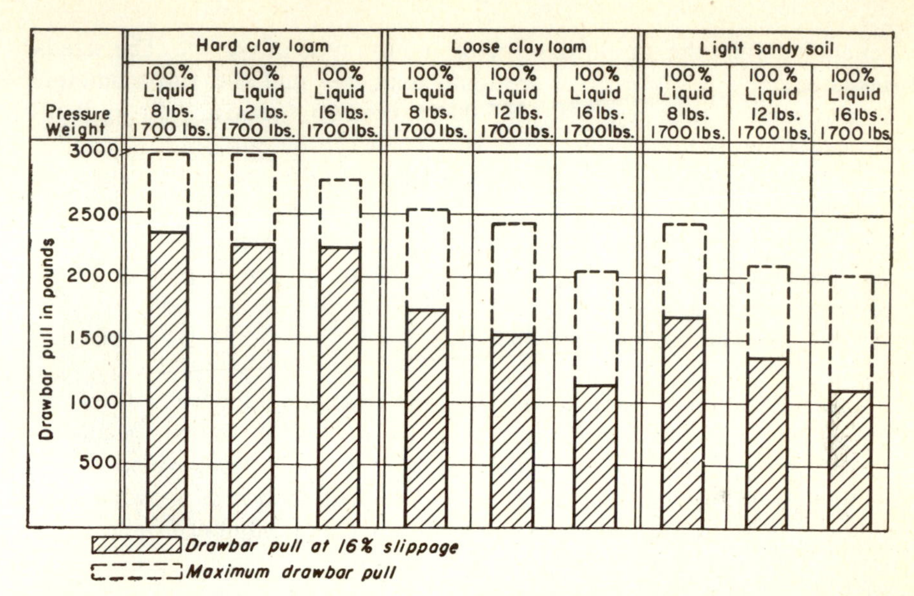

Figure 19-2 Effect of inflation pressure on drawbar pull.

Tire Size Tractor tires are classified as "front-wheel" (nonpowered) and "rear-wheel" (power driven). Nonpower driven front-wheel tires (Fig. 19-13) usually have 1, 3, or 5 concentric tread ribs to provide easy steering and to control skidding. The number of plies varies from 6 to 10

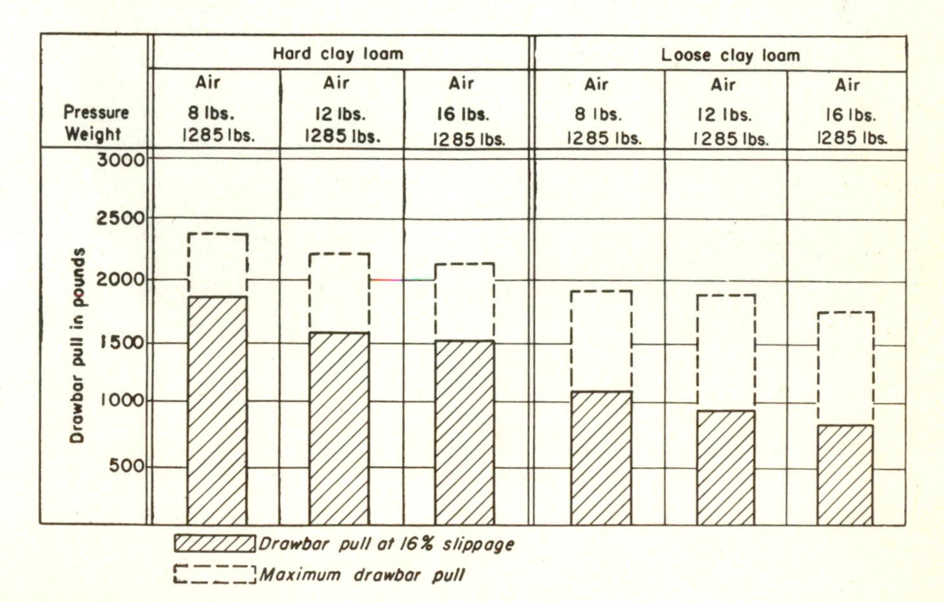

Figure 19-3 Effect of inflation pressure on drawbar pull.

depending upon the size of the tractor and the type of service. The size is designated by the tire cross-sectional diameter and the rim diameter. Some common sizes are 4.00-12, 5.00-15, 6.00-16, 6.50-16, 7.50-16, 9.50-15, and 10.00-16.

Power driven (rear) tires have special tread designs to provide traction (Fig. 19-1) under all conditions. The number of plies varies from 4 to 10, the cross-section diameter varies from 11.2 to 30.5 in (28.4 to 77.5 cm) and the rim diameter varies from 24 to 42 in (61 to 106.7 cm). Some common sizes are 12.4-24, 12.4-42, 13.6-38, 14.9-24, 14.9-38, 18.4-24, 18.4-42, 23.1-34, 24.5-32, and 30.5-32.

Identification of Agricultural Tires In order to standardize the identification of the many agricultural tires available, the Rubber Manufacturers' Association and the Tire and Rim Association developed and adopted a uniform system of coding. This identification is shown on the sidewall of each tire and consists of a letter plus a number (Table 19-1).

Table 19-1 Agricultural Tire Code Designation

Code marking	Industry tire type	Tire service
	Traction tires (rear, tractor)	
R-1	Regular agricultural	General farming
R-2	Cane and rice	Wet muck, sugar cane, rice farming
R-3	Industrial and sand	Sandy or volcanic ash soils, orchards, highway moving, golf course work, light industrial service
R-4	Industrial lug type	Light industrial service, highway mowing
LS-1	Regular tread logger	Logging operations
LS-2	Medium tread logger	Logging operations
	Steering tires (front, tractor)	
F-1	Single rib tread	Rice farming
F-2	Regular agricultural	General farming
F-3	Industrial rib	Light industrial service
	Garden tractors	
G-1	Lug type	Gardens
G-2	Universal type	Lawn mowers
	Implement tires	
I-1	Rib tread	Free-rolling wheels
I-2	Moderate traction	Drive wheels
I-3	Traction implement	Drive wheels
I-4	Smooth tread	Mower or rake, caster wheels

Radial Tires The use of radial tires on power driven wheels of farm tractors has increased within the past year. There are certain advantages in using radial instead of bias tires. Rolling resistance is slightly less, there is increased drawbar horsepower, and the ride is smoother with less jarring and vibration. Carper and Opp[1] found that the radial tire performed slightly better than the bias tire (Table 19-2), on a large four-wheel drive tractor with dual wheels. This improvement can be accounted for by the fact that a radial tire has a larger footprint than a bias tire. A greater difference was shown in the travel reduction than in maximum horsepower tests. The reason for this is that travel reduction tests were run at a lower slip range than the horsepower test. Radial tires are superior in the lower slip range but lose their advantage when the tractor approaches the higher slip ranges, where maximum horsepower occurs under field conditions.

Dual Wheels As tractor drawbar horsepower has increased, the use of dual wheels on tractors has increased (Fig. 19-4). Drawbar pull can be increased and/or soil compaction may be reduced under certain soil conditions and tractor weight, by the use of dual-power-driven wheels on tractors.

Results of tests at several locations indicate:

1 If a tractor is operating without weights on single power driven tires, the same increase in traction can be obtained by adding liquid ballast in the tires or adding duals. If soil compaction is a problem, the addition of duals is preferable.

Table 19-2 Bias Versus Radial Tire on Large Four-Wheel Drive Tractor*

	Static load		Travel reduction		Maximum drawbar horsepower		
Test comparison	Front Axle, lb	Rear Axle, lb	Sod, %	Disced Soil, %	Sod, hp	Disced Soil, hp	Clay track, hp
18.4 − 38, 6 ply, R-1, bias, dual.	19,500	11,500	12.2	12.2	171	154	154
18.4R − 38, 6 ply, R-1, radial, dual.	19,620	11,670	10.1	9.0	172	160	166

* Tire pressure for all test, inside dual 14 lb/in², outside dual 12 lb/in²

[1] R. L. Carper, and A. E. Opp, "Tire Performance on a Four-Wheel Drive Tractor," paper no. 77-1518, ASAE, St. Joseph, Mich., 49085.

Figure 19-4 Tractor with dual rear wheels.

2 If a tractor is presently operated with single tires with ballast, it will have approximately the same drawbar pull if it changes to duals without ballast, but will have less soil compaction.

3 Maximum drawbar pull can be obtained with dual wheels and liquid ballast in all wheels. Compaction will be less than that produced by the use of single tires with ballast.

Attachment of Dual Wheels There are basically four methods used to attach the outside wheel required for duals. The first method uses threaded hooks and nuts combinations to fasten the outer rim to the inner rim. A spacer is used to give small space between the wheels. No wheel or hub is used in the outside wheel.

A second method uses snap-on or over-center catches to fasten the outer tire and rim to the inner rim. No wheel is used in the outer rim. A third method uses an adapter or spacer that fastens through the wheel-weight bolts to the inner wheel. The outside dual, which includes a wheel, rim and tire, but no hub, is bolted to the adapter (Fig. 19-5).

The fourth and the most rugged method (Fig. 19-6) uses a cast-iron wheel with heavy-duty rim fastened to the axle by a tapered bushing hub. The dual wheel is fastened to the tractor axle in the same way as the original inside wheel.

There are several arrangements of dual wheels according to the way the dished wheel is turned (in or out) and if the tractor has a 96 in (2.44 m) power axle or 120 in (3.05 m). The longer axle permits the duals to be spread out to match the row spacing for the cultivation of row crops. The arrangement of duals places additional bending load on the axles whenever only the outside tire of the duals supports the tractor when

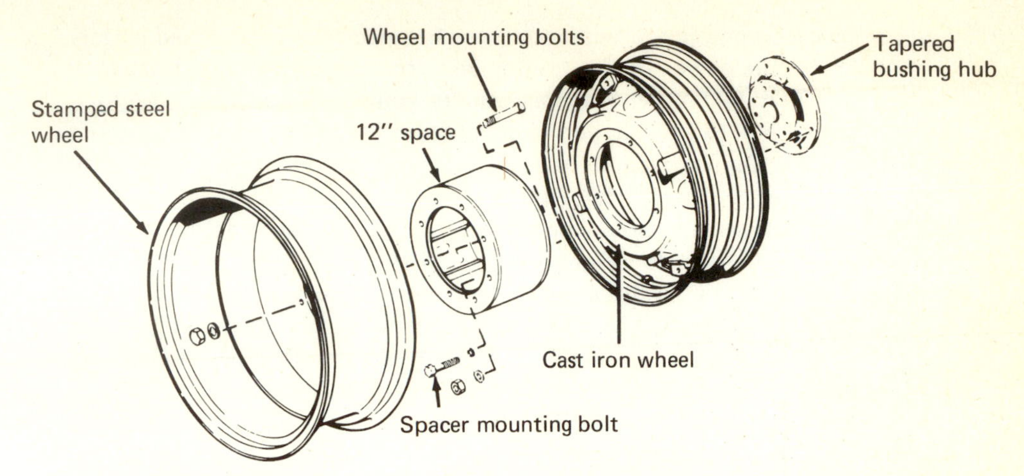

Figure 19-5 Dual rear wheel mounted by use of spacer and bolts through wheel-weight holes. (Courtesy of J I Case Company.)

passing over a high spot. Some tractor manufacturers recommend that the inner tire be operated at maximum tire pressure and the outside tire pressure should be 4 lb/in^2 (0.28 kg/cm^2) less. This enables the outer tire to deflect when passing over obstructions, thus permitting the inner tire to share some of the load and reducing stress on the axle.

Weights for Rubber Tires Since the wheels and tires alone are relatively light and hence do not provide sufficient traction under most conditions, it is necessary to add weight in some manner. This may be done by attaching special cast-iron weights to the wheels; partially filling the tire with water or some other suitable liquid; or adding dry, powdered ballast

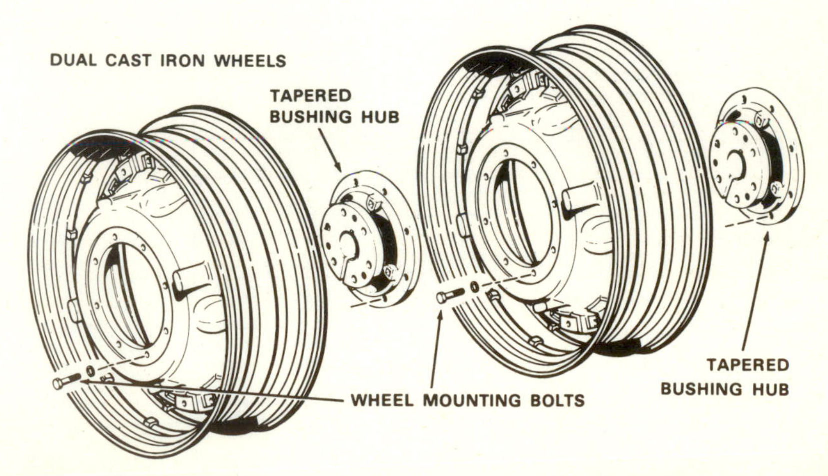

Figure 19-6 Dual rear wheels mounted by cast iron wheel fastened to the axle by tapered bushing hub. (Courtesy of J I Case Company.)

in the tire. The amount of additional weight needed is determined largely by the size and power of the tractor and traction conditions. For small tractors, 200 to 600 lb per wheel is usually sufficient. For larger tractors, as much as 1,000 to 2,600 lb per wheel may be needed to obtain maximum drawbar pull. Do not exceed load capacity of individual tires or the manufacturer's recommended total tractor weight, with ballast. In general, only sufficient weight should be added to obtain good traction without undue slippage. As little weight as possible should be used in harrowing, planting, drilling, and cultivating in order to reduce soil packing.

Figure 19-7 shows the effect upon drawbar pull of adding wheel weights. Investigations show that the increased drawbar pull will be approximately 50 percent of the weight added to the wheels.

Weight is frequently added to rubber-tired tractors by adding water to the tires until they are 75 percent full. The quantity of liquid to use depends upon tire size and the amount of extra weight desired. In areas where antifreeze protection is needed, add 3.5 lb (1.6 kg) of commercial calcium chloride flakes per gallon (3.78 liters) of water for freeze protection to −12°F (−24.4°C). For protection to −52°F (−46.6°C), use 5 lb (2.3 kg) of calcium chloride flakes per gallon water. An 18.4-38 tire filled 75 percent with water will add 917 lb (416 kg) of weight. If 5 lb (2.3 kg) $CaCl_2$ is used per gallon of water, the weight added per 18.4-38 tire is 1187 lb (538 kg).

Rolling Resistance The rolling resistance is the drawbar pull or its equivalent required to move the tractor over a given surface. The tractor in field work passes over soft traction surfaces. The wheels or tracks, in sustaining the weight of the tractor, sink into the surface. Therefore, the

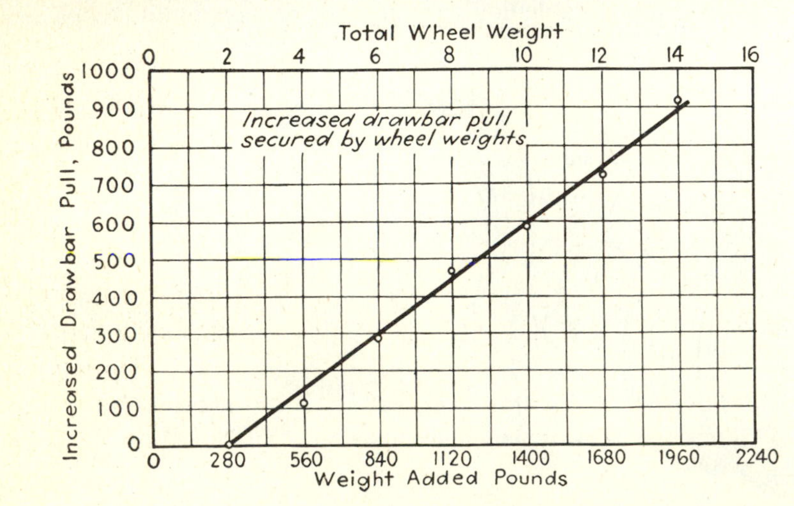

Figure 19-7 Effect upon drawbar pull of adding weight to rear wheels.

tractor is virtually climbing an incline as it moves forward. In addition, rolling resistance includes resistance due to friction in traction members and losses incurred in obtaining adhesion.

Tractor Tire Slippage and Tractive Efficiency The travel reduction or slippage of a tractor wheel may be defined as the ratio of distance traveled per wheel revolution without drawbar load to the distance traveled per wheel revolution with drawbar load.

$$\text{Travel reduction (percent)} = 100 - \frac{\text{distance traveled per wheel revolution with drawbar load}}{\text{distance traveled per wheel revolution without drawbar load}}$$

Usually tractor drawbar pull will increase as percent slippage increases (Fig. 19-8) up to 16 percent, which is considered the allowable maximum. Drawbar pull will increase after slippage is greater than 16 percent (Figs. 19-8 and 19-9), but tire wear becomes excessive.

The tractive efficiency of a tractor wheel or track may be defined as the ratio of its work output to the work applied to it at the axle. Figure 19-9 shows the results of studies of tractor tires made by Reed et al.,[1] with

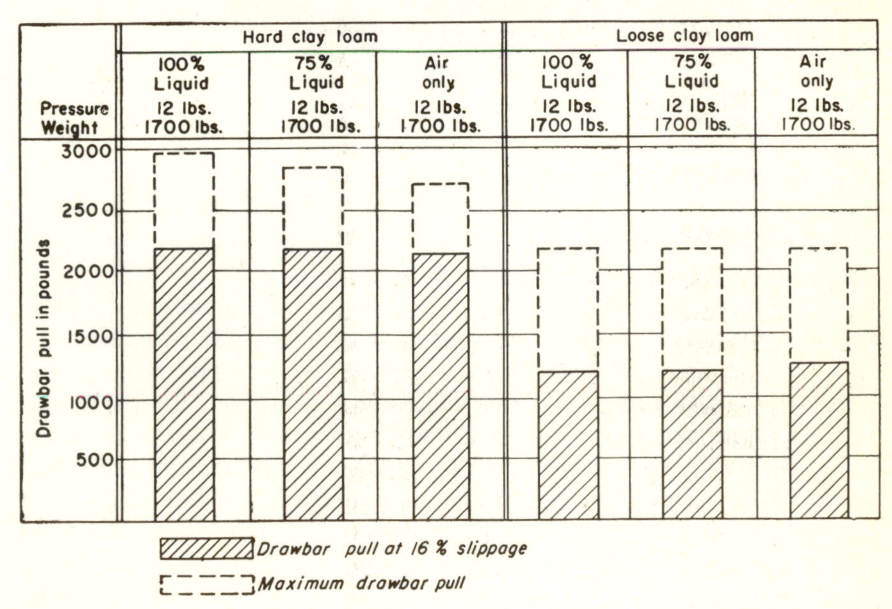

Figure 19-8 Effect of different methods of weighting tires on drawbar pull.

[1] I. F. Reed, C. A. Reaves, and J. W. Shields, "Comparative Performance of Farm Tractor Tires Weighted with Liquid and Wheel Weights," *Agr. Eng.*, **34,** (6): 391–395 (1953).

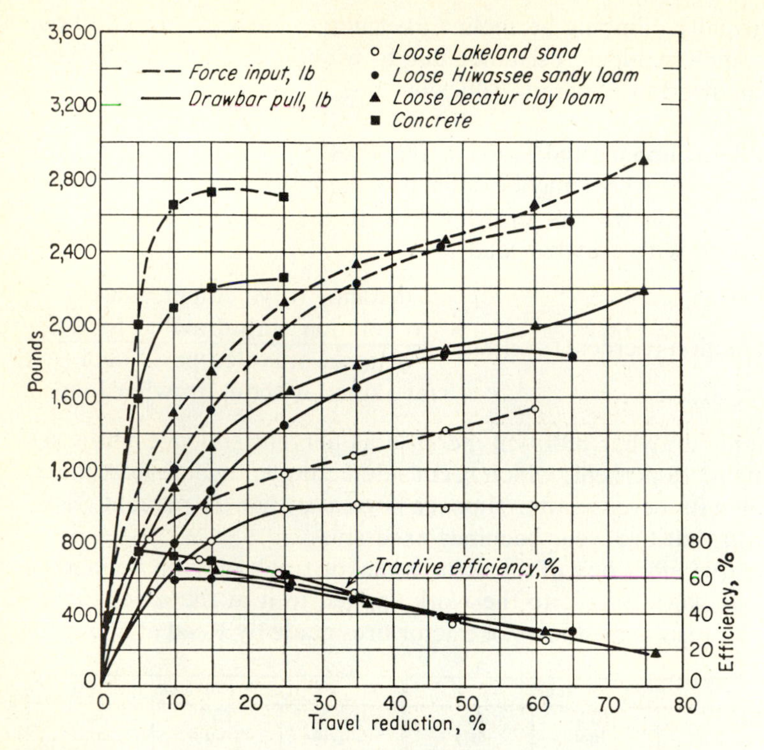

Figure 19-9 Performance of 11-38 tractor tires on various surfaces.

respect to their tractive efficiency on concrete and with different soil conditions and types.

Life of Tractor Tires The life of pneumatic tractor tires will depend on such factors as (1) abrasive wear, (2) chipping, (3) punctures and blowouts, (4) chemical decomposition, and (5) general care given. Abrasive wear is dependent upon the soil surface condition, the percent of wheel slippage, and drawbar load. Some tractors today travel many miles on paved highways between different farms that their owners operate. Tread life of tires operated on paved surfaces is reduced because the lug cannot penetrate the hard surface, causing unnatural flexing of the tire wall and excessive wear on the lugs.

McKibben and Davidson[1] report that Iowa farmers estimated the average life of tractor tires to be about seven years. Figure 19-10 shows the results of this survey and the variation in estimated life.

[1] E. G. McKibben and J. B. Davidson, "Life, Service and Cost of Service of Pneumatic Tractor Tires," Iowa State College, Agr. Exp. Sta. Bullentin 382 (1939).

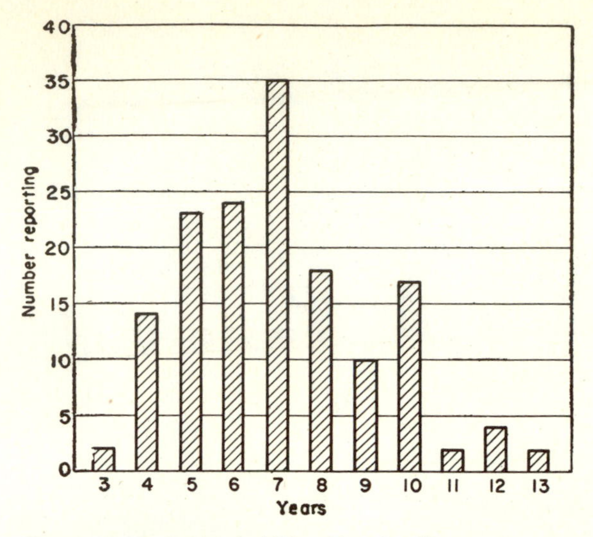

Figure 19-10 Estimated life of tractor tires.

Track Mechanisms

The traction mechanism of a track-type tractor (Fig. 19-11) consists of: (1) frame, (2) drive sprocket, (3) front idler, (4) track rollers, and (5) track. The track frames are built of structural steel. The sprocket and front idler drive and support the track. The track rollers on the underside of the track frame act as supports between the machine and the part of the track in contact with the ground. In fact, these rollers might be considered as small wheels rolling and conveying the tractor over a stationary track.

The track itself is built up of forged-steel shoes bolted to forged-steel heat-treated links, with alloy-steel heat-treated pins (see Fig. 19-12).

Since any track mechanism must have many points of wear that are continually exposed to dirt and grit, three fundamentally important considerations must be observed by the manufacturer, namely, (1) the parts themselves must be made of high-grade materials that have been carefully heat-treated and hardened; (2) any important bearings, such as track rollers, front idlers, and so on, must be well enclosed and properly lubricated; and (3) there must be some provision for taking up the slack and maintaining the proper track tension.

Wheels Versus Tracks

The overall drawbar efficiency of a tractor is the relationship of the power supplied at the drawbar and the power generated by the engine under a given set of conditions. That is,

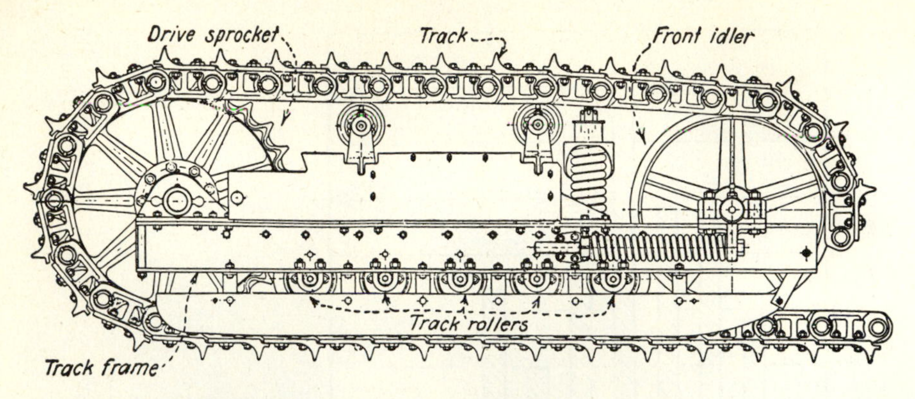

Figure 19-11 Track assembly showing principal parts.

$$\text{Drawbar efficiency (percent)} = \frac{\text{drawbar hp}}{\text{bhp}} \times 100$$

The comparative drawbar efficiency of track-type and wheel-type tractors is clearly disclosed by an analysis of tests made by the University of Nebraska.[1] For example, the maximum horsepower tests of 25 wheel-type machines show an average efficiency of 88.1 percent. Similar tests of 12 track-type tractors show an average efficiency of 84.9 percent. The average slippage in these same tests was 5.67 percent for the wheel-type tractors and 2.95 percent for the track-type machines. The higher drawbar efficiency of the wheel-type machines, in spite of their greater slippage loss, can be attributed largely to the power consumed by the track mechanism and the heavy tracks as they are rotated and move the machine. In

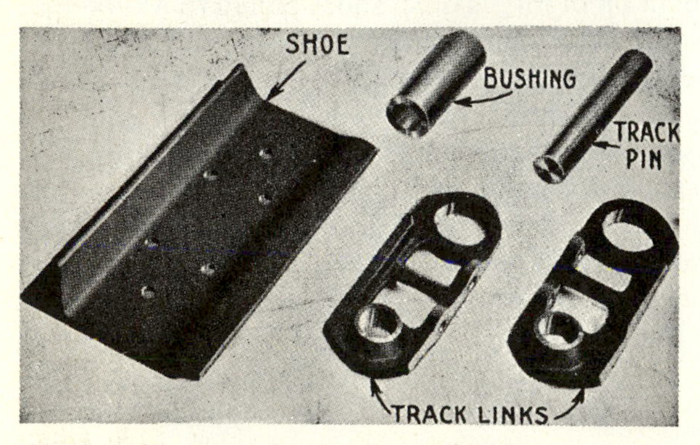

Figure 19-12 Track construction and parts.

[1] Nebraska Tractor Test, University of Nebraska, Agricultural Engineers Department, Lincoln, Nebraska 68583.

other words, the rolling resistance of wheel tractors is considerably less than that of the track types.

The outstanding advantage of the track type is, no doubt, its ability to secure traction under conditions that will not permit the wheel type to operate with any degree of success. Not only can the track tractor usually secure good footing under the most adverse conditions, but there is likely to be less loss from slippage under average operating conditions.

Front Axles

Front-axle construction for wheel tractors depends upon the tractor type. Figure 19-13 shows the axle and steering linkage for a four-wheel tractor. All front wheels are equipped with roller bearings and a convenient means of wear adjustment. Lubrication is either by packing the bearing with grease or by periodic application of the lubricant with a grease gun. Figure 19-14 shows a device for equalizing the traction of dual front wheels when traveling on ridges or over uneven ground.

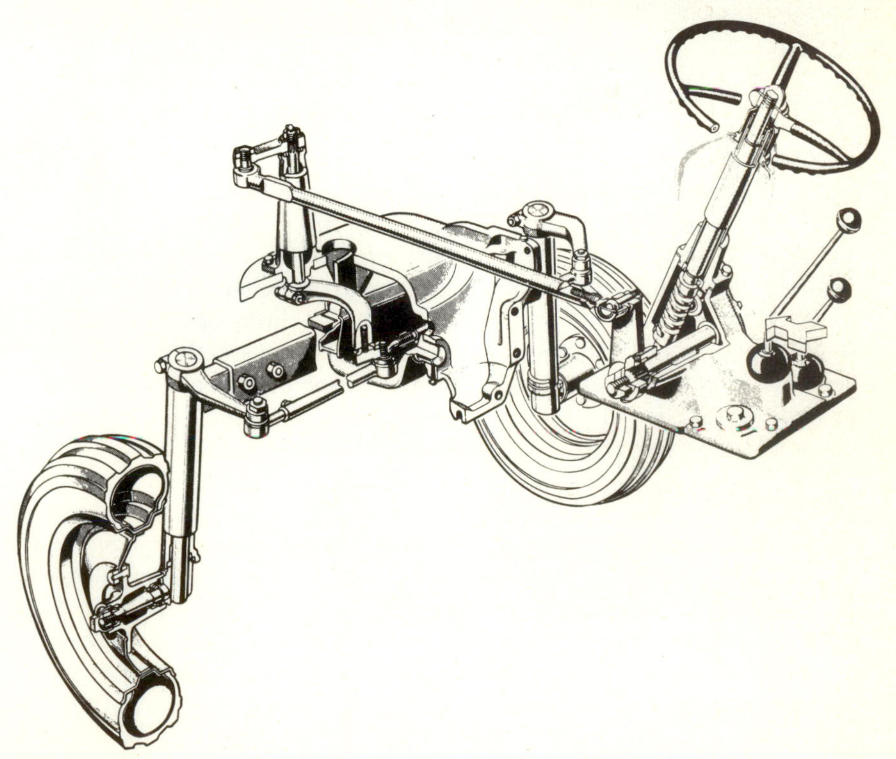

Figure 19-13 Axle construction and steering mechanism for a four-wheel tractor. (Courtesy of Massey-Ferguson, Inc.)

Figure 19-14 Special device for equalizing traction of dual front wheels. (Courtesy of Deere and Company.)

STEERING MECHANISMS

The operator of two-wheel-drive tractors controls the directional movement of the machine by means of an assembly consisting of (1) a steering wheel, (2) a steering gear, and (3) a suitable linkage to the front wheels. The steering gear is the most important part of the entire assembly because it must transmit the rotary movement of the steering wheel to the linkage at a relatively slow speed and with as little operator effort as possible. Figure 19-13 shows the complete steering assembly for a typical wheel-type tractor. Steering gears are always enclosed in a dirtproof housing and operate in oil or grease. Most have a simple and convenient adjustment for wear.

Power Steering

The medium and larger sizes of two-wheel-drive tractors, when equipped with the standard type of steering mechanism, require considerable effort on the part of the operator in steering them, particularly in loose soil, on rough ground conditions, or when the tractor is stationary or moving slowly. Hence, most of these tractors are equipped with a power-actuated steering mechanism. The necessary assembly is usually a part of the hydraulic system of the tractor and includes: (1) the pump, (2) the actuator hydraulic pressure from the pump, (3) the control valves actuated by the operator through the steering wheel, and (4) the connecting oil lines.

Obviously, the engine must be running in order to operate the pump, which forces the oil through the lines to the steering unit with full-power steering. The only force required from the operator is to turn the steering wheel in order to open the hydraulic valves. Power steering for two-wheel-drive tractors is divided into two major categories: (1) hydrostatic steering and (2) hydraulic steering with mechanical drag link.

Hydrostatic Steering There is no mechanical connection between the steering valve and the actuator that moves the wheels. Figure 19-15 shows a hydrostatic steering system with a closed-center system during a right turn.

When the steering wheel is turned to the right, the steering shaft, which is threaded through the steering valve piston, tries to pull this piston upward. Since oil is trapped in the circuit, the shaft moves the collar downward, rotating the pivot lever, and opens a pressure and return valve. With the valves open, oil under pressure enters the steering valve, forcing the piston upward. Oil then flows out of the steering valve into the right-hand steering cylinder, turning the front wheels to the right. Oil is forced out of the left-hand steering cylinder to the reservoir. When the operator stops turning the wheel, the steering shaft is moved upward by the steering-valve cylinder, forcing the collar upward, centering the pivot lever, thus closing the valve. (Left-turn valves are not shown in Fig. 19-15.)

Hydraulic Steering with Mechanical Drag Link Figure 19-16 illustrates a hydraulic steering mechanism with mechanical drag link, making a turn to the right. When the operator turns the steering wheel and shaft to the right, resistance to turning the wheel forces the shaft up out of the worm nut. This shifts the spool valve and steering shaft up, and opens the oil line to the cylinder at the front wheels. The cylinder rotates a rack and pinion device, which turns the front wheels. Oil is forced out the other side of the cylinder to the reservoir. As long as the steering wheel is turned, oil will flow into the cylinder and turn the front wheels. When the steering-wheel motion is stopped, the hydraulic pressure will turn the front wheels slightly further to the right, moving the steering link forward and pulling the valve back to the neutral position.

Steering Four-Wheel-Drive Tractors

The development of large four-wheel-drive tractors led to the development of different methods for steering. Most four-wheel-drive tractors use either (1) articulated steering or (2) four-wheel steering.

Articulated Steering Tractors that have articulated steering can pivot at a center hinge point (Fig. 19-17). The wheels are always parallel to

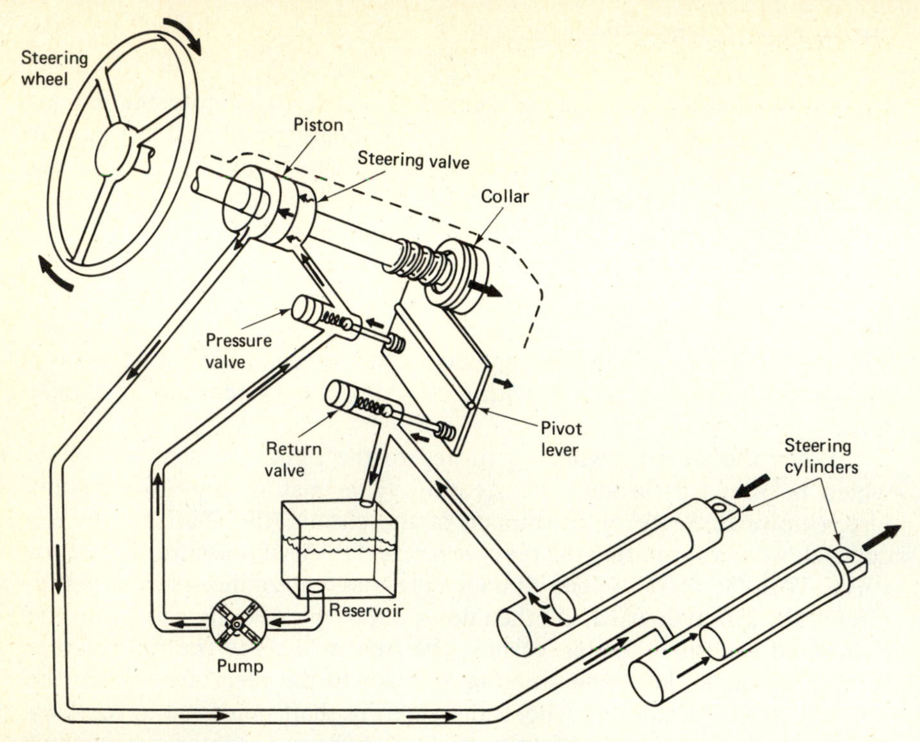

Figure 19-15 Operation of hydrostatic steering system as tractor makes right turn. (Courtesy of Deere and Company.)

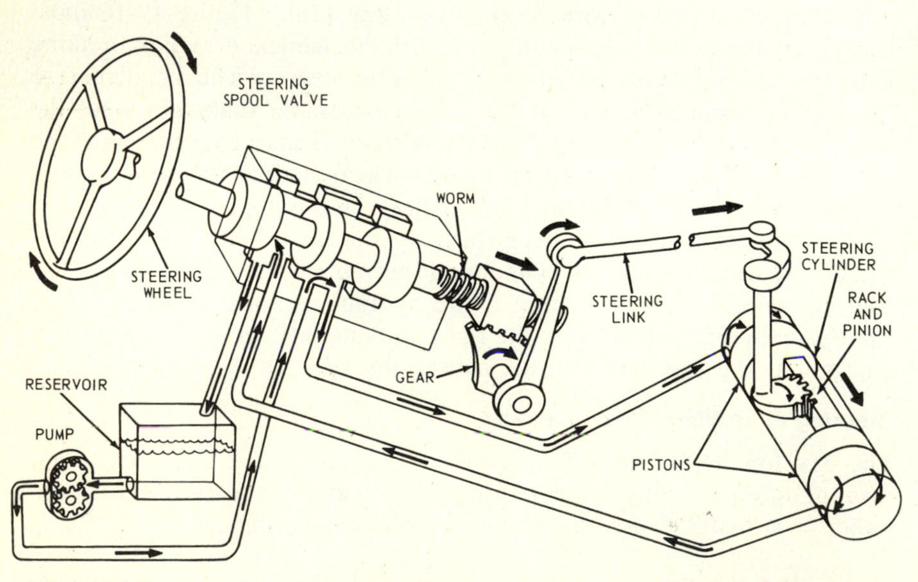

Figure 19-16 Tractor using hydraulic steering with mechanical drag link to make right turn. (Courtesy of Deere and Company.)

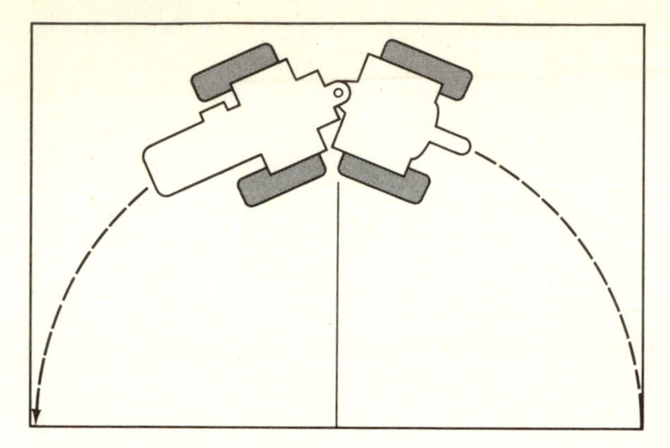

Figure 19-17 Four-wheel-drive tractor with articulated steering making left turn.

the frame and perpendicular to the axle. (With two-wheel-drive steering, the wheels turn toward the frame when the tractor makes a turn.) Steering is done by two hydraulic cylinders, which pivot the front and rear drive units around a center point. When a tractor makes a left turn, a hydraulic cylinder on the left side of the center pivot point retracts, pulling the two units together (Fig. 19-17), while a cylinder on the right side pushes. When a steering correction is made with the front frame, the rear frame is automatically turned in the opposite direction (Fig. 19-17). The rear wheels always follow the front ones with articulated steering. Tractor articulates up to 40°.

Four-Wheel Steering Four-wheel-drive tractors with four-way steering have a king pin for each wheel; thus, rear wheels can be steered like front wheels on regular two-wheel-drive tractors. Steering can be done with front and rear wheels turning in the same or opposite directions, with front wheels only or with rear wheels only (Fig. 19-18).

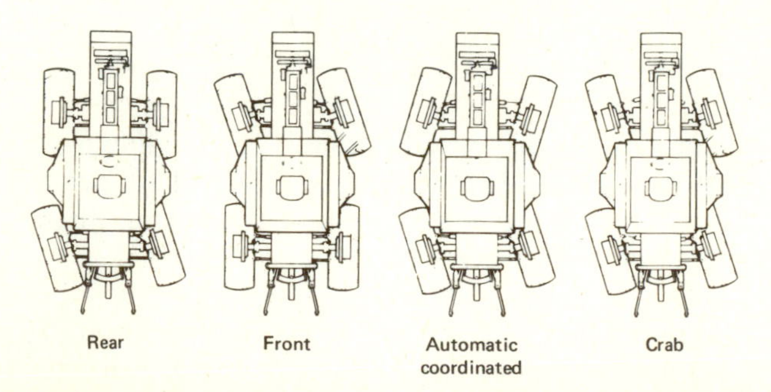

Figure 19-18 Four-wheel-drive tractor with four-wheel steering.

A schematic drawing of a four-wheel steering system is shown in Fig. 19-19. The front steering system includes the steering wheel column, hydraulic lines, and the front axle steering cylinders. It operates similar to the steering system used on two-wheel-drive tractors. The rear steering system consists of a lever-control position indicator, rear steering hydraulic control valve, hydraulic lines, and steering cylinders. A mechanical timing cable connects the two independent front and rear systems for coordinated steering. This permits automatic rear control with the steering wheel so that the rear wheels automatically follow the front wheels. Rear steering can be controlled independently by moving the rear-steering hand lever upward to turn right, or downward to turn left.

Wheel-Tread Adjustment

Row crop or all-purpose tractors must be provided with some convenient means of adjusting the wheel tread to adapt it to different machines and row spacings. In general, for plowing, the wheels must be relatively close together, while for cultivating or planting two or four rows, they must be set rather far apart.

Standard four-wheel tractors must have a means of adjusting both the front-wheel and rear-wheel treads, while tricycle-type tractors require only a rear-wheel adjustment. The front-wheel adjustment is usually taken care of by shortening or lengthening the axle on each side. A telescoping arrangement or bolts fitted through a series of holes accomplishes this.

Rear-wheel treads are changed (1) by sliding the wheel hub inward or outward on the axle; (2) by having an offset hub, or dished wheel, and reversing the wheel on the axle; (3) by shifting the rim and tire inward or outward by means of rim lugs or spacers; or (4) by a combination of these methods. Figure 19-20 shows a tread adjustment from 48 to 76 in (1.22 to 1.93 m) by 4-in (10.1 cm) intervals, by using methods 2 and 3. Figure 19-21 shows the long axle with adjustable hub and dished wheel. Figure 19-22 shows a special type of quick tread adjustment involving the use of spiral rim cleats and special locking devices. The power of the tractor itself is used to shift the wheels.

POWER TAKE-OFF

The power take-off is a special gear-driven shaft usually protruding from the backside of the tractor rear-axle housing. It is designed to power integral mounted or trailing implements such as mowers, balers, forage harvesters, cotton strippers, or corn pickers.

Types of PTOs There are three types of PTO systems used on tractors. The three systems include: (1) transmission-driven, (2) continuous-running, and (3) independent.

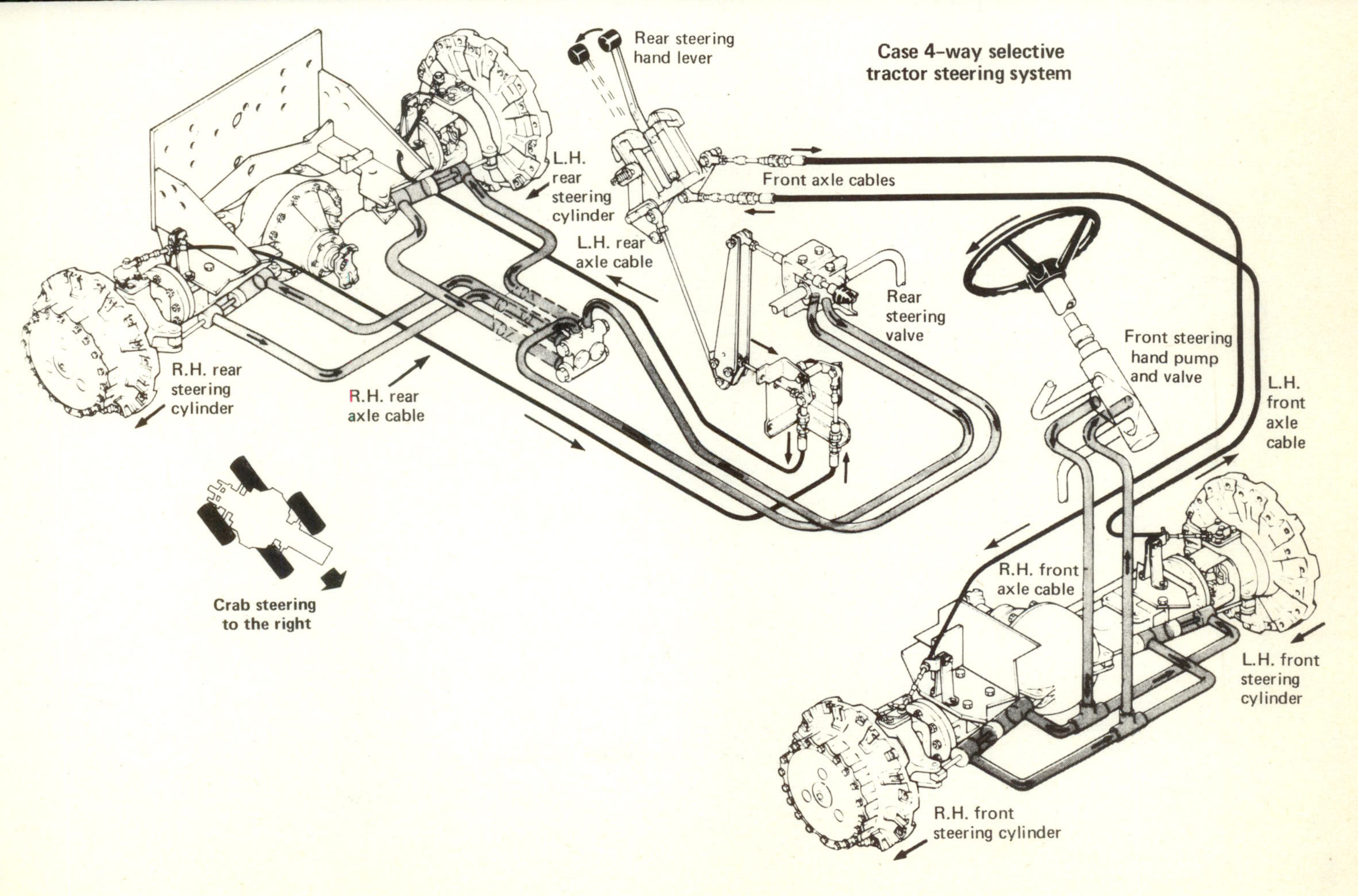

Figure 19-19 Schematic drawing of a four-wheel-steering system. (Courtesy of J I— Case Company.)

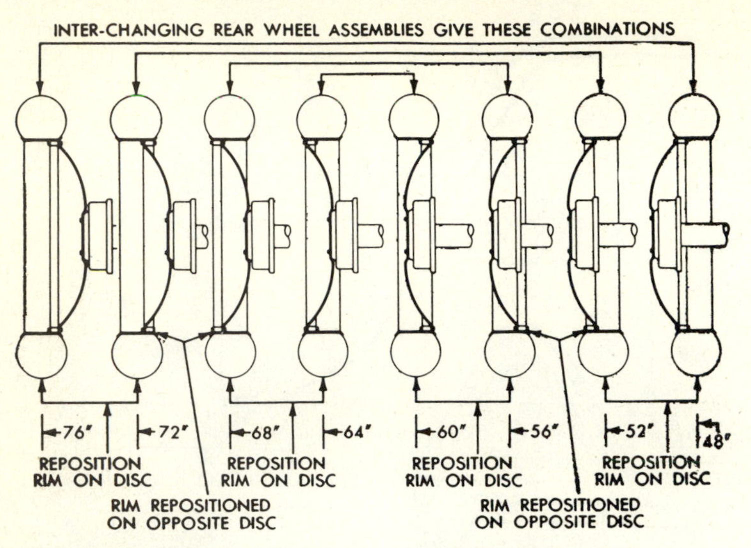

Figure 19-20 Method of spacing rear wheels.

Transmission-Driven PTO The early types of PTOs were transmission driven. They received their power through the master clutch; therefore, they operated only when the master clutch was engaged. The PTO would stop whenever the clutch was disengaged to stop or start movement of the tractor (Fig. 19-23).

Continuous-Running PTO This unit has a two-stage clutch (Figs. 19-23 and 19-24) one for the transmission and one for the PTO. Frequently both clutches are operated by one foot-opened clutch pedal. Depressing

Figure 19-21 Extension-axle method of wheel-tread adjustment.

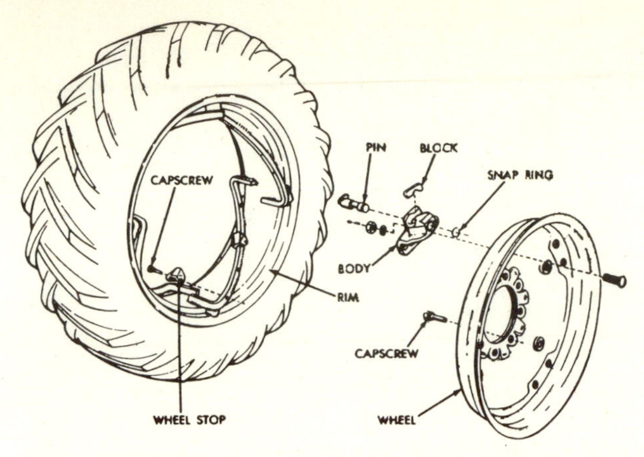

Figure 19-22 Wheel-tread adjustment by means of spiral guides and engine power. (Courtesy of Allis-Chalmers Manufacturing Company.)

the foot-controlled clutch pedal one-half of its travel distance will stop power flow to transmission, thus stopping motion of the tractor, but the PTO will continue to rotate. Depressing the clutch pedal completely stops both forward motion and PTO rotation. The rotation of the PTO cannot be stopped without stopping the forward motion of the tractor.

Independent PTO Two separate clutches are used in an independent PTO. One clutch controls travel of the tractor (Figs. 19-23 and 19-25),

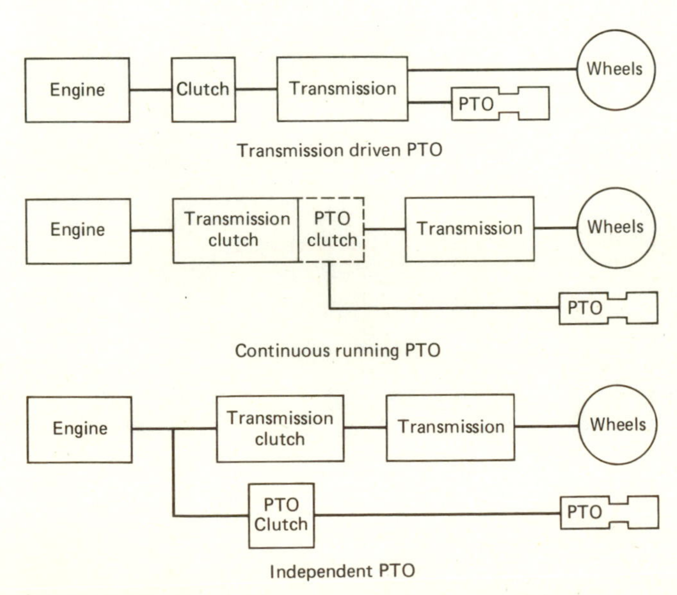

Figure 19-23 Three types of tractor power take-offs.

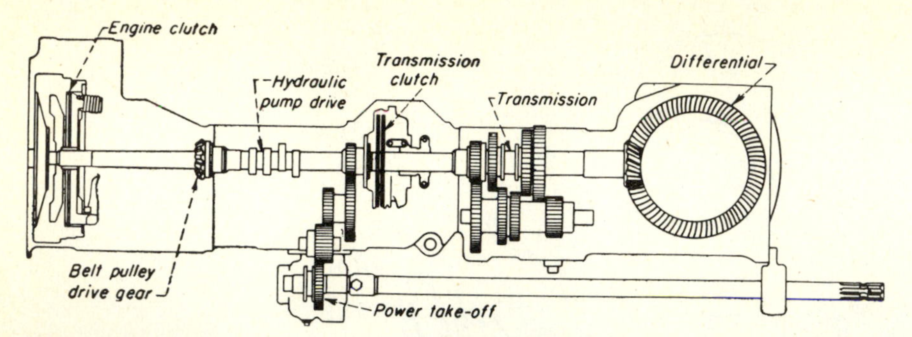

Figure 19-24 Continuous-running power take-off. (Courtesy of Allis-Chalmers Manufacturing Company.)

while the second clutch controls the PTO. Each clutch is driven independently by the tractor engine. The PTO clutch may be controlled by an over-center hand lever, or it may be actuated hydraulically by a hand-control lever. With an independent PTO, the tractor can be stopped, gears in transmission shifted, and forward travel started without stopping the PTO. The PTO can be started or stopped without stopping the forward motion of the tractor. Figure 20-15 illustrates a hydraulic control for an independent PTO.

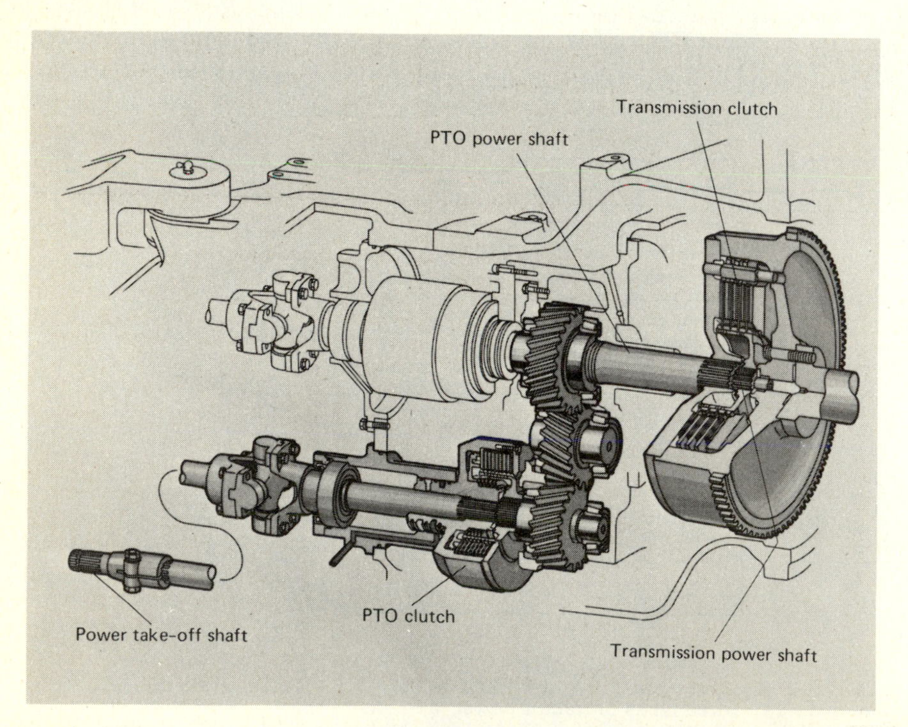

Figure 19-25 Independent power take-off. (Courtesy of Deere and Company.)

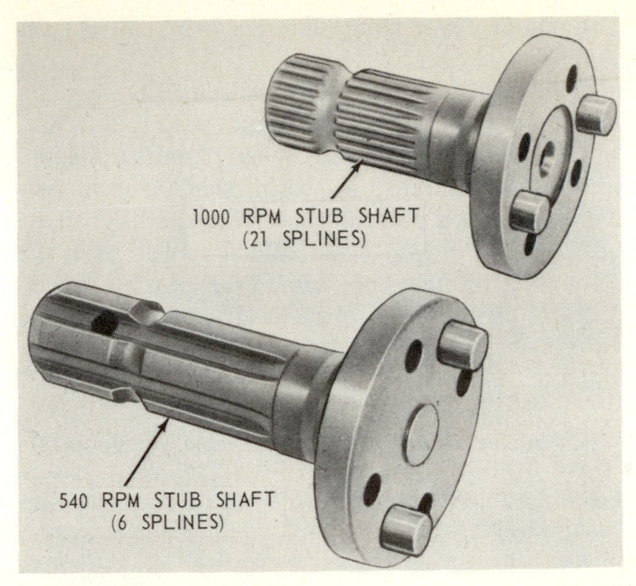

Figure 19-26 Twenty-one splines and six splines tractor PTO stub shaft. (Courtesy of Deere and Company.)

Standards for PTO Drives Standards for PTO drives have been established by the American Society of Agricultural Engineers (ASAE Standard: ASAE S203.9) for use by tractor and implement manufacturers. These standards enable the changing of equipment from one make of

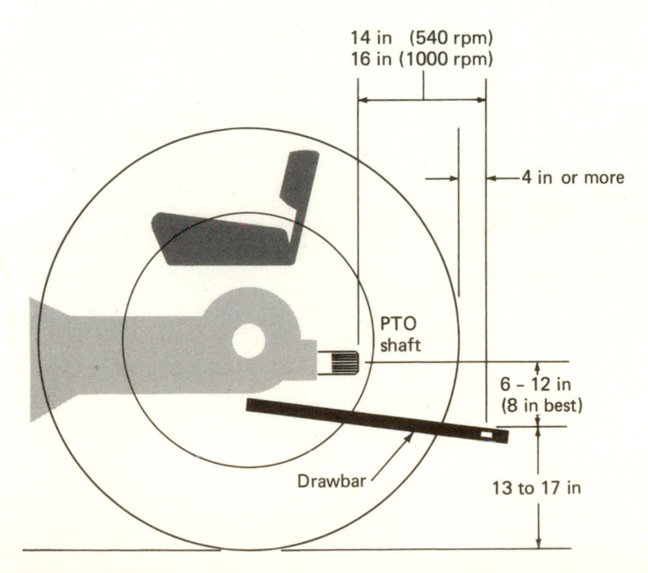

Figure 19-27 Location of PTO shaft and drawbar hitch on farm tractors with 1⅜ in PTO shaft. (ASAE–SAE Standard.)

tractor to another. The standards cover many different items in the PTO system.

The PTO shaft on all tractors has been standardized. The shaft on tractors has either 21 splines and operates at 1,000 rpm or 6 splines and operates at 540 rpm (Fig. 19-26). The higher shaft speed permits transmission of more horsepower without increase in size of shaft. The recommended location of the drawbar hitch point in relation to the PTO shaft, tires, ground height, etc. is shown in Fig. 19-27. When viewed from the rear of a tractor, the PTO shaft rotates clockwise. Tractor PTO shields provided by the manufacturer should be in place at all times.

PROBLEMS AND QUESTIONS

1 What is meant by the traction efficiency of a tractor and what are the factors affecting it?
2 Explain why dual wheels are used on some tractors. What are the advantages and disadvantages of dual wheels?
3 Explain the meaning of drawbar efficiency and compare wheel- and track-type tractors in this respect.
4 What is meant by articulated steering and four-wheel steering? Explain how each operates.
5 What is meant by tractor power take-off? Explain how each of the three tractor PTO systems operates.
6 Why was a standard developed for PTO drives? Explain the standard as it applies to farm tractors.
7 What are radial tires? Explain their advantages, disadvantages, and why they are used on some tractors.

Chapter 20

Tractor Hitches and Stability—Hydraulic Systems and Controls

Originally, all tractor-powered field operations were performed by trailing machines attached to the drawbar of the tractor at one fixed point. With the introduction of (1) certain row-crop field tools; (2) the power take-off driving mechanism; and (3) hydraulic lifting and control mechanisms, there was a transition from plain drawbar-operated machines to integrally mounted or more directly connected tools. These developments have necessitated distinct changes in tractor design and operation, which, in turn, have provided greater versatility, convenience and ease in control and operation, and better maneuverability, performance efficiency, and weight distribution.

Trailing and Mounted Machines In general, tractor-operated field machines are (1) pulled behind it by a simple flexible hitch to a drawbar; (2) mounted or directly attached to it at the rear, front, or sides; or (3) operated by a combination of these methods. Examples of trailing implements are moldboard and disk plows, disk and other types of harrows, grain drills, hay rakes, and balers. Most tools used in row-crop production such as bedders and listers, planters, cultivators, and sprayers are mounted on the tractor and thereby become more or less integral with it.

A number of basic factors must be recognized in designing tractors to pull and operate all types of field machines under the many and varying conditions encountered. These are (1) weight and weight distribution, (2) attachment of driven machine in such a manner as to obtain the most effective performance with a minimum power loss; (3) reasonable operating ease and operator safety assurance; and (4) elimination of excessive or abnormal wear and strains on both the tractor and the driven machine.

Some knowledge of the mechanics of a tractor is basic to a clear understanding of hitching principles and design and tool attachment and operation.

Principles of Mechanics A knowledge of mechanical principles is essential to a clear understanding of any analysis of the mechanics of a tractor and the reactions involved, particularly when it supplies power at the drawbar. Referring to Fig. 20-1A, a force F of 1,000 lb applied at the center P of the bar is balanced by two forces of 500 lb each and equidistant from P. Likewise, the moments (force times its distance from center of rotation) are balanced; that is,

$$500 \times 10 = 500 \times 10$$

or we may say that the moments about P are zero. In Fig. 20-1B the force F is counterbalanced by two unequal opposing forces. The moments with respect to P are likewise balanced; that is,

$$600 \times 8 = 400 \times 12$$

Figure 20-1C represents another slightly different but balanced arrangement with respect to the fixed point P; that is, taking moments with respect to P,

$$1{,}000 \times 10 = 500\,(10 + 10)$$

Figure 20-1D represents a more complex arrangement having an unbalanced system of moments which results in clockwise rotation about P; that is,

$$100 \times 10 + 200 \times 20 = 5{,}000 \text{ ft}\cdot\text{lb}$$

and

$$1{,}000 \times 4 = \underline{4{,}000} \text{ ft}\cdot\text{lb}$$

$$\text{Net unbalanced moment} = 1{,}000 \text{ ft}\cdot\text{lb}$$

Center of Gravity—Basic Tractor Mechanics A number of factors enter into the effective drawbar performance of a tractor. Some of these

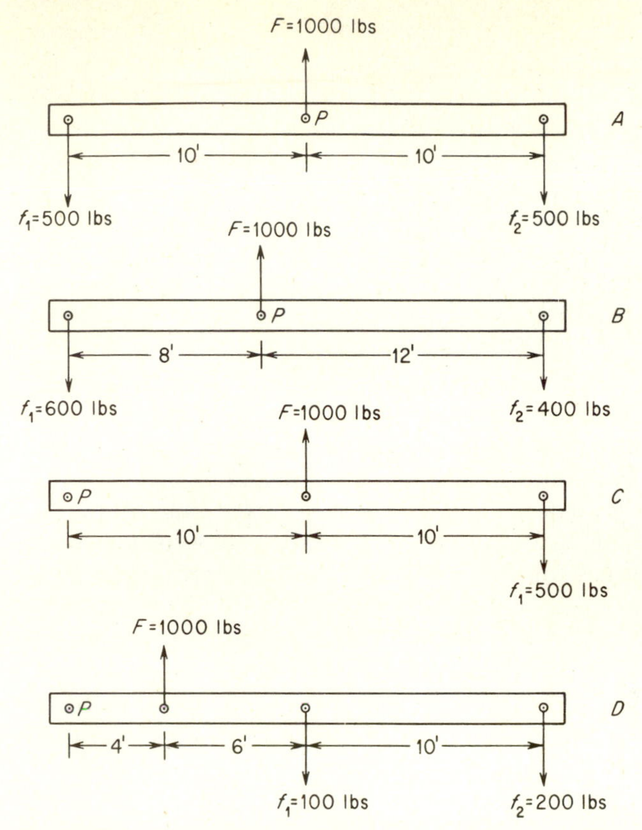

Figure 20-1 Elementary mechanics as applied to parallel force reactions.

are (1) total weight; (2) center of gravity and weight distribution, (3) traction member size and type, (4) drawbar and hitch design and location, (5) traction surface condition, (6) surface grade, (7) travel speed, and others.

The analyses which follow are based only upon the more common and simpler conditions encountered in tractor operation as follows: (1) wheel type and rear-wheel drive, (2) uniform travel in one direction on a smooth, level surface, (3) drawbar load L reacting horizontally and attached midway between drive wheels parallel to direction of travel, (4) supporting soil reaction on front and rear wheels applied vertically under centers of axles, and (5) horizontal soil reaction parallel to direction of travel and tangent to low point of wheel rim. These conditions are shown in Fig. 20-2.

The center of gravity of an object such as a tractor is that point from which it might be suspended or supported and still have it remain at rest in its normal position with respect to a level surface. In simple words, it is some point within the body of the machine where its entire weight might be considered as being concentrated.

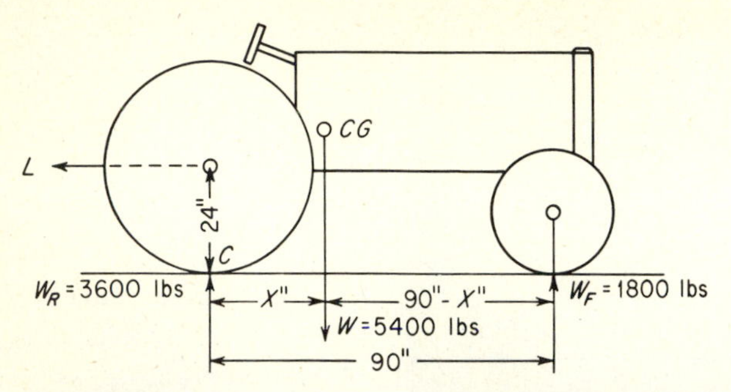

Figure 20-2 Force reactions and location of center of gravity for a tractor.

The location of the center of gravity of a tractor determines its weight distribution longitudinally with respect to the front and rear wheels, as well as transversely and vertically. In general, the center of gravity for standard rear-wheel-drive tractors is located forward of the rear axle a distance equal to about one-third of the wheel base. In track-type tractors it is located ahead of the midpoint of the tracks in order to provide uniform and maximum track pressure and soil contact.

The approximate center of gravity of a wheel tractor can be located by measuring its wheel base and weighing the loads supported by the front and rear wheels and adding these loads to get the total weight. The center of gravity is then computed by moments about the point C, as shown in Fig. 20-2. By computation

$$5{,}400x = 1{,}800 \times 90$$

$$x = 30 \text{ in}$$

Drawbar Pull and Weight Transfer Tractor stability and steering, as well as maximum traction and minimum slippage, are greatly dependent upon the drawbar design and location and the correct attachment of the implement to the tractor. Referring to Fig. 20-2, it is assumed that a load L attached directly behind the rear axle exerts a certain force in a horizontal plane. This load is counteracted by the force created by the rear wheels in contact with the ground or other surface at C. Furthermore, the moment of this force about C reacts against the moment of W_F with respect to C; that is,

$$24 \times L = 1{,}800 \times 90$$

or

$$L = 6{,}750 \text{ lb}$$

Hence, any load greater than 6,750 lb, applied as indicated, would lift the front end of the tractor off the ground and upset it. On the other hand, if L is less than 6,750 lb, some weight will remain on the front end. For example, if L is 1,200 lb then

$$1{,}200 \times 24 = F \times 90$$

or

$$F = 320 \text{ lb}$$

This means that a force of 320 lb acts against W_F, reducing it to 1,800 − 320, or 1,480 lb, and that some additional weight is transferred to the rear wheels to provide better traction. It will also be noted that attaching the hitch and load L below the rear axle reduces the weight transfer and the tendency of the tractor to rear up, while attaching it above the axle increases weight transfer but reduces front-end stability.

Rolling Resistance and Soil Reaction Thus far, we have considered the conditions existing with the tractor as a stationary body. Since, in actual practice, it is moving and rolling along a more or less level and possibly flexible surface, certain other factors must be considered that slightly alter the previous theoretical analysis. The principal factor involved is the resistance of the wheels as they roll and carry their load along. Figure 20-3 illustrates the conditions existing with a steel wheel and lugs on a somewhat loose soil surface; that is, the point of maximum resistance R is at C_1 rather than at C directly below the wheel center. However, studies[1] made of rubber-tired tractors show that on firm, level ground where lug penetration is negligible and wheel traction is at a maximum, the dynamic reaction of the soil is through a point directly below the rear axle and at the ground surface rather than somewhat forward of this point. Under less firm or loose soil conditions, the tires sink into the soil, increasing the rolling resistance and decreasing drawbar pull. Under such conditions, the point of drive-wheel reaction with the soil shifts forward of the rear axle center to a slight degree.

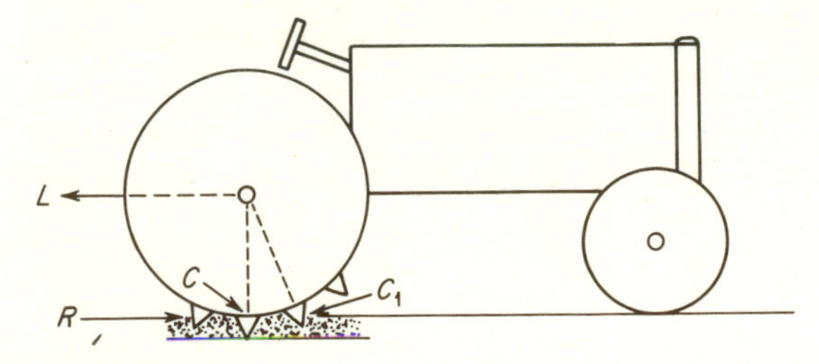

Figure 20-3 Effect of rolling resistance and soil reaction on wheel traction.

[1] W. H. Worthington, "Evaluation of Factors Affecting the Operating Stability of Wheel Tractors," *Agr. Eng.*, **30** (3, 4): 119–123, 179–183 (1949).

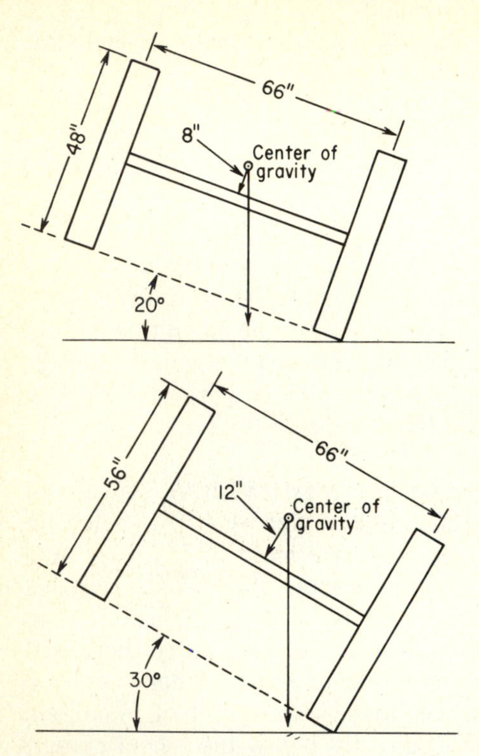

Figure 20-4 Effect of surface slope and wheel height on lateral stability.

Horizontal Reactions and Stability The performance and stability characteristics of a wheel tractor with respect to the forces acting laterally or crosswise are dependent mainly upon (1) the tractive surface whether level or sloping; (2) the hitch relationship between the load pulled and the drawbar; and (3) certain tractor design factors such as wheel diameter and spacing and location of center of gravity.

Figure 20-4 shows how lateral stability is affected by land slope, wheel height, rear-wheel tread, and the location of the center of gravity. The tractor will not tip sideways as long as the vertical projection from the center of gravity falls at some point between the wheels, as indicated. In general, for a given cross grade, the best stability is obtained with small wheels spaced far apart and with a low center of gravity. The machines pulled by the tractor and their attachment to it also affect this stability.

DRAWBARS AND HITCHES

Drawbars and hitches are provided for connecting and controlling drawn, integral (mounted), and semi-integral implements and equipment.

Drawbars, Offset Hitches, and Side Draft Figure 20-5 shows a desirable and typical drawbar arrangement. The assembly consists of (1) two

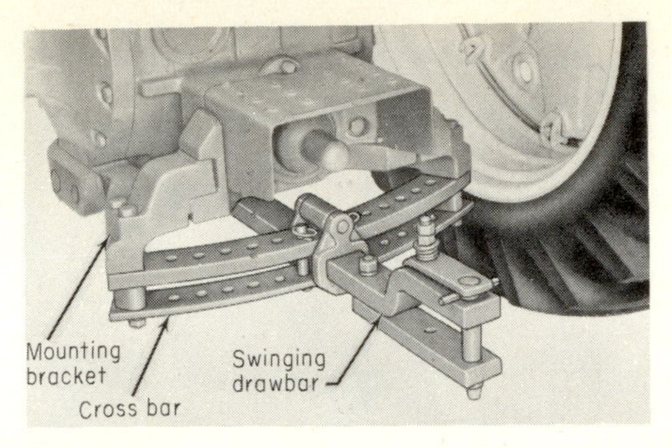

Figure 20-5 A typical tractor drawbar assembly. (Courtesy of Deere and Company.)

side members attached rigidly to the rear-axle housing and somewhat below the rear axle itself; (2) a crossbar with clevis holes; and (3) a flexible swinging drawbar attached underneath and forward of the axle housing. The pulled machine may be attached either directly to the crossbar or to the drawbar. Good drawbar design includes (1) a range of adjustability both horizontally and vertically to meet all drawbar applications, (2) proper location to give maximum drawbar power effectiveness, and (3) maximum operator safety under all conditions.

A good understanding of certain mechanical principles is important in attaching trailing machines to tractors to obtain satisfactory performance as well as safety. Referring to Fig. 20-6, the true line of pull of an ordinary tractor passes through a point in the hitch midway between the rear wheels and is parallel to the direction of travel. The true line of draft of any drawbar-operated implement passes through that point on the implement to which one can hitch and make it move squarely forward in the direction of travel. In other words, the true line of draft of an implement is a line parallel to the direction of travel and passing through its center of resistance. In all drawbar work the object should be to make the true line of pull of the tractor and the true line of draft of the implement coincide as far as possible. If they do not, more or less side draft results.

In a tractor with two drive wheels connected by the usual differential, the pull exerted by the wheels is about equal. Friction in the differential gears is all that permits one wheel to pull slightly more than the other. For all practical purposes a point *CP* (Fig. 20-6) may be called the center of pull, being on a longitudinal axis slightly ahead of and midway between the wheel centers. Side draft exists if the pull is not straight back from *CP*. For example, in Fig. 20-6A the load L_2 is attached to the right of the center hole of the drawbar. This offset reaction has a tendency to pull the front end of the tractor sideways and to the right. Hitching to the left of the center hole would pull the front end to the left. This reaction, commonly called side draft, can be counteracted only through the steering

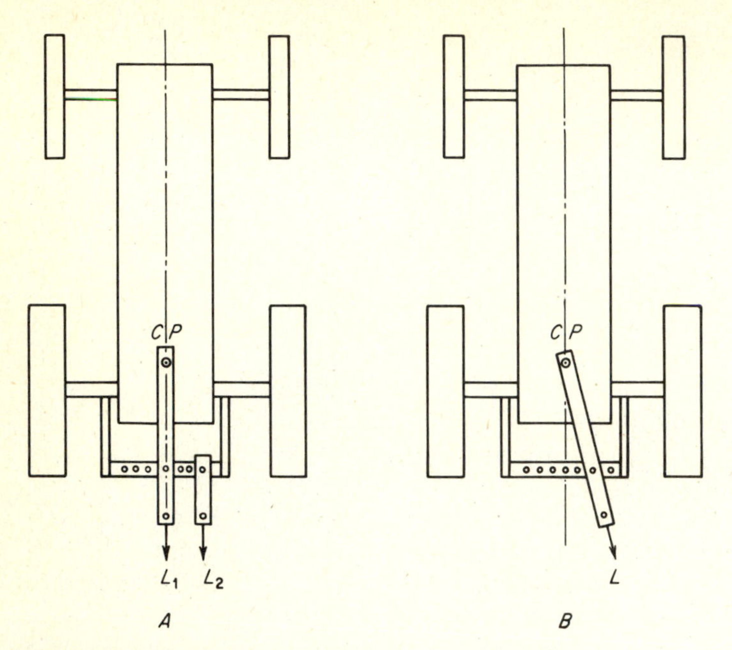

Figure 20-6 Typical load reactions for tractor drawbars.

of the tractor and may be great enough to make steering and control difficult. Another condition (Fig. 20-6B), in which the load L is pulled at an angle, also creates a similar side-draft effect. The swinging hitch bar (Figs. 20-5 and 20-7) offers the advantage of permitting a wide machine such as a disk harrow or a grain drill to trail in the proper manner in making short turns and thus reduces side draft and makes a better corner.

Side draft is encountered largely with trailing plows and, to some extent, with some harvesting machines. In these cases the path of the tractor with respect to the implement is somewhat restricted, for example, by the edge of the furrow wall or by standing grain or hay. Plows give more trouble in this respect because of their heavy draft and narrow swath. Side draft in tractor plowing is affected by a number of factors such as (1) whether the tractor runs in the furrow or entirely on the unplowed land; (2) width and tread of drive wheels; (3) type of plow; and (4) width of cut and number of plow bottoms.

In plowing, all the small and medium sizes of wheel tractors are operated with the right-hand wheels in the furrow. This simplifies steering and eliminates the use of a special guiding device. Some large-size two-wheel-drive tractors, all four-wheel-drive tractors, and track-type machines operate on the unplowed land.

Referring to Fig. 20-8, the center of resistance of a single moldboard bottom is located one-fourth of the width of the bottom to the right of the

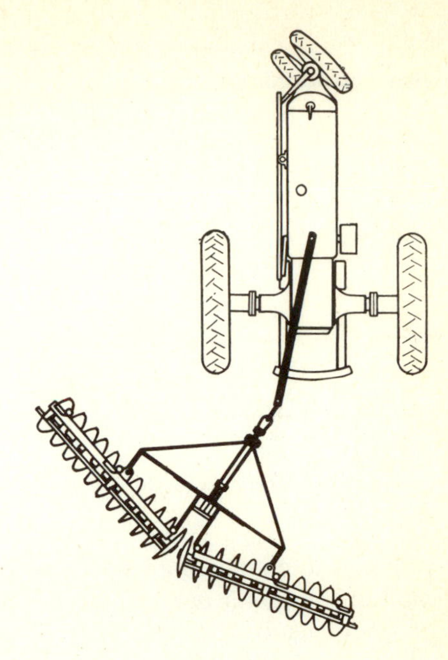

Figure 20-7 Use of a swinging drawbar with a disk harrow.

shin and about where the share joins the moldboard. For plows having two or more bottoms, the center of resistance of the entire unit is midway on a line connecting the points of resistance of all bottoms as shown by Fig. 20-8 and in Table 20-1. Assuming that one wheel of the tractor runs in the furrow, the spacing of the drive wheels (inside to inside of tires)

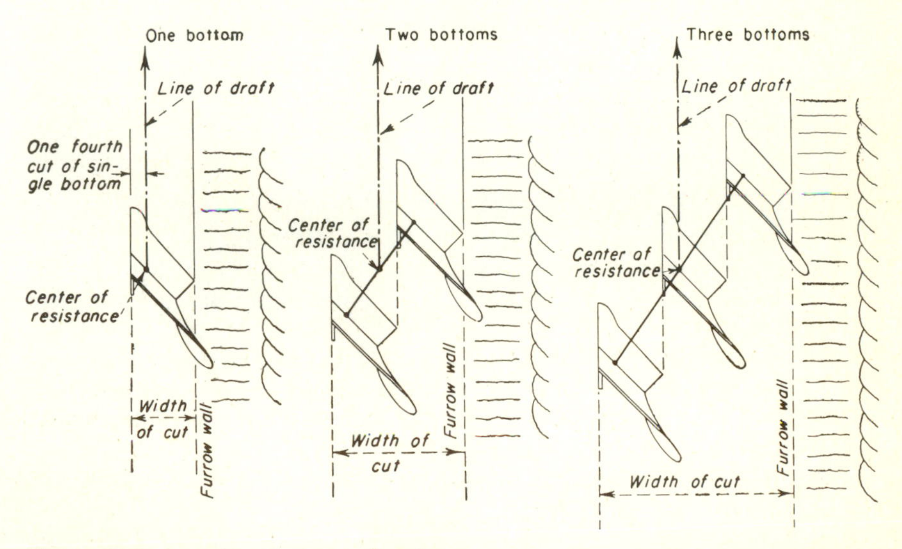

Figure 20-8 Location of center of resistance.

Table 20-1 Centers of Resistance and Tractor-Wheel Spacing for Moldboard Plows

Number of bottoms	Size of bottom, in 12		14		16	
	A[1]	*B*[2]	A	B	A	B
1	9	18	11	22	12	24
2	15	30	18	36	20	40
3	21	42	25	50	28	56
4	27	54	32	64	36	72
5	33	66	39	78	44	88

[1] *A* = distance in inches from furrow wall to true line of draft of plow and tractor.

[2] *B* = distance in inches between inside edges of drive wheels to give least side draft.

required to eliminate all side draft is twice the distance from the furrow wall to the center of resistance of the plow, as shown in Table 20-1.

Frequently it is difficult to adjust the spacing of tractor wheels to eliminate all side draft. This is particularly true when a wide-tread tractor is used with a one- or two-bottom plow or a narrow-tread tractor is used with a four- or five-bottom plow. In such cases it is recommended that any offset between the line of draft of the plow and the center line of hitch of the tractor be divided between the tractor and the plow. As the horsepower of two- and four-wheel-drive tractors increases, hitching becomes more important because of the increased use of large trailing implements such as plows, harrows, and multiple-implement units.

GENERAL RULES FOR TRACTOR HITCHES AND STABILITY

1 A long, low-mounted drawbar is preferable to a short, high one from a safety standpoint and allows for better draft and traction as well.

2 A low center of gravity and a wide tread consistent with the desired crop clearance is preferred to a high center of gravity and a narrow tread.

3 The location of the center of gravity for a two-wheel-drive tractor along a longitudinal axis should be such as to provide only sufficient weight on the front wheels to give satisfactory steering and control under normal hitch and drawbar pull conditions.

4 For four-wheel-drive tractors, the center of gravity along the longitudinal axis should be located so that when the tractor is under load part of the front weight plus implement weight is transferred to rear axle, dividing weight equally between front and rear axles.

5 The longitudinal stability of a tractor is reduced when the drive wheels dig in or drop into a ditch. Placing a timber crosswise in front of the wheels is particularly hazardous and may cause overturning.

6 Sudden engagement of the clutch of a tractor pulling a heavy load under good traction conditions may lift the front end or cause overturning.

Integral-Mounted Machines All manufacturers now provide an almost complete line of integral- and semi-integral-mounted implements for their tractors. This includes plows, harrows, planters, cultivators, earthmoving equipment, and others. Integral machines are used almost exclusively on the small- and medium-sized tractors, with semi-integral-mounted implements used on many large two- and four-wheel-drive tractors. A second and coincidental development that has given major impetus to the use of integral tools is the incorporation of hydraulic mechanisms for their operation and control.

Integral implements have the following advantages:

1 Greater maneuverability. The entire unit is compact and ease in backing and making short turns readily adapts the outfit to small irregular fields and adverse working conditions.

2 Attachment and control are relatively easy and, in most cases, can be accomplished from the tractor seat.

3 The problem of transportation about the farm or over hard-surfaced highways is simplified because the machine is not in contact with the surface, and its weight is carried entirely by the tractor.

4 The initial cost and weight of integral machines is usually less because of the elimination of transport and gauge wheels, control levers, and other parts.

5 The additional weight carried by the tractor when integral machines are used improves traction and steering control.

6 There is better visibility of machine operation and performance.

Trailing-type machines when compared with integral tools have the following advantages:

1 They are universally adaptable to all makes and models of tractors, whereas some integral tools are adapted to but one make of tractor.

2 Attachment to the drawbar by a clevis and pin is simpler.

3 The full power of the tractor can be utilized by using multiple-machine units or multiple-operation machines, thus reducing time and labor costs.

4 Trailing implements, when once adjusted, are not thrown out of adjustment when detached from the tractor.

Three-point Hitches Tractors use the three-point hitch for connecting and controlling integral- and semi-integral-mounted implements. The four basic functions of a three-point hitch are: (1) attaching implement and making it an integral part of the tractor, (2) control working depth of implement; (3) raise and lower implement; (4) transfer weight automatically to rear wheels of the tractor.

The basic parts of a three-point hitch are: (a) lower draft links, (b) upper or center link, (c) lift links, (d) lift arms, (e) rockshaft as shown in Fig. 20-9. The pull from the tractor is transmitted to the implement by the two draft links. The implement is raised when a hydraulic cylinder rotates the rockshaft lift arms thus lifting the lower draft links. Adjustments on both the side and upper links permit proper setting of the implement to give the most effective performance. A special drawbar attached between the two lower draft links readily adapts the three-point hitch for pulling ordinary trailing machines. Side movement of the draft links are restricted by sway blocks or chains attached to the underside of the tractor. This prevents the draft links or implement from swinging too widely and making contact with the tractor tires.

Weight Transfer One major advantage of the three-point hitch system is that it transfers weight automatically to the rear wheels of a tractor when using integral or semi-integral implements. The effect of this weight

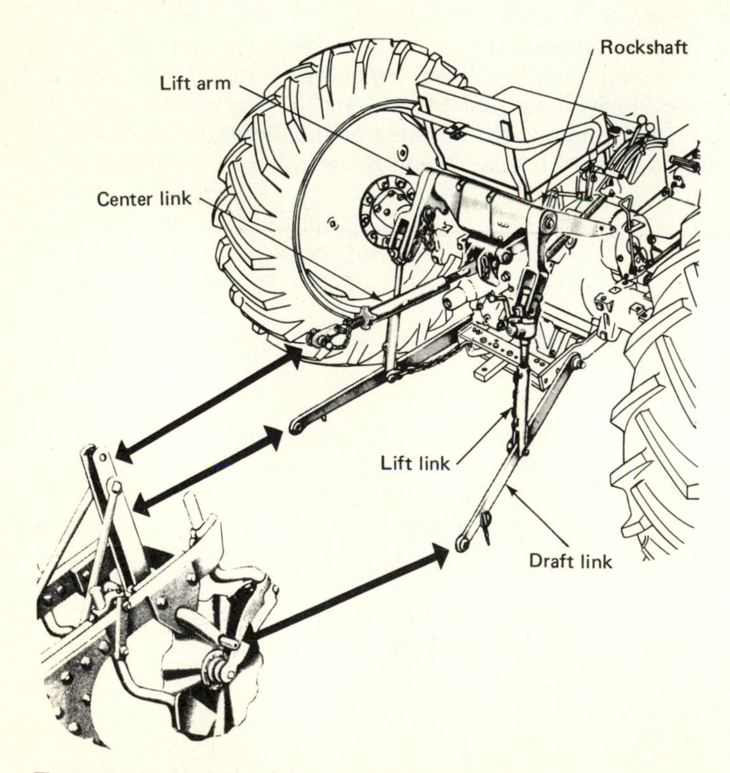

Figure 20-9 Typical three-point-hitch mechanism.

transfer was discussed earlier in this chapter under drawbar pull and weight transfer. Two general methods are used to accomplish this weight transfer. Weight transfer is the result of the resistance of the implement to forward motion, which causes the draft load to be transmitted directly to the tractor, and the hitch attempts to lift against the natural down-suck of the implement and/or the weight of the implement. The result is that weight is transferred from the front wheels of the tractor to the rear wheels. The amount of weight transfer must be controlled, otherwise the effective weight on the front wheels may be reduced so that the tractor cannot be steered safely. The addition of front-end weights increases the steering ability of tractors when transporting integral or semi-integral equipment that imposes heavy loads on the lower draft links.

Category of Three-Point Hitches There are four categories of three-point hitches for farm tractors. When the three-point hitch was first accepted by farmers, the tractors on which it was used were small (Table 20-2). Therefore, the implements used were one- or two-bottom moldboard plows, two-row planters, and other small implements. Four distinguishing dimensions that vary between the categories of three-point hitches are shown in Fig. 20-10 and include: (1) gap in top mast, (2) diameter of hole in top mast, (3) diameter of hitch pins for lower links, and (4) lower-hitch-point spread.

Depth Control of Three-Point Hitch The raising and lowering of a three-point-hitch implement and its depth setting are controlled by a hand lever connected to a sensitive valve in the hydraulic circuits. Figure 20-11 illustrates a position-controlled plow that the operator wants to raise slightly. The operator moves control lever 1 forward, which pivots the cam follower 2 forward and presses it against rod 3, which opens valve 4. Oil 9 under pressure from pump is now admitted to the cylinder, forcing piston 5 to push against shaft arm 6, rotating rockshaft 7, with lift arm 8 upward. The lift arm is attached to the lower-draft link and plow so that the plow is raised. The plow stops rising when the valve 4 is closed again,

Table 20-2 Horsepower Range of Tractors for Different Categories of Three-Point Hitches*

Category of hitch	Maximum drawbar horsepower†
I	20 to 45 (15 to 35 kw)
II	40 to 100 (30 to 75 kw)
III & III N	80 to 225 (60 to 168 kw)
IV & IV N	180 to 400 (135 to 300 kw)

* ASAE Standard: ASAE S217.1 (SAEJ715f)

† Based on ASAE Standard: ASAE S2909, Agricultural Tractor Test Code.

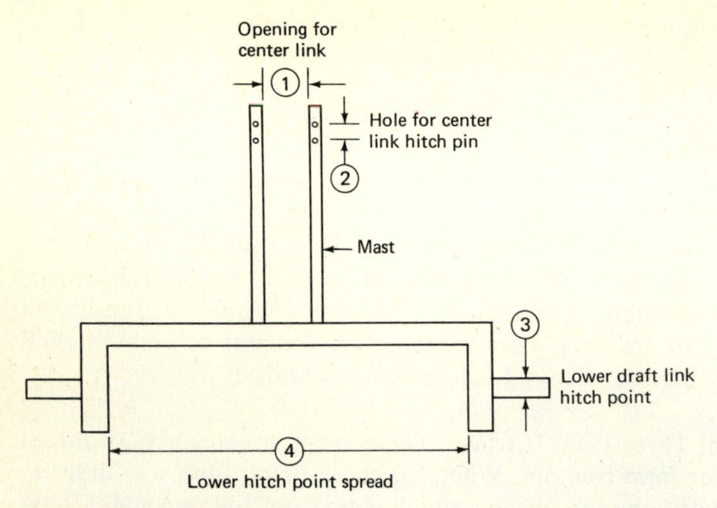

Figure 20-10 Four dimensions which vary between categories I, II, III, and IV, three-point hitches on tractors.

trapping oil in the cylinder 5. Valve is closed when the cam follower 2, working on the sloping cam of the rockshaft 7 is moved to the rear and releases the rod 3. Valve 4 is then closed by its spring. Most three-point hitches are designed to have plows operate at position-response control (fixed depth) or draft-response control (variable depth).

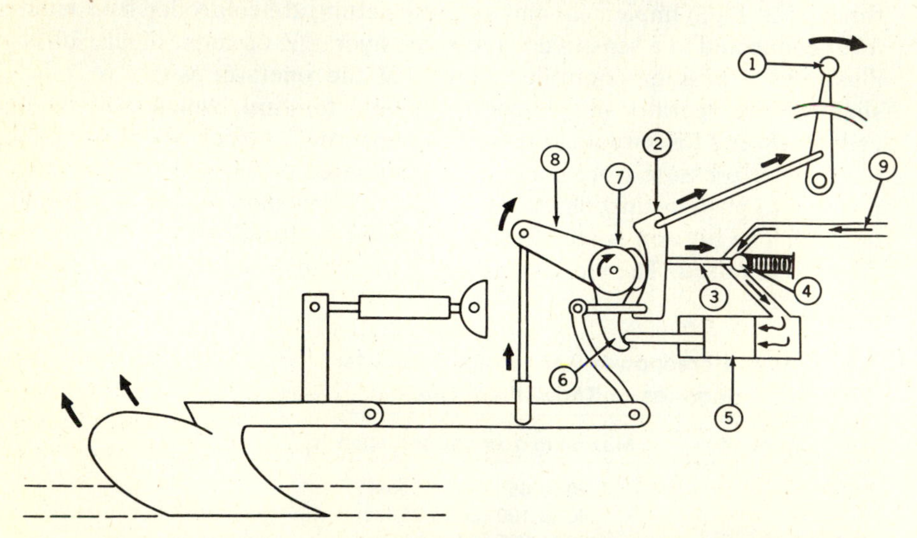

Figure 20-11 Hydraulic system for raising integral-mounted three-point-hitch plow. It consists of (1) control lever, (2) cam follower, (3) operating rod, (4) valve ball, (5) piston, (6) shaft arm, (7) rock shaft, (8) lift arm, and (9) oil from pump. (Courtesy of Deere and Company.)

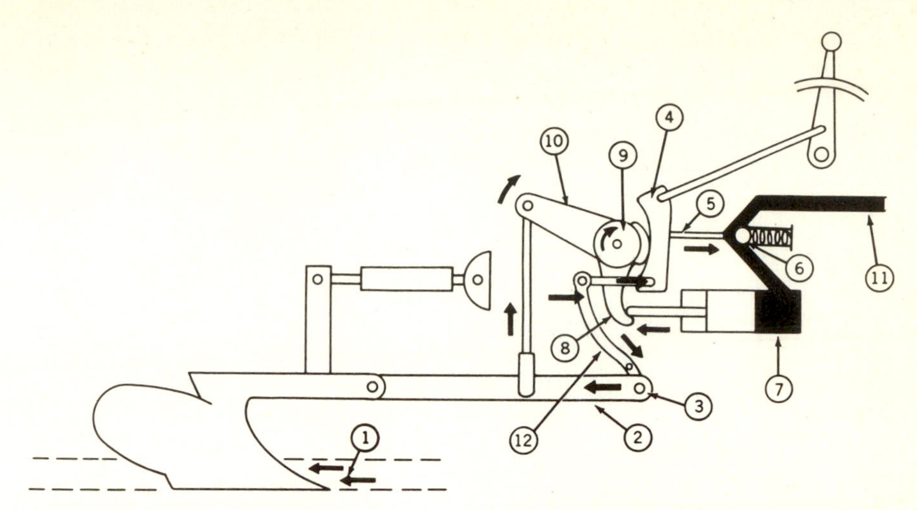

Figure 20-12 Draft response depth control for raising an integral-mounted three-point-hitch plow. It consists of (1) plow furrow, (2) draft link, load sensing shaft, (4) cam follower, (5) operating rod, (6) valve ball, (7) cylinder, (8) shaft arm, (9) rock shaft, (10) lift arm, (11) oil from pump, and (12) load control arm. (Courtesy of Deere and Company.)

Draft-Response Depth Control Draft-response control enables the implement to vary its operating depth automatically according to soil resistance, thus keeping a constant load (draft) on the implement. The operator sets the control lever (Fig. 20-12), which determines how much draft (pull) is required before the hitch lifts the implement. When the implement is raised slightly, draft load decreases and hitch holds the implement in a fixed position until soil resistance changes again.

The operation of the draft-response-depth-control system is shown in Figure 20-12. When soil resistance 1 increases, the draft on lower-draft link 2 increases. This increase in draft is monitored by load-sensing shaft 3, which moves lower end of load-control arm 12 rearward, causing it to pivot against cam follower 4. The cam follower pushes rod 5, which opens valve 6 and allows oil 11 from pump to enter cylinder 7, forcing piston against arm 8, rotating rockshaft 9 and lift arm 10 upward. The lift arm 10 is attached to lower-draft link 2 raising it and plow slightly, thus reducing the draft on the plow. When the plow is through the hard ground and the draft is reduced, the automatic load sensing will lower the plow. The regular depth of the plow can be set by the operator with the control lever (Fig. 20-12). The implement will stay at this depth unless it receives a signal from the load senser.

Position Control The depth of the implement is directly proportional to the setting of the control lever (1 in Fig. 20-11) on the tractor. The linkage of the hitch is rigid. When the tractor's front wheel lowers or raises, the implement is forced in the opposite direction. Thus, the implement will operate at varying draft depending upon how smooth the ground is.

TRACTOR HYDRAULIC SYSTEMS

The rapid growth, within the last 30 years, in the number of tractors that are equipped with the three-point hitch and the use of integral- and semi-integral mounted implements could not have been possible without the development of tractor hydraulic lifts and control mechanisms.

Since hydraulics was introduced to control integral-mounted implements, the system has grown to include: (1) power steering, (2) power brakes, (3) power shift of transmissions, (4) power differential lock, (5) operate 1 to 4 remote hydraulic cylinders, (6) power-engaged PTO, (7) power-engaged clutch, and (8) power for remote hydraulic motors. As tractor sizes and complexity increase, it becomes decidedly important that the tractor hydraulic system operate correctly. A hydraulic system failure usually makes a modern tractor inoperable.

The advantages of using hydraulics for the transmission of power include:

1 Simplifies the transmission of power to remote drives.
2 No clutch or mechanical-type overload protective device is required.
3 Motors, cylinders, etc. are not damaged if stalled.
4 Easy to reverse direction.
5 Output speed easily changed.

Principles of Hydraulics

Hydraulics is a broad subject with many different areas, but the area that applies to tractors and implements is the science of transmitting force and/or motion through the medium of a confined liquid under pressure.

The basic principle upon which all hydraulic-control mechanisms operate is known as Pascal's law. It states that the pressure applied to an enclosed fluid is transmitted equally in all directions. This is illustrated by Fig. 20-13. The hydraulic press, shown in Fig. 20-14, illustrates further the application of this principle; that is, a small pressure reacting against a surface of limited area can be multiplied many times by being applied to another surface of greater area. In fact, the pressures vary directly as the areas involved, or

$$P_2 = P_1 \frac{a_2}{a_1}$$

Also, the work done by the small piston is equal to the work done by

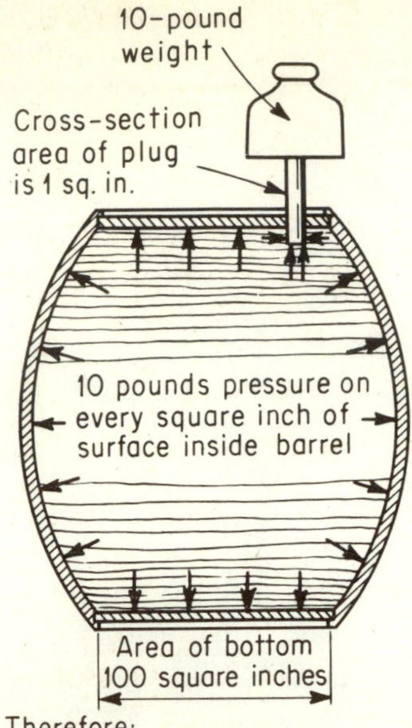

Figure 20-13 The pressure reaction of a liquid in a closed container.

the large piston; that is $P_1d_1 = P_2d_2$ and the distances through which the pistons move vary inversely as their areas.

Components of the Hydraulic System

A hydraulic system contains many parts or components as shown in Fig. 20-15. It consists of (1) pump(s), (2) actuator (cylinder or motor), (3) cooler, (4) filter, (5) control-valve, (6) necessary oil lines, and (7) reservoir.

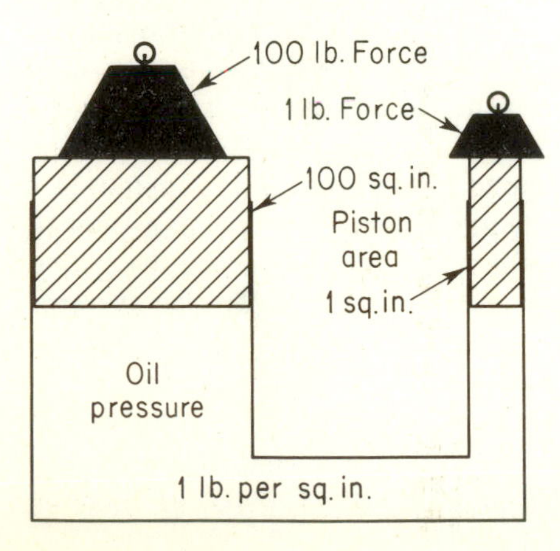

Figure 20-14 The effects and relationship of pressure and area for a liquid in a closed container.

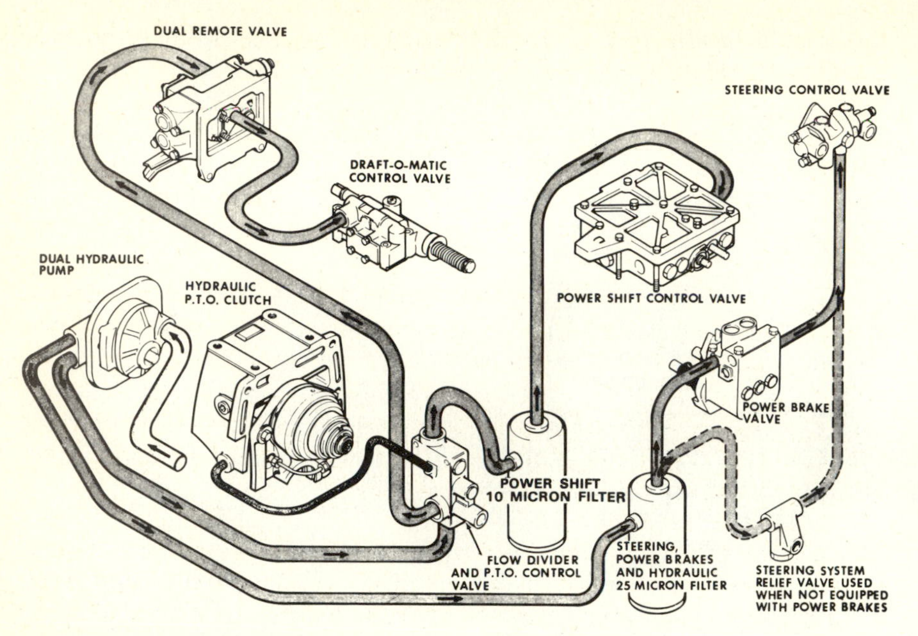

Figure 20-15 Pictorial flow diagram illustrating parts of tractor hydraulic system. (Courtesy of J I Case Company.)

Pump The pump is often referred to as the heart of the hydraulic system since none of the other actuators can function if the pump is not operating. It is driven directly by the tractor engine and operates continually. In some older tractors, the pump operated off the power-takeoff shaft; therefore, it did not operate when the clutch was disengaged. The pump converts mechanical energy from the tractor engine into pressure energy in the fluid. The actuator is operated by this pressure energy. The function of the pump is to create a flow of fluid in the hydraulic system. Pressure in the system is caused by a restriction to the fluid flow. If the resistance is a load on a hydraulic cylinder, only enough pressure is created to move the load. Pumps used in hydraulic systems are positive displacement and are designed to pump the same volume of fluid over a wide range of pressure. Pumps also may be classified as fixed or variable displacement. The amount of fluid pumped per revolution is constant in a fixed-displacement pump. A variable-displacement pump has a provision for changing the size of the pumping chamber, thus changing the gallon per minute output. The types of pumps used in tractor hydraulic systems include gear, which is fixed displacement, or vane and rotary piston, which may be fixed or variable displacement.

Because of the ever-growing application of hydraulics, many medium and large tractors utilize dual hydraulic pumps with a capacity of 16 to 20 gal/min (60.6 to 75.7 liters/min) per pump. With this capacity, application of remote hydraulic motors to agricultural tractors and implements is possible. ASAE standard: ASAE S316.1 recommends that category I tractors should handle up to a 5 hp; category II tractors, up to 10 hp; and category III tractors, up to 15 hp hydraulic motors.

Cylinder A hydraulic cylinder is a cylindrical tube that is sealed on both ends with a piston inside the tube connected to a rod that extends to the outside through an end cap. Seals on the piston prevent the hydraulic oil from leaking past the piston. When fluid is forced into the cylinder, the piston is moved.

Cylinders are available as single or double acting. Fluid enters a single-acting cylinder through a single hose and forces the piston in one direction only. Piston and rod are retracted by either gravity or a spring. A double-acting cylinder has two hoses, and fluid can be directed to either end of the piston. Thus, the cylinder is moved up or down by the force of hydraulic oil. The pounds of force that a cylinder can exert depend upon pressure in the system and diameter of the piston.

Valves There are numerous types of valves that may be used in hydraulic systems to control actuators. They can be grouped into three general categories: pressure control, volume control, and directional control with many different valves in each group.

The direction control valve that is most commonly used on tractors and farm machines is the manually operated spool valve. A simple "three-position, four-way" two-land spool valve is shown in operation in Fig. 20-16. The valve's three positions are neutral, left, and right, and it is connected to the hydraulic system in four ways: (1) pump, (2) reservoir, (3) cylinder port 1, and (4) cylinder port 2. When the spool is moved to the left (Fig. 20-16), oil is directed from the pump to cylinder port 1, expand-

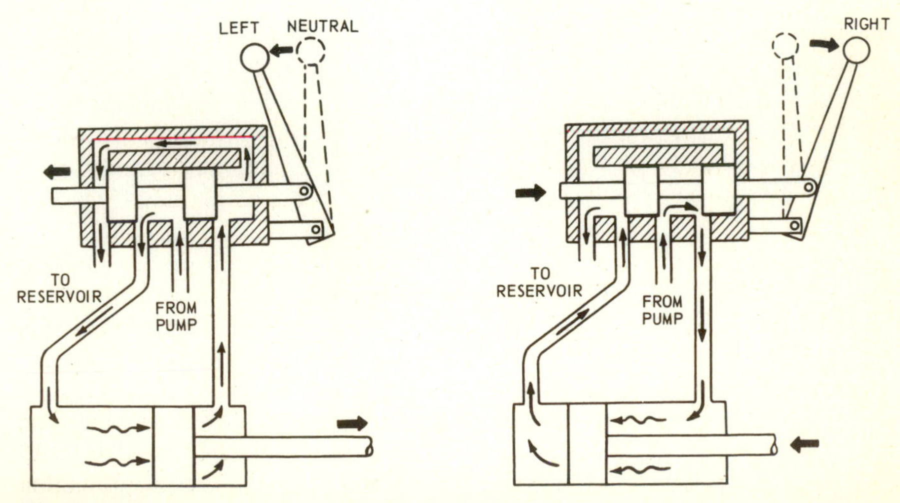

Figure 20-16 Spool valve controlling a hydraulic cylinder. (Courtesy of Deere and Company.)

ing the cylinder. At the same time, cylinder port 2 is opened so that oil from the opposite end of the cylinder flows through the valve and returns to the reservoir. When the valve spool is moved to the right the oil flow is reversed and the cylinder retracts. When the spool valve is moved to the neutral position it seals off both ports to the cylinder, trapping oil to hold the cylinder in place.

The two basic hydraulic systems used on tractors are the open center and closed center (Fig. 20-17). Each system uses a different type of spool in the valve for controlling oil flow. The open center valve (Fig. 20-18) is constructed to allow oil to flow freely from pump to the reservoir when the valve is in neutral position with cylinder ports blocked. The closed center valve uses a spool that blocks all the ports and stops the flow of oil from the pump when it is in neutral position (Fig. 20-18). Both types of valves normally have cylinder ports blocked when the spool valve is in neutral, but in some designs the ports are open to allow a cylinder to "float" or when the valve is used to control a hydraulic motor.

Reservoir The function of the reservoir or tank is to store the hydraulic oil and to provide some cooling for the system. The tank is vented to allow for expansion and contraction as the oil is heated and cooled. On many tractors, the reservoir is the case that houses the tractor transmission and differential. Reservoirs are provided with a device to indicate oil level.

Filter The filter is very important in determining the useful life of the hydraulic pump and other parts of the system. The filter removes foreign particles such as dirt, metal particles, etc. Most tractors use a full-flow filter through which all the oil passes. Many systems utilize a

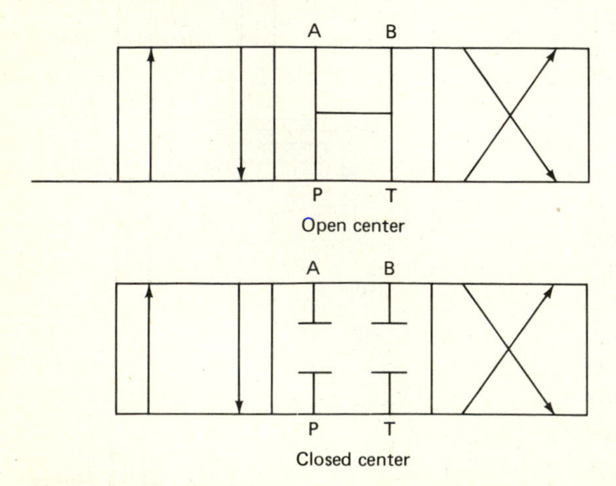

Figure 20-17 Symbols for directional control hydraulic valves.

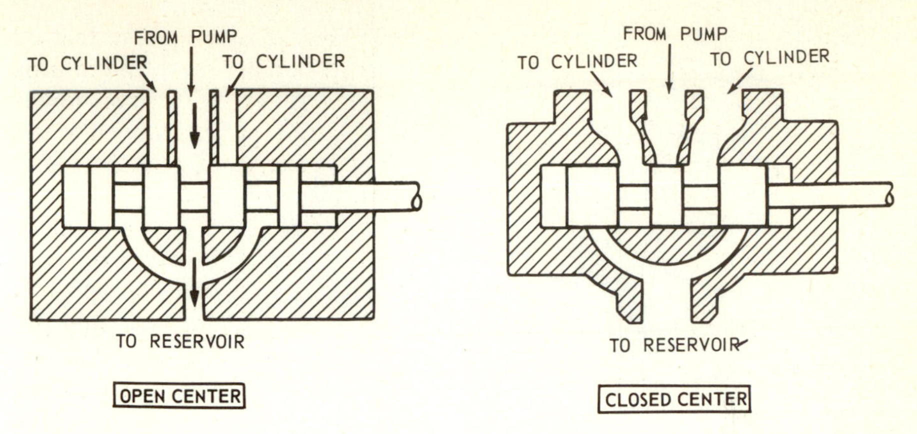

Figure 20-18 Open- and closed-center valve in neutral position. (Courtesy of Deere and Company.)

suction filter, located in the hydraulic reservoir, to filter the oil as it goes to the pump. Some systems will also have a return filter for the oil to pass through on its way back to the reservoir. It is important that the filter be replaced according to the manufacturer's instructions.

Coolers As the size of tractors increases and greater use is made of their hydraulic system, the buildup of heat is a problem. Coolers are used to remove this heat. Types of coolers used are air-to-oil and water-to-oil. The air-to-oil cooler has a coil with cooling fins in front of the radiator (Fig. 20-19). The water-to-oil uses water from the engine-cooling system to carry off heat.

Lines and Hose Hydraulic fluid is usually conducted to the various parts of the system by lines made from steel, copper, or synthetic rubber. The lines must be of sufficient size to carry the volume without causing excessive resistance to flow. Most systems on tractors have a maximum working pressure of about 2,000–2,500 lb/in^2 (140.6–175.7 kg/cm^2). Lines are rated according to working and bursting pressure. Bursting pressure of the lines is well above the working pressure.

Flexible rubber hoses are used extensively to pass hydraulic fluid from the tractor to an attached implement, since these will withstand shock and vibrations and are easy to install. Rubber hose usually has three layers. The inner layer is usually synthetic rubber of a type that will not be affected by the fluid. Fabric braid or wire braid, used to reinforce the hose, makes up the middle layer. It can be either one- or two-layer braid depending on the pressure the hose is to carry. The outer layer of the hose is a protective cover of rubber. Hose is sized according to inside diameter, tubing by outside diameter.

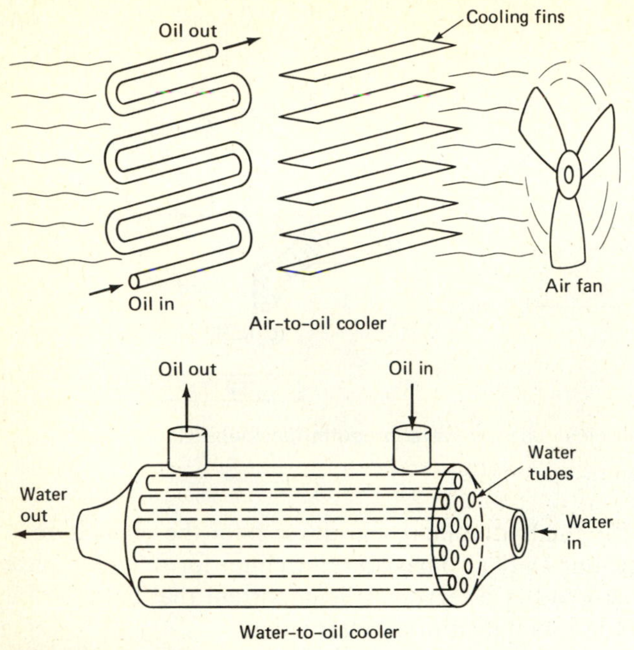

Figure 20-19 Air and water hydraulic-oil coolers.

Accumulator Accumulators, which store energy in the form of compressed gases, are used in some hydraulic systems as a means of providing an additional supply of high-pressure hydraulic fluid if the demand is greater than the capacity of the pump. An accumulator essentially increases the short-time power output of the hydraulic system without increasing the pump's capacity.

Bladder and piston accumulators are the two types used in hydraulic systems. An inert gas such as nitrogen is used as cushioning material. The gas is contained in a flexible bladder in the bladder type (Fig. 20-20), while a floating plunger keeps the gas and fluid separated in the piston type. The accumulator is designed for a specific function according to the quantity of oil it will hold and the pressure of the gas that is in it.

Hydraulic Fluid The oil in a hydraulic system serves as the power transmission medium and the system's lubricant and coolant. There are many types of oil or fluids that are used in hydraulic systems. It is important that the proper fluid be used, since most of the problems in hydraulic systems are often related to the use of improper or contaminated fluids.

The fluid must protect the system from rust and corrosion, resist foaming, and maintain proper viscosity over a wide range of temperatures.

The many different hydraulic systems found on tractors and ma-

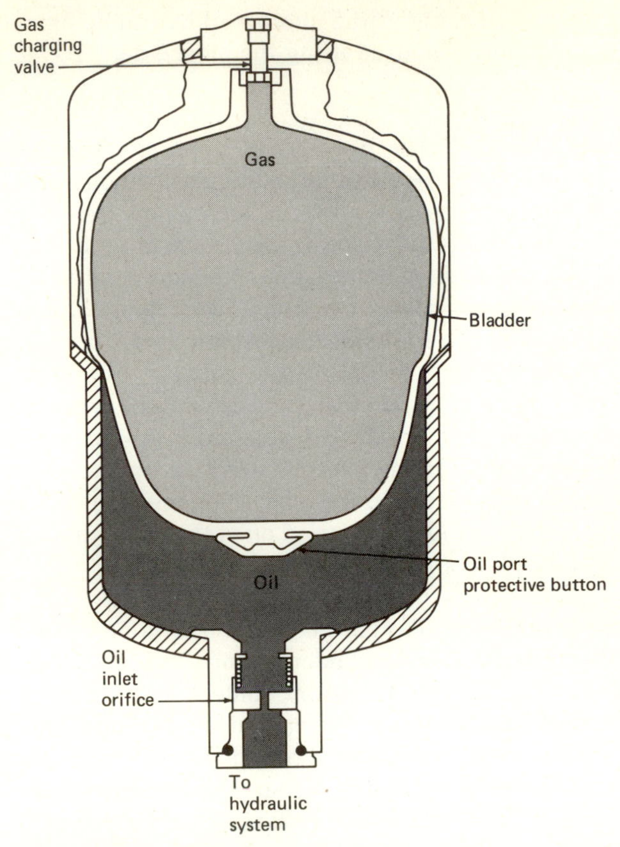

Figure 20-20 Cross section of bladder type accumulator used with some closed-center hydraulic systems.

chinery may be designed to use a wide variety of hydraulic oils. Oils from these different systems should not be mixed or interchanged. The manufacturer's recommendations should always be followed when changing or replacing oil in the system.

TYPES OF HYDRAULIC SYSTEMS

Many of the parts discussed in the previous section when assembled into a hydraulic system may perform from one to several different functions on the machine. Each function would have its own "hydraulic circuit" through the use of directional control valves. In most cases, the same pump is used for all circuits, but some large tractors use dual pumps.

The two types of hydraulic systems used on tractors are: (1) open center and (2) closed center. The basic difference between the two is the type of directional control valves that are used. Historically, tractors have used open-center hydraulic systems, but as the number of hydraulic functions increased, some manufacturers started using the closed-center system.

Open-Center Hydraulic System The open-center type is a simpler system and was used on most early three-point hitch tractors, which had a limited hydraulic demand. In the open-center system, the oil flow from the pump returns to the reservoir when the valve is in a neutral or center position. This requires that an open-center or tandem-center valve be used (Fig. 20-17), since both valves provide the free flow of fluid back to the tank when in neutral position. Fixed displacement pumps are used in open-center hydraulic systems (Fig. 20-21). These pumps provide a constant oil flow through the system, but pressure varies with the types of loads placed on the system. The peak pressure developed is only enough to move the load. Since a constant-volume pump is used, an open-center valve is required to allow the oil to return to the tank when the actuator is not in use. If two or more functions must be carried on at the same time in an open-center system, flow dividers must be used (Fig. 20-22).

Closed-Center Hydraulic System The closed-center hydraulic system uses a closed-center valve that blocks off cylinder, pump, and tank ports when in a neutral position (Fig. 20-16). Therefore some means must be used to unload the pump when the valve is in neutral position. One method is to use a relatively small constant-volume pump to charge an accumulator (Fig. 20-23), and use an unloading valve to divert pump flow to the reservoir when the accumulator pressure reaches a preset value. A check valve between the accumulator and the unloading valve prevents

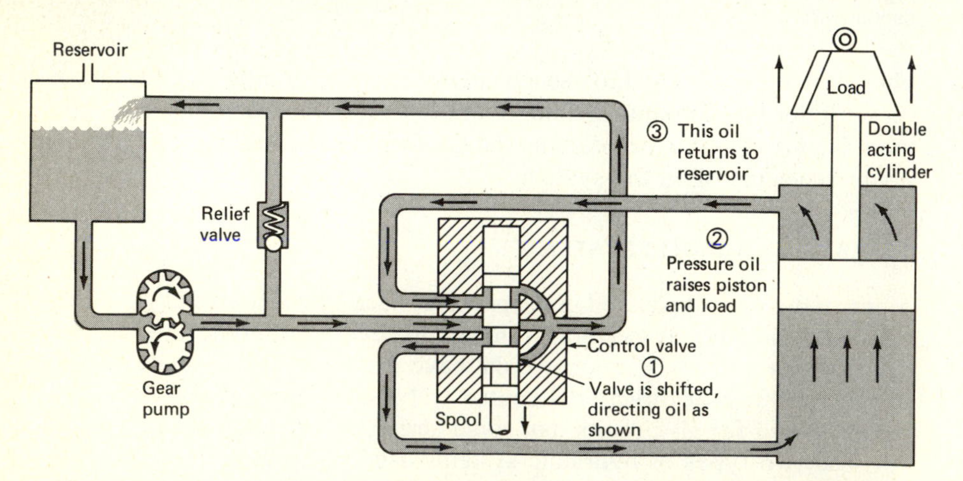

Figure 20-21 Open-center hydraulic system in operation with gear pump, relief valve, four-way control valve and double-acting cylinder.

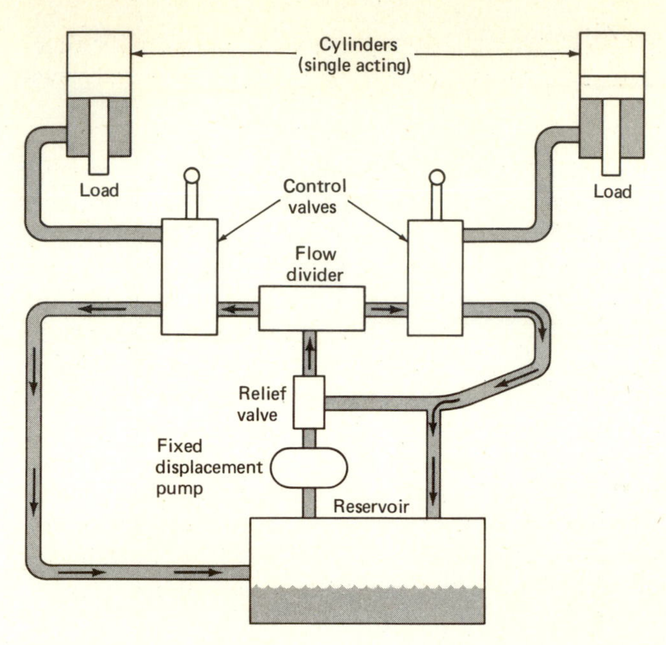

Figure 20-22 Open-center hydraulic system with flow divider.

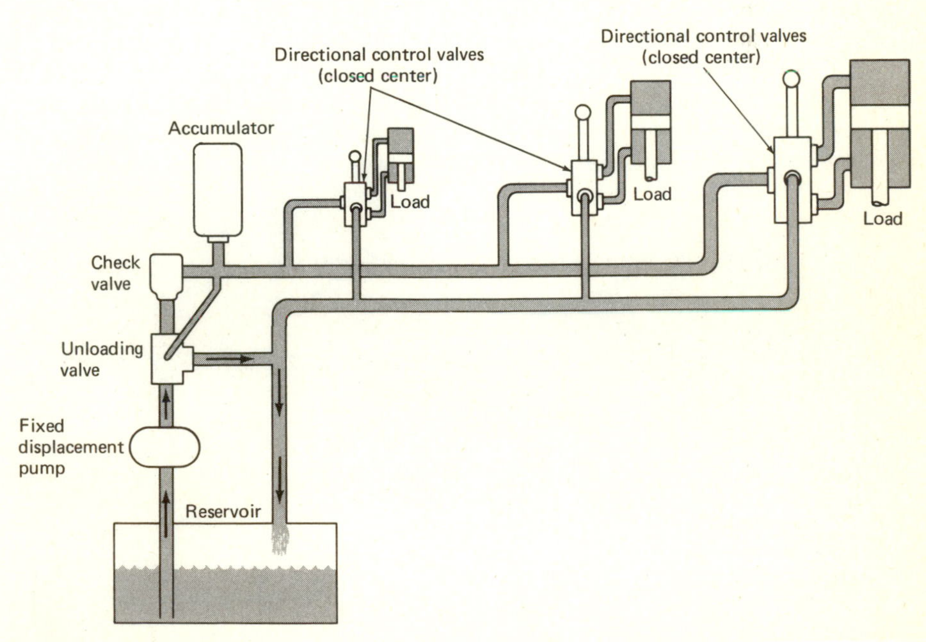

Figure 20-23 Closed-center hydraulic system with fixed-displacement pump, accumulator, and two-way directional valves.

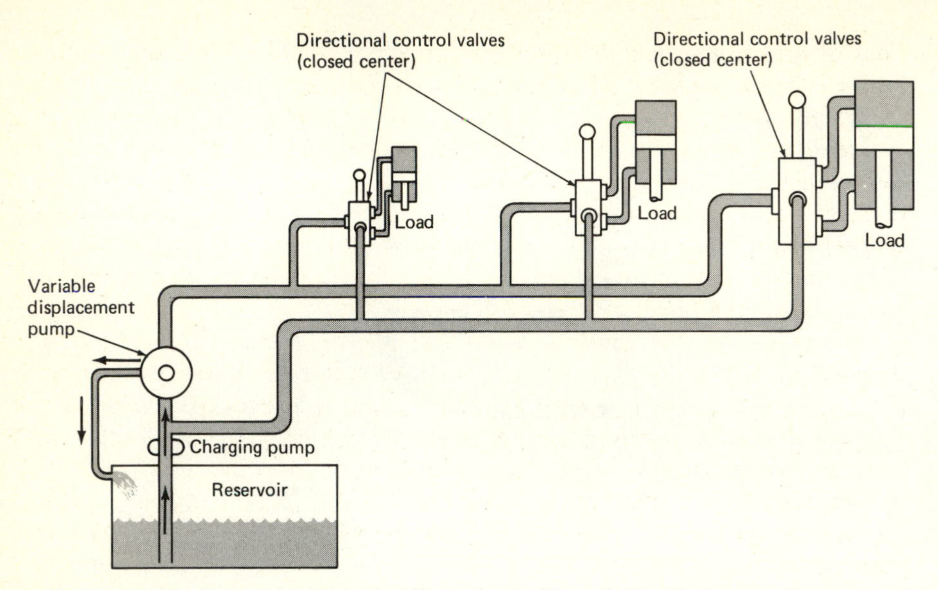

Figure 20-24 Closed-center hydraulic system with variable-displacement pump and two-way directional control valves.

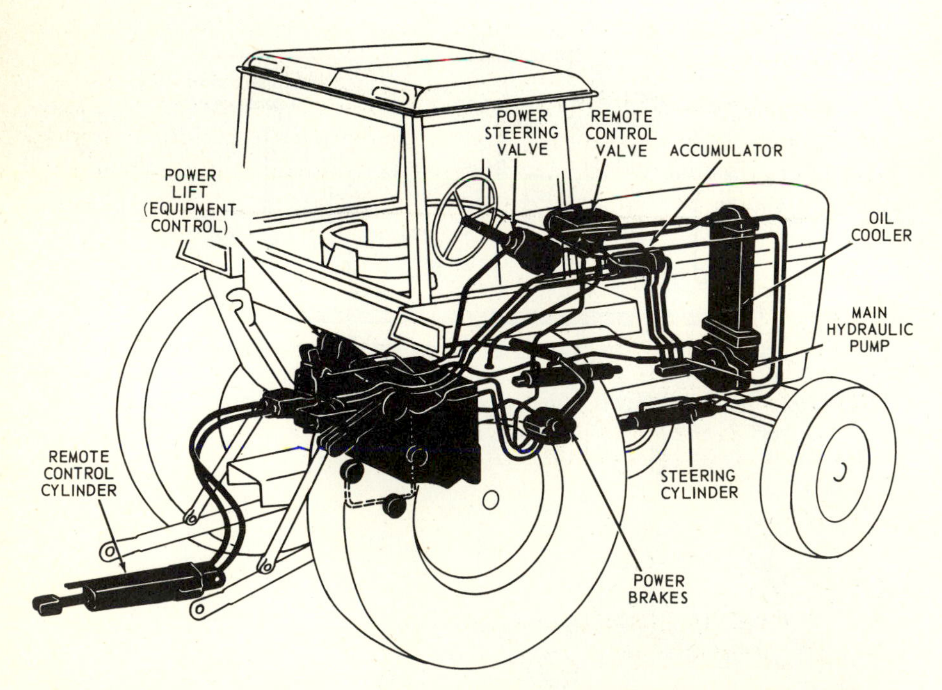

Figure 20-25 Complete closed-center hydraulic system. (Courtesy of Deere and Company.)

loss of pressure and fluid stored by the accumulator. When the control valve is open, stored fluid under pressure will flow into the circuit being operated. When system pressure drops to a preset level, the unloading valve diverts pump flow into the system to operate hydraulic components and refill the accumulator. This system depends on most tractor hydraulic system operations requiring a high flow rate of oil for a relatively short period of time. The accumulator can supply the required volume of oil. The primary advantage of an accumulator closed-center system is the reduced volume of oil required from the pump.

A second type of closed system uses a variable-displacement pump, which automatically changes flow to meet the demands of the system (Fig. 20-24). Even when no functions are in use pressure is maintained, but flow is reduced to a rate just sufficient to replace fluid lost through leakage, plus that quantity required to keep the system cool. The pump must have the capacity to meet the combined demand created if all functions on the tractor were put into operation simultaneously. The primary advantage of the variable-displacement pump is its ability to supply maximum oil capacity for sustained periods. This enables the tractor to operate equipment such as front loaders and backhoes for extended periods. The fluid pressure in the working part of the closed-center hydraulic system is held at a constant pressure; however, the flow rate varies with the demand of the circuits that are operating. A complete closed-center tractor hydraulic system is shown in Fig. 20-25.

Trailing-Implement Control Certain drawbar-pulled trailing implements such as plows, disk harrows, and grain drills utilize remote hydraulic cylinders to raise and lower or adjust them to operating conditions, as shown by Fig. 20-26. Remote-lift cylinders are connected to the

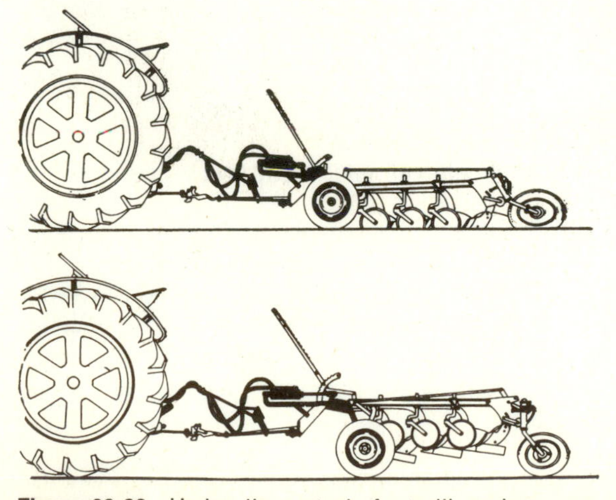

Figure 20-26 Hydraulic control of a trailing plow.

pressure pump and control mechanism by heavy-duty flexible hose lines, with convenient leakproof push-pull quick-connecting couplings. Lift cylinders may be either single or double acting. Single-acting cylinders produce only a lifting action and depend upon the weight of the implement or load and oil-flow rate for the downward movement. Double-acting cylinders provide positive action in either direction and must have two oil lines and connections. Hydraulic cylinders may be equipped with stroke-limiting stops to limit the piston travel and working depth of the implement. These stops are either a simple mechanical device on the cylinder or a piston-actuated valve that controls the oil flow. Other machines such as hay balers and cotton strippers utilize hydraulic cylinders for controlling certain machine parts such as the pickup on a hay baler and the stripper unit on a cotton stripper.

Care, Lubrication, and Servicing of Hydraulic Transmissions and Implement Controls Hydraulic-type tractor transmissions and hydraulic mechanisms for controlling and operating various machines and attachments powered by the tractor will perform properly, provided certain precautions are observed and followed. These are (1) use the exact type and viscosity of oil recommended by the manufacturer; (2) maintain the correct oil level at all times; (3) avoid mixing of oils of dissimilar characteristics; (4) drain the system and refill with the specified lubricant and service the filter according to instructions; (5) avoid getting water, dust, grit, or foreign material in the system or on the oil line connections; and (6) check frequently for leaks.

There is considerable variation in the lubricant types and service recommendations for hydraulic transmissions and hydraulic-control mechanisms with respect to the different makes, types, and sizes of tractors. Hence it is most important that the specific instructions for any tractor be observed.

HYDRAULIC FACTS*

Here are some key facts that will help you understand hydraulics:

1 Hydraulic power is nearly always generated from mechanical power. Example: A hydraulic pump driven by an engine crankshaft.

2 Hydraulic power output is nearly always achieved by converting back to mechanical energy. Example: A cylinder which raises a heavy plow.

3 There are three types of hydraulic energy: (a) potential or pressure energy; (b) kinetic energy, the energy of moving liquids; and (c) heat energy, the energy of resistance to flow, or friction.

* Courtesy of Deere and Company

4 Hydraulic energy is neither created nor destroyed, only converted to another form.

5 All energy put into a hydraulic system must come out either as work (gain) or as heat (loss).

6 When a moving liquid is restricted, heat is created and there is a loss of potential energy (pressure) for doing work. Example: A tube or hose that is too small or is restricted. Orifices and relief valves are also restrictions, but they are purposely designed into systems.

7 Flow through an orifice or restriction causes a pressure *drop*.

8 Oil must be confined to create pressure for work. A tightly sealed system is a must in hydraulics.

9 Oil takes the course of least resistance.

10 Oil is normally *pushed* into a pump, not drawn into it. (Atmospheric pressure supplies this push. For this reason, an air vent is needed in the top of the reservoir.)

11 A pump does not pump pressure; it creates flow. Pressure is caused by *resistance* to flow.

12 Two hydraulic systems may produce the same power output—one at high pressure and low flow, the other at low pressure and high flow.

13 A basic hydraulic system must include four components: a reservoir to store the oil; a pump to push the oil through the system; valves to control oil pressure and flow; and a cylinder (or motor) to convert the fluid movement into work.

14 Compare the two major hydraulic systems:
Open-center System = pressure is varied but flow is constant.
Closed-center System = flow is varied but pressure is constant.

15 There are two basic types of hydraulics:

(a) Hydrodynamics is the use of fluids at high speeds "on impact" to supply power. Example: a torque converter.

(b) Hydrostatics is the use of fluids at relatively low speeds but at high pressures to supply power. Example: most hydraulic systems, and all those covered in this book.

PROBLEMS AND QUESTIONS

1. Referring to Fig. 20-1*D*, compute the force F which is required for equilibrium if f_1 and f_2 are 900 lb and 450 lb respectively.
2. A tractor has a wheel base of 90 in and weighs 7,800 lb. The static weight on the front wheels is 2,300 lb. Calculate the location of the center of gravity longitudinally with respect to the rear axle.
3. Referring to the tractor in Prob. 2, compute the drawbar load L which would be required to upset the tractor if the drawbar is attached below the rear axle and 16 in vertically above point C.
4. What force will be supported by the large piston shown in Fig. 20-14 if the pressure is 25 lb/in^2 and the area is 60 in^2.
5. What are the advantages of integral-mounted machines.
6. Explain what is meant by weight transfer when using integral-mounted machines.

7 Explain what is meant by draft-response depth control. How does it operate and what are its advantages?
8 Name the principal components of a tractor hydraulic system and explain the function of each.
9 Name the two basic hydraulic systems used on farm tractors, explain the difference in the two systems, and give the advantages and disadvantages of each system.

Chapter 21

Tractor Safety and Comfort

The modern farm tractor offers a level of operator comfort, convenience, and safety that could not have been contemplated by the farmer of a decade or more ago. An increased awareness of the types of injuries that occur and improved tractor technology have led to this improved work environment.

Safety in the agricultural environment differs in several important respects from safety in other contexts. The operator of an agricultural tractor works in an area generally removed from direct supervision and often completely isolated. With no foreman present to supervise, and, often, without a co-worker to assist, the operator must rely upon the design of the machine, his or her own knowledge and experience, and the information provided by safety signs and operator's manuals to perform field tasks.

ROLLOVER PROTECTION

Rollovers are the greatest single cause of tractor-related fatalities, accounting for more than one-half of such deaths. An upset or rollover results from an unstable condition usually attributable to operating conditions. Operating on steep slopes or on extremely rough terrain is a common cause of rollovers. This type of accident is also likely to occur when front-end loaders are used improperly.

Specific conditions that may lead to instability and a possible rollover include:

1 Operating near steep or slippery slopes.
2 Operating around holes, stumps, large rocks or other obstructions.
3 Driving at too great a speed for existing conditions.
4 Improperly loading, hitching, or ballasting.
5 Turning too abruptly for speed and conditions.

Rollover Protective Structures (ROPS) The most effective way to prevent injuries from rollover is to operate a tractor in such a manner that this cannot occur. In the event that such an accident does occur, however, the operator is almost certain to receive a serious and possibly fatal injury unless adequate protection is provided. A rollover protective structure (ROPS) is a steel unit intended to provide operator protection in the event of a rollover, when used in conjunction with a seat belt. These units have been designed and built as: (1) two-post ROPS frame (Fig. 21-1); (2) two- or four-post ROPS with canopy (Fig. 21-2); (3) fully enclosed cab with ROPS (Fig. 21-3).

In general, an operator sustains injury when the body receives a blow (or a number of these), of sufficient intensity, in a short period of time. In the case of a rollover accident, the force involved is that associated with overturning, with possibly the tractor pinning the operator under it. The ROPS, then, protects the operator by absorbing a substantial portion of

Figure 21-1 A two-post rollover protective structure (ROPS). (Courtesy of J I Case Company.)

Figure 21-2 A two-post rollover protective structure (ROPS) with a canopy. (Courtesy of J I Case Company.)

this force and channeling it into the deformation of the ROPS structure itself. In other words, during a rollover, the ROPS structure will no doubt become deformed, bent, or crumpled. If this were not the case, a ROPS would not absorb the impact but rather would transmit it to the frame of the tractor and to the operator.

Though it may appear that the design and construction of a ROPS would be a straightforward matter, this is not the case. In fact, a ROPS is a sophisticated engineering device that must meet exacting standards. The structure must absorb as much of the impact as is feasible without its deforming enough to endanger the operator. It must accomplish this after

Figure 21-3 A fully enclosed tractor cab providing operator with rollover protection and controlled environment for comfort. (Courtesy of J I Case Company.)

having been subjected for an indefinite period to the severe conditions of agricultural operations.

For example, the structure must be capable of withstanding the shocks and vibration of field work without exhibiting significant metal fatigue. It must be subjected to temperature extremes, including possible sub-zero operation, without the metal becoming brittle. Finally, it must withstand rain, dust, dirt, and exposure to corrosive chemicals without deterioration.

Care and Maintenance of ROPS Although a ROPS is a carefully designed unit, its effectiveness may be compromised if it is not properly maintained. A list of maintenance guidelines includes:

1 Periodically inspect the structure for visible signs of damage or other deterioration.

2 Visually inspect all welds to ensure that cracks have not formed.

3 Never hammer, weld, bend or otherwise tamper with a ROPS as this may weaken the structure.

4 Never attempt to straighten or repair a damaged ROPS. Even though it may appear to have been restored to original condition, its structural strength has been compromised.

Seat Belt Usage Whenever a ROPS-equipped tractor is being operated, a seat belt must be worn. The obvious reason for this is that in the event of an overturn, the operator who is not wearing a seat belt will be thrown from the seat and thereby not receive the benefit of a ROPS. Another benefit of seat belt usage is that the operator will not fall from the tractor during its operation. The National Safety Council estimates that twenty percent of all tractor-related farm fatalities result from being run over by a tractor. In roughly two-thirds of these accidents, the person first fell from the tractor.

TRACTOR CABS

The fully enclosed ROPS cab provides the highest available level of operator comfort and safety. In addition to rollover protection, the operator is sheltered from temperature extremes, dust, dirt, insects, and chemicals (Fig. 21-4). By creating a more comfortable environment in which to work, operator fatigue is reduced and attention may be more closely focused on operating the machine. In this section, some aspects of the design of the operator station and controls will be discussed. Although included under the heading of tractor cabs, many of these topics apply also to tractors that are not equipped with an enclosed cab.

Figure 21-4 The interior of a modern agricultural tractor. Note symbols on controls. (Courtesy of J I Case Company.)

Operator Seating Since the operator may remain seated for long periods of time, the type of seat (Fig. 21-5) used is of some significance. Individuals of both sexes and various ages and sizes must be accommodated. The following principles allow for a high level of operator comfort:

1 The seat should be adjustable, preferably for forward and backward movement and height, in order to accommodate as wide a range of operators as possible.

2 The seat back should provide adequate back support.

3 The seat should be adjusted so that legs can be comfortably extended, reach the floor or other support, and can readily reach all foot-activated controls.

4 The seat should be positioned so that there is a minimum amount of pressure on the back of the thighs.

5 The seat should be adjustable so the operator can have a good, unobstructed vision of the area around the tractor.

Controls and Displays The various operator controls such as those for ground speed, direction, engine speed, power take-off, and others, must be designed in such a way as to allow easy access and operation. The primary requirement is that all these controls should be located where they can readily be reached by the operator. Since the size and shape of operators vary, it is important that controls be located where they are

Figure 21-5 Swivel-suspension operator's seat with adjustable (1) arm rest, (2) back-rest, (3) seat tilt, (4) forward-rearward, (5) swivel, (6) weight and (7) height. (Courtesy of J I Case Company.)

accessible to most. The adjustable seat helps in this regard. As a general rule, the most frequently used and most important controls should be located in the most easy-to-reach locations.

To efficiently and safely operate a tractor, the operator must be able to rapidly identify the proper control for the task to be performed. The experienced operator will, of course, automatically select the appropriate control without having to search for it. Not all tractor operators, however,

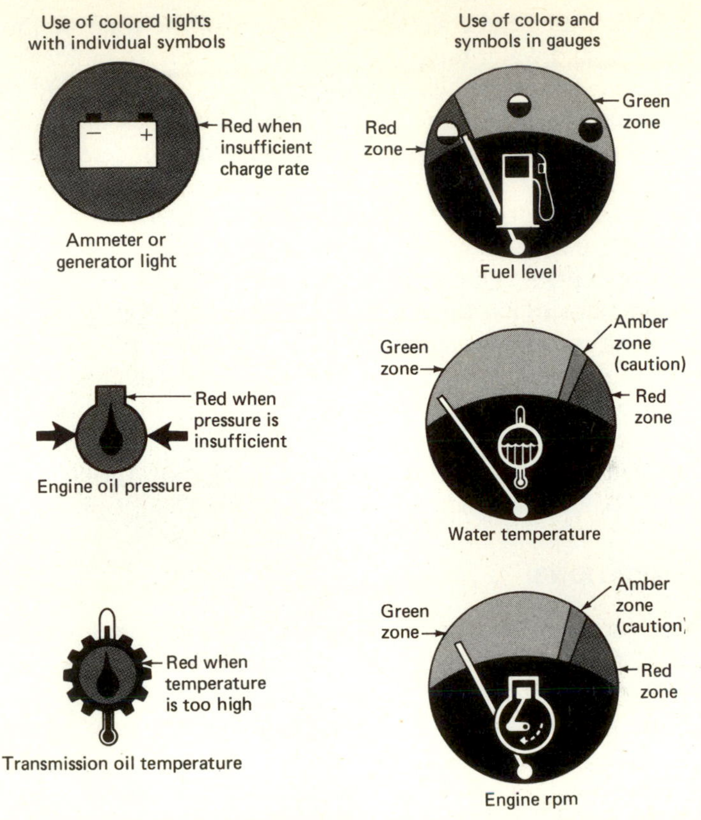

Figure 21-6 Typical illustrations of the use of color with universal symbols for operator's controls on agricultural equipment. (Reproduced with permission of the American Society of Agricultural Engineers from the 1978–1979 *Agricultural Engineers Yearbook.*)

are experienced, and even those who are may become confused occasionally, for example, when operating a machine with which they are unfamiliar. The most common means of identifying controls uses labels with words or symbols specifying function.

In addition to being readily identified and reached, the controls on modern agricultural tractors are designed to be easily and correctly operated. A properly designed control does not require an excessive amount of force to operate, nor should the result of its use deviate from the operator's expectations. As a simple example, clockwise rotation of the steering wheel should effect a right turn, while counterclockwise rotation will effect a left turn. Similarly, a hand-operated engine speed control would be operated in a forward or upward direction to effect a decrease in foward speed.

Displays and gauges provide the means by which the operator may obtain information regarding the condition and status of the tractor. Engine rpm, oil pressure, and water temperature are all important indices of machine performance. Although these instruments vary considerably in importance and frequency of use, the same general principles of layout given for controls apply to displays. All displays should be located such that they are clearly visible from the operator's normal seated position, with the more important and more frequently used ones in prime locations. Each display should be identified by words or symbols.

The principles of control and display design outlined above are helpful in allowing an operator to use the machine efficiently and safely. Most agricultural tractors of recent vintage incorporate these and other guidelines. Most manufacturers follow voluntary industry standards that allow for standardization of many of these guidelines across product brands. Figure 21-6 illustrates some of the symbols for operators controls on agricultural equipment as outlined in the American Society of Agricultural Engineers (ASAE) Standard S304.5.

SAFETY COMMUNICATIONS

There are numerous sources of information available to the operator that are intended to assist in safe operation. Among the primary sources are the operator's manual and the safety decals or signs affixed to the machine. Jointly, these provide the fundamental guidelines for safe tractor usage.

The operator's manual is a relatively detailed document that conveys not only safety information but also general specifications for operating, servicing, and maintaining the unit. It is intended to be used by all operators of the machine and, hence, it is written for a rather general audience. It must be recognized, however, that no manual can be complete enough to thoroughly educate the novice operator. Sound instructions and supervised practice with the machine are important.

The manual should always be read thoroughly prior to operating a unit for the first time. Subsequently, the manual should be stored on the tractor. It should be placed in a secure location so as not to be damaged or lost.

While the operator's manual provides a reference source for the operator, the safety signs or decals on the unit give brief, precise warnings related to safety. Their purpose is not to provide detailed instructions but rather to prevent those unsafe actions which, due to carelessness or lack of experience, could lead to injury. If further information is needed, the operator's manual should be consulted.

There is an unfortunate tendency to regard safety communications of all types as being elementary and "meant for the other person." The operator's manual and safety signs, however, provide the most direct means of instructing and warning an operator, and for this reason these are important to all who work with tractors.

OTHER TRACTOR SAFETY CONSIDERATIONS

Although tractor overturns and accidents from being run over account for more than two-thirds of all tractor-related fatalities, there are other potential hazards. For example, moving belts or shafts can cause serious injury if one becomes entangled in them. Loose-fitting clothing can easily draw the body into contact with a belt or shaft. The power take-off (PTO) shaft is of particular importance in this regard. Industry standards prescribe the use of shields to minimize the likelihood of accidental contact or entanglement with this shaft. These shields should be kept securely in place and replaced promptly if missing or damaged. The engine should be turned off prior to servicing or unclogging a PTO-driven implement.

Another source of potential injury is starting a tractor while in gear. Many manufacturers utilize a neutral-start switch, which prevents the engine from being started except when the transmission is in neutral. Older machines may not be equipped with this feature, and thus may start while in gear. It is also important to note that in jump-starting a tractor, the neutral-start switch may be bypassed thereby allowing the tractor to start while in gear.

Riders other than the operator should never be allowed on the tractor while in motion unless the manufacturer has specifically provided for them. Without such provision, riding a tractor may result in a fall or fatality.

It is customary in many farming communities that children begin to work with agricultural equipment at a relatively young age. It is most important that a child not be allowed to operate a tractor or other type of equipment until he or she is both intellectually and physically capable of handling it. Intellectually, the child should not only be fully aware of the operating procedures and safety rules for the machine, but must have sufficiently mature judgment to make proper decisions in a panic situation. Physically, the child must be of sufficient strength and stature to reach and activate all the controls easily and have enough visibility. If there is ever any doubt as to the capability of a child or anyone else to operate a tractor safely, that person should not be allowed access to it.

SAFETY EMBLEM

The National Safety Council recommends that all slow-moving vehicles operated or traveling on highways have a safety emblem attached to their rear so that it can be easily seen. This unique identification emblem should be used only on vehicles that are designed for and travel at rates of speed

Figure 21-7 Standard safety emblem for slow moving vehicles when operated on highways. This emblem color is fluorescent orange in the center and reflector red on the borders.

less than 25 mph. The emblem consists of a fluorescent yellow-orange equilateral triangle with a 4-in dark red reflective border, positioned with a point of the figure up (Fig. 21-7). The reflective border defines the shape of the fluorescent color in daylight and appears as a hollow red triangle at night, when in the path of a motor vehicle's headlights. This emblem does not replace warning devices such as reflectors, taillights, or flashing lights, but supplements them.

PROBLEMS AND QUESTIONS

1. Briefly describe the design and function of a rollover protective structure (ROPS).
2. What is the most common cause of tractor-related fatalities? State conditions which may contribute to the accidents.
3. Differentiate between the safety information which is provided by decals and signs affixed to the machine and that contained in the operator's manual.
4. What principles are important in the positioning and design of tractor controls?
5. How does safety in an agricultural environment differ from safety in other industries?
6. Explain what should be included in maintenance guidelines for ROPS. Why is maintenance of ROPS important?
7. Explain what is meant by a SMV emblem; describe its construction and shape and explain how and where it should be displayed.

Chapter 22

Economics of Tractor Operation and Utilization

The farm tractor has come to be recognized as a major source of power in both American and foreign agriculture. Mechanical power is now being applied successfully to practically all the operations involved in the production of the major field crops as well as fruits, vegetables, and similar farm commodities.

There is a vast amount of information available concerning the subject of farm power and its utilization under the great variety of existing agricultural conditions. Yet, because of the rapid changes in tractor design and equipment and in crop-production methods, much of this information is more or less obsolete. It is not the purpose of the author to present the detailed results of all the published material on the subject of tractor economics but merely to give such information as will be of general application and answer the more common questions along this line.

ECONOMICS OF THE TRACTOR

Factors Determining Kind of Power to Use The principal factors to be considered in choosing the most suitable kind of power for a given farm setup are:

1 Size of the farm
2 Topography of the land
3 Crops and kind of farm
4 Soil characteristics
5 Size of fields

The smallest size of farm on which a tractor can be used profitably cannot be stated definitely because other factors, such as crops raised, size of fields, and so on, also enter into the problem. Before the introduction of the cultivating type of tractor, farms devoted largely to row crops were dependent upon animal power for planting and cultivation. Consequently, the size of the farm on which a tractor could be used profitably was seldom less than 80 acres, and some investigations showed that farms considerably larger than this could be operated about as profitably with horses alone. Now that small tractors are available that will successfully perform all operations in row-crop production and thereby eliminate the need of a single horse or mule, it seems reasonable to say that the size of farm on which a tractor can be used with profit might be as low as 10 to 20 acres.

The topography of the land is of minor consideration except in extreme cases. A tractor will not operate satisfactorily on steep hillsides because of the tendency to slip and slide downhill. In going directly up the hill most of the power may be consumed in propelling the tractor without a load. Land that is terraced to control soil erosion can usually be handled satisfactorily with tractors.

The particular crops to be grown on a given farm and the acreage of each must be considered along with the other factors mentioned. However, the introduction of the all-purpose tractor has removed certain limitations in the utilization of mechanical power for producing a number of crops and greatly simplified the problem from this standpoint.

The size of fields is no longer an important factor in economical tractor operation. It is true that for most operations larger fields mean somewhat less loss of time and more efficient results, but the smaller tractors are now capable of being handled easily in small or irregular fields containing as few as 5 to 10 acres.

In selecting the proper kind of power for a farm, one should consider carefully the numerous jobs, both tractive and stationary, that are likely to develop. If a tractor of a certain type is well suited to handling the majority of these jobs, it should prove a profitable investment. In other words, the greater the amount of time the machine is kept busy, the lower the cost per horsepower-hour of power developed. The possibility of doing a certain amount of custom work for neighbors, provided it is done at a profit, often solves the problem of whether to buy a tractor or not. Frequently certain special tractor jobs develop, such as clearing land, dragging roads, moving buildings, operating snowplows, and the like, which assist in reducing the annual power cost.

Choice of Type and Size As a rule the acreage and kind of crops raised determine the type of tractor best adapted to a given farm. If such row crops as corn, cotton, or grain sorghums are grown either alone or with wheat, oats, hay, and similar broadcast crops, the all-purpose tractor can probably be utilized to best advantage and with greater efficiency. This type is particularly well adapted to dairy and general grain and livestock farms because of the variety of power jobs which arise. For wheat, rice, or all-grain farming the ordinary general-purpose tractor of the wheel or track type is most suitable.

The choice of the correct size of tractor is important, particularly if the farm setup justifies the use of an all-purpose tractor. In general, for small farms or for large farms made up of small fields, the small or two-row size will prove most satisfactory. On the other hand, for large farms having large fields, a larger tractor capable of handling large machines will likely prove more economical.

Make of Tractor In selecting a tractor, construction and design should be observed closely. This is important from the standpoint of durability, service, accessibility, and adaptability to the kinds of work to be done. Parts requiring frequent adjustment should be accessible, and lubrication should be simplified but positive. Convenience and safety in operation are especially desirable. This includes ease of steering, control levers that are readily manipulated, good rear-wheel fenders, comfort and rollover protection, a convenient and properly protected power take-off, and suitable hydraulic control equipment.

The past reputation and future stability of the manufacturer and the character and dependability of the local dealer are factors of prime importance in selecting the most suitable make of tractor. The tractor business is highly competitive and necessitates constant alertness and persistent effort on the part of the manufacturer to place upon the market a reliable machine at a reasonable price. He must ever be on the lookout for weaknesses in design and possibilities of improvement in construction and operation. He knows that his best salesman is the satisfied customer.

The success of a tractor often lies with the local dealer. The successful tractor dealer must have a thorough knowledge of the merits, adaptability, construction, and operation of the machines that he sells. Furthermore, he must be able to provide his customers with prompt, reliable, and competent service. This includes a well-equipped shop, trained mechanics, and a good stock of staple repair parts.

TRACTOR POWER COSTS

Factors Affecting Power Costs The cost of the use of a tractor or of any other kind of mechanical power-generating device is dependent upon a number of factors, as follows: (1) probable life in hours, days, or years, (2) annual use in hours or days, (3) interest on investment, (4) housing, (5) insurance, (6) taxes, (7) fuel consumption, (8) lubricants, and (9) repairs. Specifically, for a tractor these items may be classified as follows:

1 Fixed or overhead costs
- **a** Depreciation
- **b** Interest on investment
- **c** Housing
- **d** Insurance
- **e** Taxes

2 Operating costs
- **a** Fuel
- **b** Lubricants
- **c** Repairs

The fixed or overhead cost items are those which remain relatively constant whether the machine is used or not. Of course depreciation is affected both by obsolescence and by use; hence, strictly speaking, it may vary to some extent in accordance with use. The variable-cost items are those created only by use of the machine and vary directly as the days or hours of use. One of the most important factors affecting the cost of operation of a tractor or any other machine is the number of days or hours it is used per year. The reason for this is that the total annual fixed costs remain about the same whether the machine is used 100 h or 1,000 h per year. Therefore, the greater the annual use, the lower the daily or hourly fixed cost and hence the lower will be the total daily or hourly operating cost.

Tractor Life and Depreciation Depreciation is the decrease in value and service capacity of a machine as a result of natural wear, obsolescence, damage, corrosion, and weathering. Obviously the longer the service life of a tractor in hours, days, or years, the lower its annual depreciation rate and cost. A machine wears out with use, but the rate of wear depends upon the skill of the operator, lubrication and general maintenance, design, quality of materials, and so on. Obsolescence is an important factor in depreciation but is difficult to evaluate.

Studies and surveys indicate that tractor life varies greatly, depending upon the factors mentioned and other conditions. However, a distinct majority of farmer estimates fall within a range of 10 to 16 years. These estimates are based upon an annual use of 500 to 700 h. Hence, in terms of total hours, it can be said that farm tractor life varies from 6,000 to 10,000 h.

METHODS OF DETERMINING DEPRECIATION

According to Fenton and Fairbanks,[1] there are three methods of computing depreciation: (1) estimated-value; (2) straight-line; and (3) constant-percentage. They state further that the three important needs for depreciation estimates are (1) determination of resale, trade-in, or appraisal values of used machines; (2) determination of depreciation charges to be used in computing operating costs; and (3) for income tax purposes. The method used in the first case should give, as nearly as possible, throughout the machine's life, values that represent values on a used-equipment market. The second objective is the one commonly encountered in calculations to determine the cost of operation of a machine, such as cost per hour or per acre. The method of calculating depreciation for this purpose should assume that the equipment is to remain in use on the farm and perform its particular job throughout its useful life. It should give results that are uniform throughout the machine's life. Of the three methods enumerated, the estimated-value and constant-percentage methods are more suitable for determining resale values, and the straight-line method is better for calculating the cost of use of equipment.

Estimated-Value Method The owner's estimate of the value of a used machine may be used to determine its depreciation. When estimates are obtained from a sufficient number of owners of similar machines ranging in age from new to worn-out, the data may be used to determine the rate of depreciation of that particular type of machine.

Straight-Line Method The straight-line method of depreciation reduces the value of a machine by an equal amount each year during its useful life. As shown by Fig. 22-1, the rate per year is constant and therefore results in a straight-line graph. It is the simplest method of calculating depreciation and is widely used with farm machinery. While it is true that a machine depreciates less during the first few years by this method than its resale value would indicate, it is also true that farm machines are not bought for resale purposes but are bought to perform a given service on the farm. As long as the machine will perform this service satisfactorily, there is no legitimate reason for charging larger amounts for depreciation during the early years of its life.

In view of the large number of variables that are not, and cannot be, taken into account in any system of calculating depreciation of farm ma-

[1] F. C. Fenton and G. E. Fairbanks, *Kansas State Coll. Eng. Expt. Sta. Bull.* No. 74, pp. 16–32, 1954.

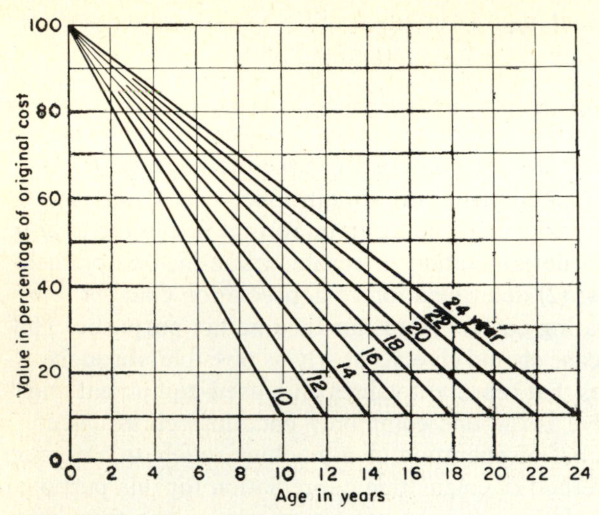

Figure 22-1 Depreciation curves based on straight-line depreciation.

chines, the accuracy of the straight-line method is all that can be expected. Many of the other methods that have been proposed are claimed to be more accurate; however, they do not offer anything in the way of accuracy that will offset their difficulty and complexity of handling. In calculating the annual depreciation by the straight-line method, if the resale value is not considered or it is assumed the machine has no value at the end of its service life, then the annual depreciation charge is simply the original cost divided by the life in years. For many purposes this gives a satisfactory answer, but since the machines usually have a trade-in value at the end of their service life and since this may be a considerable amount, it should generally be included. The present value of a machine can be found by obtaining its value in percentage of original cost at its present age from Table 22-1 or Fig. 22-1 and multiplying this by the original cost.

Constant-Percentage Method The constant-percentage method depreciates the value of the machine at the same percentage of the value remaining each year. It results in a larger reduction of value during the early life of the machine and a decreasing amount of depreciation in its later life. It takes into consideration the original cost, the trade-in value, and the estimated service life of the machine, as does the straight-line method, but calculates the percentage that will reduce the value to the trade-in value at the end of the machine's useful life. It may be used where the value of a machine is desired for resale purposes or for making farm-machinery appraisals.

To determine the rate or percentage necessary to reduce the original

Table 22-1 Values of Machines, by Straight-Line Depreciation, in Percentage of Original Cost* (10 percent trade-in value)

Age, years	10-year service life	12-year service life	14-year service life	16-year service life	18-year service life	20-year service life	22-year service life	24-year service life
0	100.00	100.00	100.00	100.00	100.00	100.00	100.00	100.00
1	91.00	92.50	93.57	94.37	95.00	95.50	95.91	96.25
2	82.00	85.00	87.14	88.75	90.00	91.00	91.82	92.50
3	73.00	77.50	80.72	83.12	85.00	86.50	87.73	88.75
4	64.00	70.00	74.29	77.50	80.00	82.00	83.64	85.00
5	55.00	62.50	67.86	71.87	75.00	77.50	79.54	81.25
6	46.00	55.00	61.43	66.25	70.00	73.00	75.45	77.50
7	37.00	47.50	55.00	60.62	65.00	68.50	71.36	73.75
8	28.00	40.00	48.58	55.00	60.00	64.00	67.27	70.00
9	19.00	32.50	42.15	49.37	55.00	59.50	63.18	66.25
10	10.00	25.00	35.72	43.75	50.00	55.00	59.09	62.50
11	—	17.50	29.29	38.12	45.00	50.50	55.00	58.75
12	—	10.00	22.86	32.50	40.00	46.00	50.91	55.00
13	—	—	16.43	26.87	35.00	41.50	46.82	51.25
14	—	—	10.00	21.25	30.00	37.00	42.73	47.50
15	—	—	—	15.62	25.00	32.50	38.63	43.75
16	—	—	—	10.00	20.00	28.00	34.54	40.00
17	—	—	—	—	15.00	23.50	30.45	36.25
18	—	—	—	—	10.00	19.00	26.36	32.50
19	—	—	—	—	—	14.50	22.27	28.75
20	—	—	—	—	—	10.00	18.18	25.00
21	—	—	—	—	—	—	14.09	21.25
22	—	—	—	—	—	—	10.00	17.50
23	—	—	—	—	—	—	—	13.75
24	—	—	—	—	—	—	—	10.00

* *Source:* F. C. Fenton and G. E. Fairbanks, *Kansas State Coll. Eng. Expt. Sta. Bull.* No. 74, pp. 16–32, 1954.

value to trade-in value at the end of the estimated service life, the following formula is used:

$$r = 1 - \sqrt[L]{\frac{S}{C}}$$

where r = percentage annual rate of depreciation
L = total service life, years
S = trade-in value
C = original cost of machine

The value V at any age n during the useful life is equal to $C(1 - r)^n$. The curves (Fig. 22-2) were prepared by using this formula.

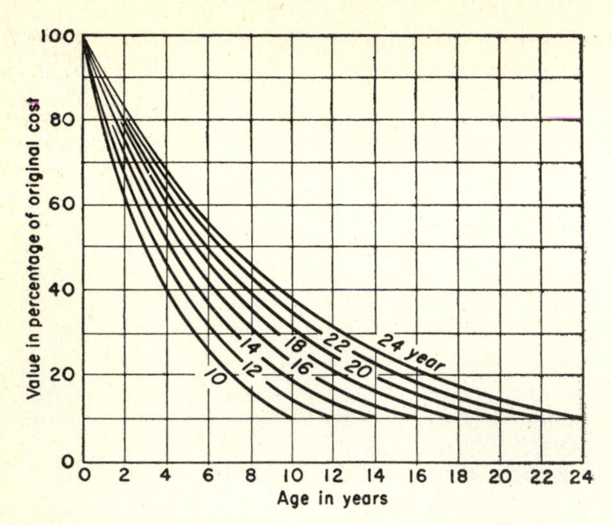

Figure 22-2 Depreciation curves based on the constant-percentage method of depreciation.

OTHER FIXED COSTS

Interest on Investment Interest on the investment in farm equipment is usually included in operational cost estimates, since money used to buy a machine cannot be used for other purposes such as the purchase of land, livestock, bonds, or other productive enterprises.

The amount invested in a machine is greater during its early life than during the later years, since an amount is written off each year as depreciation. This is true regardless of the method used in calculating depreciation. Interest charges are usually desired when operating costs are being determined but not when depreciated values for resale or trade-in are being determined. In calculating interest charges where operating costs are concerned, it is desirable to use a method that results in constant or equal yearly charges throughout the machine's life; that is, the sum of the interest charge and the depreciation should be constant, for reasons that have already been discussed. When the straight-line method of depreciation is used, this is accomplished by making an annual interest charge on the average investment in the machine over its full life. The average investment is equal to one-half the sum of the first cost and the trade-in value. The annual interest charge will then be the product of the interest rate and the average investment.

Insurance—Shelter—Taxes Insurance and housing expense and taxes, if incurred, must be included in any tractor operation costs. Although insurance and tax rates vary according to local conditions and other factors, Fenton and Fairbanks[1] suggest an annual tax charge of 1 percent and an annual insurance charge of 0.25 percent of the original cost of the machine.

Regarding shelter for farm equipment, Fenton and Fairbanks state that numerous attempts have been made by various investigators to obtain conclusive evidence of the value of a shelter for farm machinery. Most of these have failed to find such evidence but from observation have stated that, although monetary savings are not apparent, there are indeterminate values such as better management, better appearance of the farmstead, and ease of making repairs during a slack or stormy season, and have thereby justified the expense of a machine shed.

The average annual estimated repair expense is also consistently smaller for sheltered machines. Sheltering usually goes along with better care and management, and this factor alone could account for the difference in favor of sheltering.

The cost of shelter will depend upon how expensive or elaborate the structure is. Machine sheds should be built as simply and inexpensively as possible. A fair annual charge recommended for adequate shelter facilities for tractors and farm equipment is 1 percent of the original cost of the machine.

OPERATING COSTS

Items making up the operating cost of a tractor are (1) fuel; (2) lubricants, including motor oil, transmission lubricant, and grease; and (3) repairs, including the labor required to do the work. These costs are incurred only through the use of the machine and vary to a large extent in direct proportion to hours or days of use per year.

Fuel and Lubricants Costs Fuel is one of the major items affecting tractor operating costs. Factors affecting fuel costs are (1) the prevailing market price, (2) carburetor adjustment, (3) engine condition, and (4) load factor or ratio of used power to available power. A tractor uses less fuel at light loads, but the fuel consumption per unit of power developed is high; therefore, the most economical power, so far as fuel consumption is concerned, is obtained when the tractor load is at least 50 percent of its available capacity. The load may be adjusted to the rated power of a tractor by (1) increasing the width or size of the equipment; (2) pulling more than one piece of machinery at the same time; (3) using a higher gear and doing the work faster; or (4) using a higher gear and throttling the engine down to the required speed.

[1] F. C. Fenton and G. E. Fairbanks, op. cit.

Proper lubrication and the use of good-quality lubricants are very important in reducing the wear and repair expense of a tractor. The annual cost of all lubricants for a tractor is relatively small compared to other costs, but neglect with respect to the recommended lubrication requirements for a tractor can be very expensive.

Repair Costs The cost of the repairs for a tractor is an important item of expense and depends largely upon proper everyday maintenance. A survey made among Kansas tractor users indicates a considerable variation in annual repair costs, but 3.50 percent of the first cost is suggested as being representative.

Operating Costs of Tractors According to Size and Annual Use Table 22-2 has been prepared by the authors in order to present a simple but representative operating cost analysis for farm tractors according to both size and annual use in hours. The values are based as nearly as possible on 1979 costs and conditions. A careful examination of the information given in this table will disclose the following general deductions relative to tractor operating costs:

1 The larger the tractor, the greater its total hourly and yearly operating costs.

2 The fixed costs remain relatively constant, regardless of the total hours of annual use.

3 The operating costs vary directly as the total hours of annual use.

4 Depreciation and fuel are the two most important cost items in tractor operation.

5 The greater the annual use in hours, the lower the total operating costs per hour.

6 The cost per horsepower-hour remains relatively constant (depending upon drawbar horsepower rating used) regardless of the size of the tractor but under any conditions is greatest for a low annual use and lowest for a high annual use.

Table 22-2 Operating Costs of Various Sizes of Tractors per Year and per Hour*

	Used	
Size of tractor and cost items	600 h/year	1,200 h/year
A. Tractor costing $8,000, 35–45 PTO hp		
Fixed costs:		
Depreciation, 10% of first cost	$800.00	$800.00
Interest, 8% of average investment ($4,000)	320.00	320.00
Shelter, taxes, and insurance, 3% of original cost	240.00	240.00
Total	$1,360.00	$1,360.00

Table 22-2 (*Continued*)

Size of tractor and cost items	Used	
	600 h/year	1,200 h/year
Operating costs:		
Fuel, 3.5 gal/hr	$1,470.00	2,940.00
Engine oil	40.00	75.00
Hydraulic fluid, grease, and transmission lubricant	25.00	40.00
Repairs	250.00	450.00
Maintenance service labor	75.00	150.00
Total	$1,860.00	$3,655.00
Grand total per year	$3,220.00	$5,015.00
Cost per hour	$5.37	$4.18
B. Tractor costing $11,000, 60–70 PTO hp		
Fixed costs:		
Depreciation, 10% of first cost	$1,100.00	$1,100.00
Interest, 8% of average investment ($5,500)	440.00	440.00
Shelter, taxes, and insurance, 3% of original cost	330.00	330.00
Total	$1,870.00	$1,870.00
Operating costs:		
Fuel, 5.2 gal/hr	$2,184.00	$4,368.00
Engine oil	50.00	90.00
Hydraulic fluid, grease, and transmission lubricant	35.00	50.00
Repairs	400.00	750.00
Maintenance service labor	80.00	100.00
Total	$2,749.00	$5,358.00
Grand total per year	$4,619.00	$7,228.00
Cost per hour	7.70	6.02
C. Tractor costing $21,000, 90–110 PTO hp		
Fixed costs:		
Depreciation, 10% of first cost	$2,100.00	$2,100.00
Interest, 8% of average investment ($10,500)	840.00	840.00
Shelter, taxes, and insurance, 3% of original cost	630.00	630.00
Total	$3,570.00	$3,570.00
Operating costs:		
Fuel, 7.6 gal/hr	3,192.00	6,384.00
Engine oil	75.00	140.00
Hydraulic fluid, grease, and transmission lubricant	45.00	60.00
Repairs	800.00	1,500.00
Maintenance service labor	150.00	250.00
Total	$4,260.00	$8,394.00
Grand total per year	$7,832.00	$11,964.00
Cost per hour	$13.05	$9.97
D. Tractor costing 30,000, 140–160 PTO hp.		
Fixed costs:		
Depreciation, 10% of first cost	$3,000.00	$3,000.00
Interest, 8% of average investment ($15,000)	1,200.00	1,200.00

Table 22-2 (*Continued*)

Size of tractor and cost items	Used 600 h/year	Used 1,200 h/year
Shelter, taxes, and insurance, 3% of original cost	900.00	900.00
Total	$5,100.00	$5,100.00
Operating costs:		
Fuel, 10 gal/hr	$4,200.00	$8,400.00
Engine oil	120.00	240.00
Hydraulic fluid, grease, and transmission lubricant	50.00	65.00
Repairs	1,200.00	2,500.00
Maintenance service labor	200.00	400.00
Total	$5,770.00	$10,405.00
Grand total per year	$10,870.00	$15,505.00
Cost per hour	$18.12	$12.72
E. Tractor costing $50,000, 200–225 PTO hp		
Fixed costs:		
Depreciation, 10% of first cost	$5,000.00	$5,000.00
Interest, 8% of average investment ($25,000)	2,000.00	2,000.00
Shelter, taxes, and insurance, 3% of original cost	1,500.00	1,500.00
Total	$8,500.00	$8,500.00
Operating costs:		
Fuel, 15 gal/hr	$6,300.00	12,600.00
Engine oil	150.00	300.00
Hydraulic fluid, grease, and transmission lubricant	60.00	80.00
Repairs	2,000.00	4,000.00
Maintenance service labor	250.00	500.00
Total	$8,760.00	$17,480.00
Grand total per year	$17,260.00	$25,980.00
Cost per hour	$28.77	$21.65

* In preparing these cost data, certain assumptions were made as follows: average tractor life is 12 years, fuel cost (diesel fuel) .70 cents per gal, motor oil cost, $4.00 per gal., hydraulic fluid, grease and transmission lubricant, repairs and maintenance labor costs are estimates made by the authors as a result of information obtained from various sources.

Appendixes

Appendix One

Conversion Factors

English-to-metric and metric-to-English conversion factors

Multiply	by*	to obtain
Acres	0.404687	Hectares
Acres	4.04687×10^{-3}	Square kilometers
Acres	1076.39	Square feet
Centimeters	3.28083×10^{-2}	Feet
Centimeters	0.3937	Inches
Cubic centimeters	3.53145×10^{-5}	Cubic feet
Cubic centimeters	6.102×10^{-2}	Cubic inches
Cubic feet	2.8317×10^{-4}	Cubic centimeters
Cubic feet	2.8317×10^{-2}	Cubic meters
Cubic feet	6.22905	Gallons, British Imperial
Cubic feet	28.3170	Liters
Cubic feet	2.38095×10^{-2}	Tons, British Shipping
Cubic feet	0.025	Tons, U.S. Shipping
Cubic Inches	16.38716	Cubic centimeters
Cubic meters	35.3145	Cubic feet
Cubic meters	1.30794	Cubic yards
Cubic yards	0.764559	Cubic meters
Degrees, angular	0.0174533	Radians

Table A1-1 (*Continued*)

Multiply	by*	to obtain
Degrees, Fahrenheit (less 32°F.) .	0.5556	Degrees, centigrade
Degrees, centigrade	1.8	Degrees, Fahrenheit (less 32°F.)
Foot-pounds	0.13826	Kilogram-meters
Feet	30.4801	Centimeters
Feet	0.304801	Meters
Feet	304.801	Millimeters
Feet	1.64468×10^{-4}	Miles, nautical
Gallons, British Imperial	0.160538	Cubic feet
Gallons, British Imperial	1.20091	Gallons, U.S.
Gallons, British Imperial	4.54596	Liters
Gallons, U.S.	0.832702	Gallons, British Imperial
Gallons, U.S.	0.13368	Cubic feet
Gallons, U.S.	231.0	Cubic inches
Gallons, U.S.	3.78543	Liters
Grams, metric	2.20462×10^{-3}	Pounds, avoirdupois
Hectares	2.47104	Acres
Hectares	1.076387×10^{-5}	Square feet
Hectares	3.86101×10^{-3}	Square miles
Horsepower, metric	0.98632	Horsepower, U.S.
Horsepower, U.S.	1.01387	Horsepower, metric
Inches	2.54001	Centimeters
Inches	2.54001×10^{-2}	Meters
Inches	25.4001	Millimeters
Kilograms	2.20462	Pounds
Kilograms	9.84206×10^{-4}	Long tons
Kilograms	1.10231×10^{-3}	Short tons
Kilogram-meters	7.233	Foot-pounds
Kilograms per meter	0.671972	Pounds per foot
Kilograms per square centimeter	14.2234	Pounds per square inch
Kilograms per square meter ...	0.204817	Pounds per square foot
Kilograms per square meter ...	9.14362×10^{-5}	Long tons per square foot
Kilograms per square millimeter .	1422.34	Pounds per square inch
Kilograms per square millimeter .	0.634973	Long tons per square inch
Kilograms per cubic meter	6.24283×10^{-2}	Pounds per cubic foot
Kilometers	0.62137	Miles, statute
Kilometers	0.53959	Miles, nautical
Liters	0.219976	Gallons, British Imperial
Liters	0.26417	Gallons, U.S.
Liters	3.53145×10^{-2}	Cubic feet
Meters	3.28083	Feet
Meters	39.37	Inches
Meters	1.09361	Yards
Miles, statute	1.60935	Kilometers
Miles, statute	0.8684	Miles, nautical
Miles, nautical	6080.204	Feet
Miles, nautical	1.85325	Kilometers
Miles, nautical	1.1516	Miles, statute
Millimeters	3.28083×10^{-3}	Feet
Millimeters	3.937×10^{-2}	Inches

Table A1-1 (*Continued*)

Multiply	by*	to obtain
Pounds, avoirdupois	453.592	Grams, metric
Pounds, avoirdupois	0.453592	Kilograms
Pounds, avoirdupois	4.464×10^{-4}	Tons, long
Pounds, avoirdupois	4.53592×10^{-4}	Tons, metric
Pounds per foot	1.48816	Kilograms per meter
Pounds per square foot	4.88241	Kilograms per square meter
Pounds per square inch	7.031×10^{-2}	Kilograms per square centimeter
Pounds per square inch	7.031×10^{-4}	Kilograms per square millimeter
Pounds per cubic foot	16.0184	Kilograms per cubic meter
Radians	57.29578	Degrees, angular
Square centimeters	0.1550	Square inches
Square feet	9.29034×10^{-4}	Acres
Square feet	9.29034×10^{-6}	Hectares
Square feet	0.0929034	Square meters
Square inches	6.45163	Square centimeters
Square inches	645.163	Square millimeters
Square kilometers	247.104	Acres
Square kilometers	0.3861	Square miles
Square meters	10.7639	Square feet
Square meters	1.19599	Square yards
Square miles	259.0	Hectares
Square miles	2.590	Square kilometers
Square millimeters	1.550×10^{-3}	Square inches
Square yards	0.83613	Square meters
Yards	0.914402	Meters

* The expressions $\times 10^{-2}$, $\times 10^{-3}$, $\times 10^{-4}$, $\times 10^{-5}$, and $\times 10^{-6}$, following certain multipliers, indicate that the decimal point in the product—of left-column value times multiplier—is to be moved respectively 2, 3, 4, 5, or 6 places to the left.

Appendix Two

Tractor Development

The Nebraska Tractor Test has been publishing reports on power output, fuel consumption, and design integrity for tractors since 1920. Highlights of the world-renowned tests are listed in the following table.

Major Milestones in Tractor Development

Year	Highlights	Test	Make and model
1919	The Nebraska Tractor Test Law passed by the Nebraska legislature		
1920	Completion of first official test	1	Waterloo Boy "N"
1920	Cast-iron unit-frame construction and water-bath air cleaner	18	Fordson
1920	Early garden-type tractor	28	Beeman "G"
1920	Earliest practical approach to a general-purpose tractor-storage battery, starter, lights, electric governor, high-tension battery ignition system, turning brakes, differential lock and high-speed (1800 rpm) engine	33	Moline Universal "D" 9-18
1920	First crawler tractor	45	Cletrac "W" 12-20

Major Milestones in Tractor Development (*Continued*)

Year	Highlights	Test	Make and model
1920	One-piece boiler plate unit-frame construction	49	Wallis 15-25
1920	Last tractor to retain the outward appearance of a steam traction engine	63	Townsend 15-30
1920	Mechanical lift with direct engine drive	66	Square Turn 18-35
1922	Articulated four-wheel drive with hydraulic power steering	84	Rogers Four-Wheel Drive
1922	A practical power take-off (PTO) and one-piece cast-frame construction	87	International 15-30
1925	Successful all-purpose row-crop tractor	117	McCormick-Deering "Farmall"
1926	Distillate fuel replaces kerosene	128	Hart-Parr 18-36
1927	Certified performance	134	Wallis 20-30
1930	Imported tractor (from Ireland)	173	Fordson "F"
1930	"Tip-Toe" wheels with adjustable spacing on splined axles	176	Oliver Hart-Parr ROW CROP
1931	Two-speed belt pulley which could also operate in reverse	192	Bradley "General Purpose"
1932	Early diesel tractor (crawler)	208	Caterpillar "Diesel"
1934	Hydraulic lift available	222	John Deere GP "A"
1934	Pneumatic tires on a tractor	223	Allis-Chalmers "WC"
1934	Last test using kerosene	229	McCormick-Deering W-12
1935	Diesel engine in wheel-type tractor	246	McCormick-Deering "WD-40"
1936	High-compression engines for tractors	249	Minneapolis-Moline Twin City KTAHC Oliver Hart-Parr 70 HC
1937	Hesselman (fuel injection with spark ignition)	285	Allis-Chalmers WK-O Dsl.
1938	Compact or "baby" tractor (rubber tires)	302	Allis-Chalmers B
1940	Three-point hitch and hydraulic draft control	339	Ford-Ferguson System 9N
1940	Two-cycle diesel engine in a crawler	360	Allis-Chalmers HD-7W
1947	Independent PTO	382	Cockshutt 30 Gasoline
1948	Torque converter (crawler)	397	Allis-Chalmers HD-19
1948	Engine mounted behind rear axle	398	Allis-Chalmers G
1948	Power spacing wheels	399	Allis-Chalmers WD Tractor Fuel
1949	Tractor using LPG	411	Minneapolis-Moline Universal Standard
1953	LPG fuel tank mounted in front of radiator	512	Allis-Chalmers WD-45 LP
1954	Hydraulic power-assist steering	528	John Deere 70 Diesel
1955	Partial-power-shift transmission (TA)	532	McCormick Farmall 400
1955	A supercharger used in a crawler	550	Allis-Chalmers HD-21
1955	Turbocharger used in a crawler	584	Caterpillar—D-9
1956	Concrete test course constructed at Nebraska		

Major Milestones in Tractor Development (*Continued*)

Year	Highlights	Test	Make and model
1956	Record fuel economy (10-h run)		
	8.87 hp·h/gal	590	John Deere 520 LPG
	16.56 hp·h/gal	594	John Deere 720 Diesel
	11.25 hp·h/gal	598	John Deere 620 Gas
1956	Last use of distillate fuel during a test	606	John Deere 720 All Fuel
1958	Torque converter with lock-out	679	Case 811-B
1959	Starting PTO testing at Nebraska	684	Fordson Dexta Diesel
1959	Air-cooled diesel engine in a tractor	699	Porsche Diesel Jr. L108
1959	Cab and air conditioner (four-wheel drive)	700	Wagner TR-14A Diesel
1959	Full-power-shift transmission	701	Ford 881
1959	Radial tractor tire	707	Fiat 411 R Diesel
1960	Hydrostatic power steering	759	John Deere 4010 Gas
1960	Front-mounted fuel tank (rubber-tired tractor)	759	John Deere 4010 Gas
1962	Turbocharger used in wheel-type tractor	811	Allis-Chalmers D-19
1962	Largest wheel tractor tested (crab steering)	815	International 4300
1963	Alternator charging system	855	Allis-Chalmers D-21
1964	Vacuum advance distributor	874	Oliver 1650
1965	Rear-mounted fuel tank (general purpose tractor)	918	Case 930 GP Diesel
1965	Driver's seat raised and lowered hydraulically	923	Massey-Ferguson 1100 Diesel
1965	Saddle-mounted fuel tanks	923	Massey-Ferguson 1100 Diesel
1966	Roll guard protection for operator	931	John Deere 4020 S.R. LPG
1967	Hydrostatic drive	967	International 656 Hydro Diesel
1968	Fender tanks for fuel	986	Oliver 2150 Row Crop Diesel
1969	Transistorized ignition system	1007	International 544 Hydro Gas
1969	Cab and roll guard protection for operator	1026	Ford 8000 Diesel
1969	Nonmetallic fuel tank	1030	Case 870 P.S. Diesel
1970	Official sound testing	1034	Case 970 Diesel
1975	Record low sound level [76.5 dB(A)]	1195	Allis-Chalmers 7000 Diesel
1977	Power take-off ratings above 250 hp	1241	Case 2870 Diesel
1978	Hydrostatic partial-power PTOs	1272	Steiger Bearcat III PT 225
1979	Drawbar Power over 350 hp	1282	Steiger Tiger III ST 450

Source: Nebraska Tractor Test Lab, University of Nebraska, Lincoln, Nebraska.

M. D. Zimmerman, "Farm Machinery: Tools of Substenance and Survival", *Machine Design* **51**(12):18–53 (1979).

Appendix Three

Bibliography

Accident Facts, National Safety Council, 1978.

American Society of Agricultural Engineers, *Annual Agricultural Engineers Yearbooks.*

Armstrong, L. V., and J. B. Hartman: *The Diesel Engine,* The Macmillan Company, New York, 1959.

ASTM Standards on Petroleum Products, issued annually by the American Society for Testing Materials, Philadelphia.

"Automotive Engine Oil," *Lubrication,* **51**(5):45–61 (1965).

Battery Service Manual, 8th ed., Battery Council International, Chicago, 1978.

Bickerton, R. G.: "Some Aspect of Synthetic Lubricants," *Scientific Lubrication,* **15**(3):82 (1963).

Borgman, Donald E., Everette Hainline, and Melvin E. Long: *Fundamentals of Machine Operation,* Deere and Company, Moline, Illinois, 1974.

Carver, S. E.: Rolling Bearings Lubrication—II, *Lubrication,* **60**(4):61–88 (1974).

Charging Circuits—Alternator, Booklet L-661., The Prestolite Company, Toledo, Ohio 43601.

Charging Circuits—Generator, Booklet L-660. The Prestolite Company, Toledo, Ohio 43601.

Cranking Motors, Booklet DR5133C. Delco Remy Division of General Motors, Anderson, Indiana 46011, 1977.

Crouse, W. H.: *Automotive Electrical Equipment,* 8th ed., McGraw-Hill Book Company, New York, 1976.

———: *Automotive Mechanics,* 7th ed., McGraw-Hill Book Company, New York, 1975.

——— and D. L. Anglin: *Automotive Engines,* 5th ed., McGraw-Hill Book Company, New York, 1976.

Doss, H. J., R. H. Bittner, R. E. Childers, et al.: *Agricultural Machinery Safety,* Deere and Company, Moline, Illinois, 1974.

Dupuis, Heinrich: "Effect of Tractor Operation on Human Stress," *Agricultural Engineering,* **40**(9):510–519, 525 (1959).

Engine Service Classifications and Guide to Crankcase Oil Selection, American Petroleum Institute Publication 1509, 7th ed., 1979.

Farm Equipment Red Book, Intertec Publishing Corp., Kansas City, Missouri, 1979.

Forrest, P. J., et al.: "Effects of Improper Inflation Pressures on Farm Tractor Tires," *Agricultural Engineering,* **35**(12):853–858, 860 (1954).

Fundamentals of Electricity and Magnetism, Booklet DR5133A, Delco Remy Division of General Motors, Anderson, Indiana, 1967.

Fundamentals of Electricity and Magnetism, Booklet L-657, The Prestolite Company, Toledo, Ohio.

Fundamentals of Service—Engines, 2d ed., Deere and Company, Moline, Illinois, 1972.

Fundamentals of Machine Operation: Tractors, Deere and Company, Moline, Illinois 61265, 1974.

Fundamentals of Service: Power Trains, Deere and Company, Moline, Illinois 61265, 1972.

Fundamentals of Service Hydraulics, 2d ed., Deere and Company, Moline, Illinois, 1972.

Generators, Booklet 5133E., Delco Remy Division of General Motors, Anderson, Indiana 46011, 1977.

Gray, R. B.: *Development of the Agricultural Tractor in the United States,* Parts I and II, American Society of Agricultural Engineers, St. Joseph, Michigan.

Gunderson, R. C., and A. W. Hart: *Synthetic Lubricants,* Reinhold Publishing Corp., New York, 1962.

High Energy Ignition Systems, Booklet OR5133H, Delco Remy Division of General Motors, Anderson, Indiana 46011, 1973.

Hingest, R. A.: "Synthetic Oil—Is It Really Super?," *Automotive News,* August 26, 1974, p. 8.

Hollinghurst, R., and A. Singleton: "Viscosity and Engine Performance," *Journal of the Institute of Petroleum,* **55**(4):205 (1969).

The Ignition Circuit, Booklet L-659, The Prestolite Company, Toledo, Ohio.

The Ignition System, Booklet DR5133D, Delco Remy Division of General Motors, Anderson, Indiana, 1968.

Implement and Tractor, Annual Statistical Issue, Intertec Publishing Corp., Kansas City, Missouri.

Integral Charging System, Booklet DR51335. Delco Remy Division of General Motors, Anderson, Indiana 46011, 1973.

Know Your Motor Oil, 5th ed., American Petroleum Institute Publication 1507, 1977.

Lichty, Lester C.: *Combustion Engine Processes,* McGraw-Hill Book Company, New York, 1967.

Lincoln, C. W.: "Hydraulic Power Steering," *Agricultural Engineering,* **38**(1):18–21 (1958).

Magneto Ignition—Fundamentals of Design and Service, Bulletin SI 3-23, Colt Industries, Engine Accessories Operation, Beloit, Wisconsin.

McCoy, R. R., and D. S. Taber: "A Cold Look at Lubricants," *SAE Transactions,* vol. 80, paper no. 710716 (1971).

Mobile Hydraulic Manual, Sperry Rand Corporation, Troy, Michigan, 48084, 1967.

Nelson, G. S.: "Texas Farm and Ranch Accident Costs," *Texas Agricultural Progress,* **24**(2):15–18 (1978).

Obert, E. F.: *Internal Combustion Engines,* 3d ed., International Textbook Company, Scranton, Pennsylvania, 1968.

Reed, I. F., and M. O. Berry.: "Equipment and Procedures for Farm Tractor Tire Studies Under Controlled Conditions," *Agricultural Engineering,* **30**(2):67–70 (1949).

———, C. A. Reaves, and J. W. Shields, "Comparative Performance of Farm Tractor Tires Weighed with Liquid and Wheel Weights," *Agricultural Engineering,* **34**(6):391–395 (1953).

Relyea, K. D.: "Diesel, Dual Fuel and Gas Engine," *Lubrication,* **55**(3):49–76 (1973).

Schulz, Erich J.: *Diesel Mechanics,* McGraw-Hill Book Company, New York, 1977.

Simpson, H. W.: "Evolution in Tractor Transmissions," *Agricultural Engineering,* **40**(6):326–331, 335, 1959.

Smith, H. P., and L. H. Wilkes: *Farm Machinery and Equipment,* 6th ed., McGraw-Hill Book Company, New York, 1976.

Starting Circuits, Booklet L-658., The Prestolite Company, Toledo, Ohio 43601.

Stewart, H. L., *Practical Guide To Fluid Power,* 2nd ed., Theo. Audel and Company, Indianapolis, 1968.

Storage Batteries, Booklet DR5133B, Delco Remy Division of General Motors, Anderson, Indiana, 1967.

Taylor, Charles F.: *The Internal Combustion Engine in Theory and Practice,* 2nd ed., M.I.T. Press, Cambridge, 1966.

Taylor, J. H.: "Comparative Traction Performance of R-1, R-3, and R-4 Tractor Tires," *Transactions of the ASAE,* **19**(1):14–16 (1976).

———, E. C. Burt, and A. C. Baily: "Radial Tire Performance in Firm and Soft Soils," *Transactions of the ASAE,* **19**(6):1062–1064 (1976).

Tepper, Marvin: *Electronic Ignition Systems,* Hayden Book Company, Rochelle Park, New Jersey, 1977.

Toboldt, William K.: *Diesel Fundamentals, Service and Repair,* The Goodheart-Wilcox Company, South Holland, Illinois, 1973.

Turner, J. H.: *Small Engines,* American Association for Vocational Instructional Materials, Athens, Georgia, 1974.

Turnquist, P. K., and J. C. Thomas: "The Subjective Response of Males to Comfort Under Controlled Tractor Cab Conditions," *Transactions of the ASAE,* **19**(3):402–404 (1976).

"Viscosity," *Lubrication,* **52**(3):21–48 (1966).

Yeaple, F. D.: *Hydraulic and Pneumatic Power and Control,* McGraw-Hill Book Company, New York, 1966.

Index

Index